AKADEMIYA NAUK UKRAINSKOI SSR. GLAVNAYA ASTRONOMICHESKAYA
OBSERVATORIYA

Academy of Sciences of the Ukrainian SSR. Main Astronomical Observatory

V. P. Tsesevich

RR LYRAE STARS

(Zvezdy tipa RR Liry)

Izdatel'stvo "Naukova Dumka"
Kiev 1966

Translated from Russian

Israel Program for Scientific Translations
Jerusalem 1969

NASA TT F-562
TT 69-55025

Published Pursuant to an Agreement with
THE NATIONAL AERONAUTICS AND SPACE ADMINISTRATION
and
THE NATIONAL SCIENCE FOUNDATION, WASHINGTON, D.C.

Translated by Z. Lerman

Printed in Jerusalem by IPST Press
Binding: Wiener Bindery Ltd., Jerusalem

Available from the
U.S. DEPARTMENT OF COMMERCE
Clearinghouse for Federal Scientific and Technical Information
Springfield, Va. 22151

In memoriam
L.P. TSERASSKAYA
and
S.N. BLAZHKO,
the founders of the Soviet school
of variable stars

PREFACE

At the International Astronomical Congress held in Moscow in 1958 it
was resolved to resume the regular observations of the short-period
cepheids — RR Lyrae stars. That year marked the beginning of an annual
revision program of data on RR Lyrae stars. Initially only the behavior of
the 60% of the brightest stars of this type could be reconstructed for the
20 odd years during which the systematic observations were discontinued.
During the five years that followed, various doubtful and disputed cases
were successfully settled. The present book in fact summarizes the results
of no less than half a million magnitude observations of RR Lyrae stars.
During the 40 years of my active scientific life, I myself performed over
fifty thousand observations, using Soviet collections of star sky photographs
(Moscow, Simeise, Odessa, and Dushanbe). These collections are the
cumulative result of the combined efforts of numerous astronomers. The
invaluable Simeise collection was mainly founded by S.I. Belyavskii,
G.N. Neuimin, and P.F. Shain. The magnitude estimates from these
photographs were mainly carried out by myself. In certain cases other
astronomers assisted me in this ardous task: G.S. Filatov, A. Filin,
V. Satyvaldyev, T. Nikulina, Yu.E. Migach, V.F. Karamysh, B.A.
Dragomiretskaya, R.I. Chuprina, and others. To all of them, my heartfelt
thanks. The archives of Prof. S.N. Blazhko at the Shternberg State
Astronomical Institute were also used.

The observations listed in the monograph are accompanied by detailed
bibliographical references. My own new data or the data of other observers
who carried out the measurements at my request are given without
reference. Highly valuable results were obtained by G.A. Lange and
A.A. Batyrev. Romanian astronomers under G.G. Kisc were also highly
active in this direction.

I was greatly assisted in the preparation of this manuscript by
O.P. Nikushina, E.S. Kheilo, and M.S. Kazanasmas. It is my pleasant duty
to acknowledge their help.

After the manuscript had been completed, I was offered the opportunity
to examine the Harvard photographs. As a result, the data for the principal
stars covered in this book were improved by some 10%. Appropriate
corrections were introduced in the text, but the twenty thousand observations
made in Harvard were not listed. I would like to acknowledge the kind
assistance of Prof. D. Menzel, W. Liller, L. Goldberg, and C. Payne-
Gaposchkin during the 4 months of my stay at the Harvard Observatory.

TABLE OF CONTENTS

INTRODUCTION

RR Lyrae stars — variables with a short light-variation period — are one of the most interesting objects in astronomy. As new data on variables are published and new objects are discovered, it becomes progressively clearer that almost 25% of the pulsating variables (if not more) are represented by RR Lyrae stars. So far, RR Lyrae variables have been mainly used in various problems of stellar astronomy. As typical representatives of the class of regular pulsating variables, however, they have been studied most inadequately, although numerous individual studies on the subject are available.

We will start with a brief enumeration of the various applications of the RR Lyrae stars, which in my opinion are of the greatest significance.

The discovery of RR Lyrae variables in globular star clusters made it possible for the first time to determine the photometric parallaxes and to construct the first model of the Galaxy (Shapley). It was established that the globular clusters showed a distinct tendency to concentrate around the (then hypothetical) nucleus of the Galaxy, the distance of the Galactic nucleus was estimated, and its probable direction was indicated.

Baade /184/ investigated the dwarf galaxy in Draco, where he found a considerable number of RR Lyrae stars; this enabled him to determine the distance of this extragalactic system and its size. In the course of Baade's observations, a hitherto unexplained phenomenon was discovered. The frequency of the light-variation periods of RR Lyrae stars was essentially different from the corresponding frequency for globular clusters and the general Galactic background.

Kinman and Wirtanen /295/ searched for weak RR Lyrae variables in the parts of the galaxy free from absorbing interstellar gas. In two parts of the sky, centered at $\alpha_{1960} = 12^h16^m.6$, $\delta_{1960} = +33°7$, and $\alpha_{1960} = 13^h06^m.8$, $\delta_{1960} = +29°6$, each with an area of some 80 sq. degrees, they discovered 32 RR Lyrae variables up to 18^m using the 20-in. Lick astrograph. Variables of this type were found within distances of up to 25 kpc, i.e., they belong to the galactic halo. This discovery should clearly lead to a revision of our notions of the galactic structure on the whole. Figure 1 compares the actual frequency of RR Lyrae stars with the frequency predicted by Kukarkin (the solid curve).

Finally, we should mention another use of RR Lyrae stars for the study of intergalactic space. Statistical counts of extragalactic nebulae show that a large accumulation of light-absorbing matter lies in the direction of Microscopium. RR Lyrae stars played a fundamental role in establishing the exact position, distance from the Sun, and extent of this cloud. Hoffmeister /273/ surveyed that part of the sky and discovered a number of fairly weak RR Lyrae variables, which were projected onto the dark cloud.

This immediately gave the lower-bound value for the distance of the cloud. It was found to exceed 5000 pc. The cloud may thus lie in the intergalactic space. The area of this dark cloud should now be studied with a high-power instrument in order to establish with more precision its minimum distance.

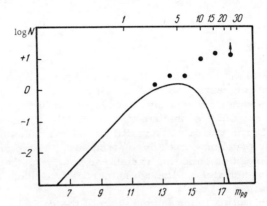

FIGURE 1. Log frequency of the apparent stellar magnitude of RR Lyrae stars (the solid line plots Kukarkin's
results, the points are the Lick observations; the num-
bers at the top of the figure give the distance in kpc).

Many examples can be cited of the use of RR Lyrae stars in the study of the structure and the kinematics of the Galaxy. However, each of these stars was regarded as a member of a large group, and the individual properties of the different members were never investigated. The aim of this book is to study the individual properties of the RR Lyrae stars, without going into their group properties. I am concerned mainly with problems relating to stability and regularity of the oscillations, without directly touching on the physics of the phenomenon.

Fairly extensive observational material has been accumulated by now pertaining to the constancy of the light-variation periods, to periodic and often non-periodic changes in the shape of the light curve, and to some "extraordinary" stars, which I call "strange" variables. The information presented in this book was mainly assembled in two periods of intensive research. In the 1930's, the interest that the RR Lyrae stars aroused among the Soviet astronomers led to the establishment of what we then called the "antalgol service" (using the old name for the RR Lyrae variables). Regular observations continued for nearly five years. Extensive observation series were built up, which gave the maxima of numerous bright variables. Since then the number of known RR Lyrae stars has greatly increased, mainly as a result of the great work done by Harvard and Sonneberg astronomers. It is of course practically impossible to stage regular observations completely covering such an extensive group of objects. In 1958, however, it was decided to resume systematic observations of RR Lyrae stars.

The catalogue of RR Lyrae stars with maximum magnitudes of less than $11^m.5$ is revised annually. The Krakow Observatory is engaged on a calculation and publication of the ephemerides, based on the revised catalogue. This work naturally requires systematic observations. They are being mainly carried out visually in Odessa and Rostov in the USSR and in Sonneberg (East Germany). Photoelectric observations are carried out in Budapest (Hungary) and Tucson (USA). Valuable photoelectric observations of selected stars were also carried out at the Lick Observatory (USA). The steady accumulation of star sky photographs in Odessa, Dushanbe, and Sonneberg further extends the project in the direction of weaker stars and leads to the discovery of highly interesting objects. Comparative analysis of the observational findings provided the basis for this book, which is mainly devoted to the "secular" stability of oscillations of RR Lyrae stars. A number of problems, such as the position of the RR Lyrae stars on the Hertzsprung—Russell diagram, determination of luminosities from a comparison of light curves and radial velocities, etc., is not touched upon in the book. These problems are treated in the original publications of M.S. Frolov and G. Preston.

Chapter 1

GENERAL INFORMATION ON RR LYRAE STARS

RR Lyrae stars are pulsating variables with rapid, short-period light variations. Observations show that RR Lyrae variables may have periods of from 56 min to 40 hrs.

The light-variation curves of RR Lyrae stars have furnished the conventional division of the entire group into three subtypes, designated *a, b,* and *c*. The light curves of variables of subtypes *a* and *b* are not much different from one another. A prolonged minimum is followed by a rapid increase in brightness — the steep rising branch. Approximately in one-tenth of the whole period the brightness reaches its maximum, and then the fairly steep descending branch gradually merges into the next wide minimum. Stars of subtypes *a* and *b* differ mainly in the duration of the rise branch: for stars of subtype *a* the rise branch is steeper than for stars of subtype *b*.

Stars of subtype *c* show an almost symmetric light-variation curve, not unlike a sine curve. The light-variation amplitude of these stars is substantially less than for stars of subtypes *a* and *b*. Their periods are generally about 0.3 — 0.4 days, whereas stars of subtypes *a* and *b* show longer periods.

The light curve is characterized to a certain extent by its skewness, defined as the ratio $\frac{M-m}{P} = \varepsilon$, where M is the time of the maximum, m the time of the preceding minimum, and P is the light-variation period. For stars of subtype *a*, the skewness is close to 0.1, for subtype *b* it is 0.3, and for subtype *c* 0.4 — 0.5.

Some of the objects discovered in recent years do not fit in any of the three subtypes. These are primarily the RR Lyrae stars of ultrashort periods. Their brightness fluctuates in a more complex fashion. The light curve is not stable. The brightness maxima wax and wane. These variations are also periodic, with a longer secondary period.

The RR Lyrae variables are also limited from the direction of long periods. If the light-variation period is close to $1^d.5$, the star is generally classified as a classical cepheid. I have shown, however, that the periods of some (not many) typical RR Lyrae stars are as long as the periods of the classical cepheids, although in all other respects they retain the characteristic features of the RR Lyrae variables. There thus seems to be no sharp boundary between RR Lyrae stars and the classical cepheids.

RR Lyrae stars reveal another characteristic effect which is not observed among the classical cepheids, although an analogous phenomenon is observed for variables of a different type, namely the β Canis Majoris type stars. This is the so-called B l a z h k o e f f e c t .

At the onset of the 20th century, Prof. S.N. Blazhko carefully studied the variables which L.P. Tserasskaya was discovering at that time at the Moscow Observatory. Two of these variables — XZ Cygni and RW Draconi — Blazhko investigated in particular detail. He found a certain instability in the periods of these stars, and the light curve was found to change its shape regularly and periodically. Subsequently this phenomenon was also observed for RR Lyrae itself, the prototype of the entire group.

This effect essentially amounts to the following. If the variable has a constant light-variation period, all the observed maxima can be described by the equation

$$M_E = M_0 + PE,$$

where M_E is the ephemeris of the maximum, M_0 is the time of the initial (zero) maximum, P is the period, and E is an integer, often called the epoch, although a more proper name would be simply the number or the index of the observed maximum. M_0 and P are found from observations, and the ephemeris is then calculated using this relation. The differences between the observed and the calculated maxima are known as the ephemeris corrections and are generally designated by the letters $O-C$, $O-A$, $O-B$, etc. The correction $O-C$ is the result of observation errors, as well as various processes which take place in the star. The study of the residues $O-C$ is of great importance in the theory of variables.

The astronomer generally plots a graph of $O-C$ vs. E, which provides an indication of the constancy of the star's period. For some stars this plot is made up of several linear segments. In this case, the light-variation period changes discontinuously, in jumps, and one can find the magnitude of the jump and the exact time when the period changed. In some cases, on the other hand, the $O-C$ plot is continuous, gradually increasing with time. The curve is sometimes reminiscent of a parabola. This implies that the light-variation period is proportional to time, and a "parabolic formula" with a quadratic term has to be used to calculating the maximum magnitude times:

$$M_E = M_0 + PE + aE^2.$$

Sometimes the variation of the period is irregular and the $O-C$ plot has a strange form.

Blazhko was the first to show that some RR Lyrae stars reveal periodic variation in the $O-C$ residues. These variations are also plotted graphically: the period of the effect Π is then found and finally the observer attempts to reduce the $O-C$ curve to one common period of "inequality". The periods Π are also sometimes seen to vary between wide limits.

The resulting conclusion is that the maximum periodically lags behind or precedes the ephemeris maximum calculated from the linear relation. The deviation may sometimes reach 1 hr for periods of 12 hrs ! The effect can be described by the equation

$$M = M_0 + PE + \alpha \sin\left(\frac{2\pi}{\Pi} E + \varphi\right).$$

Blazhko also discovered another side effect. The shift of the maximum was found to be accompanied by substantial changes in the light curve.

The amplitude of the maximum and the steepness of the rise branch change. Careful studies carried out at the Budapest Observatory have shown that the Blazhko effect is not an exception but a rule for numerous objects.

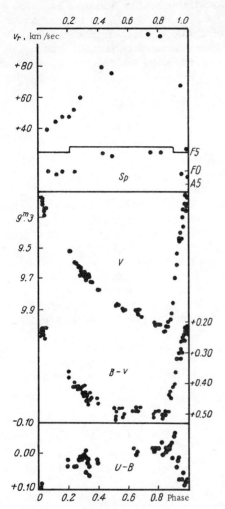

FIGURE 2. Variation curves of radial velocities v_r, spectral type Sp, visual magnitude V, and color indices $B-V, U-B$ of RX Eridani /294/.

Some objects reveal a modification of the Blazhko effect. The position of the maximum hardly changes, but the shape of the light curve and its amplitude change considerably. A typical example is provided by RZ Lyrae, which I studied in some detail /117/.

What is the theoretical background for the Blazhko effect? Some authors tried to interpret it as a superposition of two independent oscillations, as a form of interference. Unfortunately, this effect is much more complex and is not amenable to this simple interpretation.

PHYSICAL PROPERTIES

The most obvious and easily detected property of the RR Lyrae stars is the periodic variation of their magnitude. This light variation is accompanied by fluctuations of spectral type and color index, synchronized with brightness variations: at the maximum the star is "bluer" than in the minimum. Both these phenomena are correlated with changes in radial velocities. The periodic curve of the radial velocity variation is similar on the whole to the light variation curve. Figure 2 shows the variation of the radial velocities, magnitude, and color indices $U-B, B-V$ (U is ultraviolet light, B blue light, V visible light) and changes in the spectral type of RX Eridani according to Kinman /294/. All these effects are applicable to RR Lyrae stars to the same extent as to the classical cepheids.

If the radial velocities are interpreted from the standpoint of the pulsation theory as recession and approach velocities of the reversing layer, we have $v_r = \frac{dR}{dt}$, where R is the radius of the reversing layer. Integration of the radial velocity curve will clearly give a plot of the star's radius as a function of time. Then the times when $v_0 = \frac{v_{r\,max} + v_{r\,min}}{2}$ correspond to the times of maximum compression and expansion of the star. Maximum compression is observed when the velocity goes through v_0, while the differences $v_r - v_0$ change from negative to positive.

We thus have a tool for checking the pulsation theory. Indeed, using the bolometric correction, the visual stellar magnitude m_v can be converted to the bolometric magnitude $m_b = m_v + \mathrm{BC}$. Let R be the radius of the photosphere and T its temperature; the stellar surface is $4\pi R^2$, and the energy radiated every second per unit surface is σT^4. The bolometric luminosity of the star is thus expressed by the relation $L_b = 4\pi\sigma R^2 T^4$. Hence

$$\frac{L_1}{L_2} = \left(\frac{R_1}{R_2}\right)^2 \left(\frac{T_1}{T_2}\right)^4 \quad \text{and} \quad \frac{L_1}{L_2} = 2.512^{m_{b2} - m_{b1}},$$

so that

$$m_{b2} - m_{b1} = 5\log\frac{R_1}{R_2} + 10\log\frac{T_1}{T_2}.$$

If T is identified with the temperature which is determined from the known color index, the radius of the star can be calculated from the above expression. This argument, however, involves certain simplifying assumptions.

1. The color index is assumed to determine the stellar temperature. This is not quite so. The point is that the color index is affected by interstellar absorption which, as we know, causes definite reddening of the stars. This effect is very difficult to take into consideration. Moreover, the color index can be intrinsically altered by absorption in spectral lines (and also by emission lines and bands in the spectra of variables). We see that the temperature entering the theoretical expression is greatly distorted.

2. The bolometric corrections are known for non-variable stars only.

3. The radius of the photosphere is not necessarily equal to the radius of the reversing layer. Our formula gives the radius of the photosphere, whereas the radius of the reversing layer is determined by integration of the radial velocity curve.

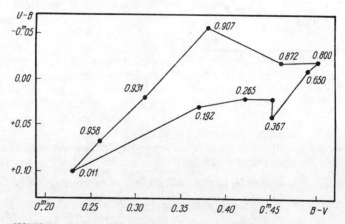

FIGURE 3. The "oval" in the ($U-B$, $B-V$) plane for RX Eridani.

Calculations carried out along these lines thus do not lead to conclusive results at this stage.

Our doubts are further confirmed by the following fact. For those odd cases when the two color indices $U-B$ and $B-V$ are known, we can plot the corresponding diagram in the $(U-B, B-V)$ plane. The instantaneous state of a star is then determined by a point on this diagram. When the magnitude and the color index change, the representing point moves tracing a closed line during one complete cycle. Had there been single-valued correspondence between $U-B$ and $B-V$, all the points would have remained on one straight line. The observations, however, lead to an entirely different result: the diagram represents a complex closed line, a character-istic "oval". This points to substantial changes in the energy distribution in the spectrum during each cycle (Figure 3).

Using Kinman's data for RX Eridani /294/, we plotted the corresponding "oval" in the $(U-B, B-V)$ plane. The numerical data are listed in Table 1 (φ is the age after maximum in fractions of period, n is the number of averaged observations).

TABLE 1. Variation of the color index of RX Eridani

φ	$B-V$	$U-B$	n	φ	$B-V$	$U-B$	n
$0^{P}.011$	$+0.23$	$+0.10$	3	$0^{P}.832$	$+0.50$	-0.02	3
.192	$+$ .37	$+$ .03	2	.872	$+$ .46	$-$ .02	5
.265	$+$ .42	$+$ .02	6	.907	$+$ .38	$-$ .06	2
.313	$+$ .45	$+$ .02	6	.931	$+$ .31	$+$ .02	3
.367	$+$ .45	$+$ .04	5	.958	$+$ .26	$+$ .07	3
.650	$+$ .49	$-$ .01	3	.985	$+$ .22	$+$ .09	3
.774	$+$ .50	$-$ .02	4				

THE BLAZHKO EFFECT

Let us consider in more detail the Blazhko effect. The main feature of this effect is the periodic variation in the shape of the light curve, and not in its period. In RZ Lyrae (Tsesevich) the amplitude of light variation markedly changes, whereas the times of the maxima accurately follow the linear relation $M = M_0 + PE$. RZ Lyrae shows another peculiar feature. Its minimum magnitude hardly changes, whereas the maximum magnitude is subjected to very pronounced variation with a period $\Pi = 116.803$ days ! Figure 4 shows two extreme light curves, and Figure 5 illustrates the variation in the "height" of the maximum of this star. Romanov has recently

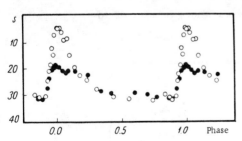

FIGURE 4. Extreme shapes of the light-variation curves of RZ Lyrae:

$\circ - \psi = 0^{\Pi}.557$; $\bullet - \psi = 0^{\Pi}.998$.

shown that the periodic fluctuation of amplitude goes on to this very day, and my original assumption that the period Π also varies (though extremely slowly) has not been confirmed.

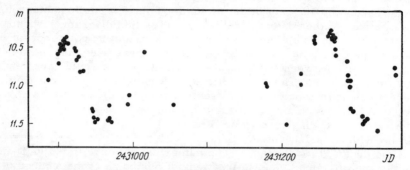

FIGURE 5. Variation of the height of the maximum of RZ Lyrae.

RZ Lyrae, however, is a unique star: no other of this kind has ever been found. In all the rest, the Blazhko effect involves both a change in the height of the maximum and in its "position"; the time of the maximum shows periodic lag relative to the time calculated from the linear relation, or alternatively precedes the predicted figure. To study this effect, we clearly require a very large amount of observations.

Let us consider the available results in some detail.

V 365 Herculis /161/. Period $0^d.613$, maximum calculated from

$$\text{Max hel JD} = 2436047.522 + 0.6130535 \cdot E; \; P^{-1} = 1.6311790. \tag{1}$$

However, if the average light curve is plotted using this relation, the points will show very large scatter (Figure 6). This is so because the shape of the light curve changes with a period $\Pi = 40^d.64$. If in addition to the age of observation used in (1) we also introduce the "inequality" age ψ, calculated from

$$T_0 (O-C) = 2436350.7 + 40^d.64 \cdot n; \; \Pi^{-1} = 0.0246063, \tag{2}$$

we can obtain for any time T two ages φ and ψ to be used in analyzing the entire observation series.

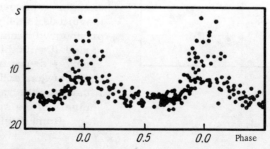

FIGURE 6. Scatter of observations of V 365 Herculis reduced to one period ignoring the Blazhko effect.

6

If all the observations are classified into different groups according to the age ψ, we obtain the average curves of Figure 7 which illustrate the variation in the shape of the light curve, in the height of the maximum, and in the "age" φ. This plot makes it possible to determine the corresponding O−C value. By plotting the variation of O−C and the maximum height M against the age ψ, we obtain a clear picture of the Blazhko effect (Figure 8).

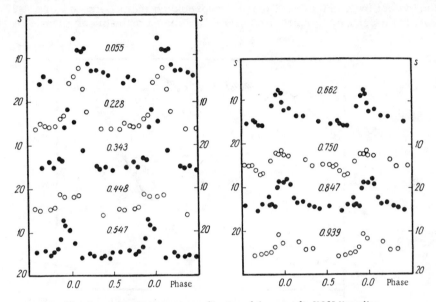

FIGURE 7. The shape of the light curve as a function of the age ψ for V 365 Herculis.

When our analysis had been completed, C. Hoffmeister published his observations of this star, which also revealed a secular variation of the period (see Ch. 4).

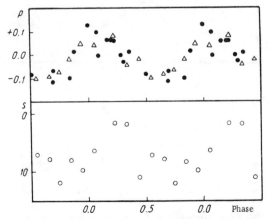

FIGURE 8. The position (top) and the height (bottom) of the maximum vs. the age ψ for V 365 Herculis.

RW Draconis. Basic period $P = 0^d.4429001$. The period is variable, however, and the deviations from the linear equation

$$\text{Max hel JD} = 2426610.242 + 0.442905 \cdot E \qquad (3)$$

have built up to substantial values over a period of 55 years. This star was observed by numerous astronomers. Active observations were first begun by Blazhko and continued by Detre; an extensive observation series was carried out by Klepikova. Figure 9 reproduced from Detre's work /222/ reveals the peculiar variation of the O−C residue for the linear equation (3) (top).

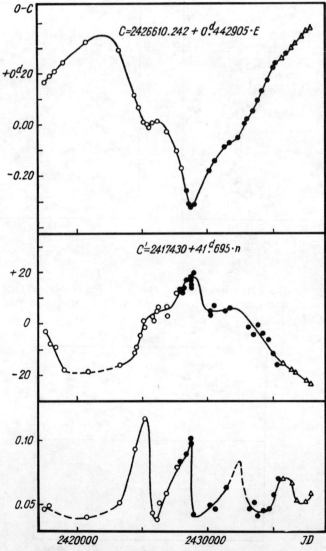

FIGURE 9. The variations of the periods P and Π of RW Draconis:

O visual, ● photographic, △ photoelectric observations.

The abscissa gives Julian dates; the star was regularly observed over a period of 20,000 days. The maximum magnitude is also variable, but its variation is periodic, not secular; the period of the Blazhko effect is $41^d.695$. The deviation of the maximum from (3), corrected for the slow secular variation of the period, changes with the same period of $41^d.695$. The period of the Blazhko effect is also variable. The middle curve in Figure 9 plots the deviations from the equation

$$T_0 = 2417430 + 41.695. \tag{4}$$

We see that in those cases when the maximum precedes the ephemeris (3), the Blazhko effect lags relative to (4).

These two effects — the primary rapid fluctuation of luminosity and the secondary variation of period — are interrelated. The lower curve in Figure 9 shows the amplitudes of the O—C variation curves, observed at different times. We see that the Blazhko effect gradually decays. In the past, the O—C values fluctuated by $0^d.12$, whereas the current fluctuations have dropped to $0^d.06$.

Analogous effects were observed for RR Lyrae. The main light-variation period is $0^d.5668$. The deviations from the linear equation

$$T_0 = 2414856.487 + 0.56683957 \cdot E \tag{5}$$

are quite substantial, and their variation is shown in Figure 10 (top).

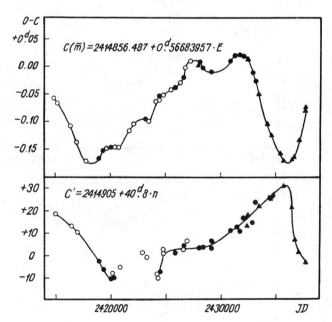

FIGURE 10. The variations of the periods P and Π of RR Lyrae:

○ visual, ● photographic, △ photoelectric observations.

9

The Blazhko effect has a primary period of $\Pi = 40^d.8$. This period is not stable, however. The deviations from the equation

$$T_1 = 2414905 + 40.8 \cdot n$$

are shown in Figure 10 (bottom). The two effects are also closely inter-related.

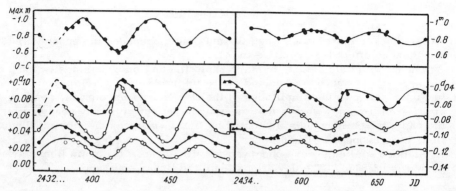

FIGURE 11. Variation of the maximum magnitude in 1947 and 1953 (top). Variation of the O—C residues for the maximum and three different points on the rise branch in 1947 and 1953 (bottom):

O visual, ● photographic, △ photoelectric observations.

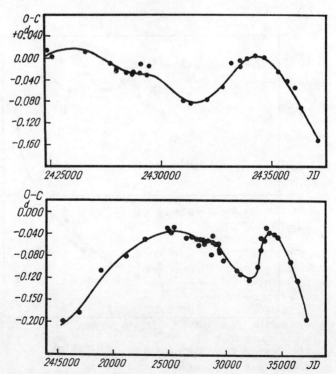

FIGURE 12. Secular variation of the period of AR Herculis (top — deviation of the time of the maximum, bottom — deviation of the midpoint of the rise branch from the linear elements).

10

The "height" of the maximum of the fundamental oscillation varies synchronously. Figure 11 shows that near the times of maximum positive deviations $O-C$ from (5) the oscillations of the maximum magnitude die out. These effects were also observed by Tsesevich /117/ and later by Almar /180/ for AR Herculis. Its period is $0^d.4700228$, the Blazhko effect period $31^d.5489$. The period P undergoes "secular" variations (Figure 12), and the inequality period Π is also variable (Figure 13).

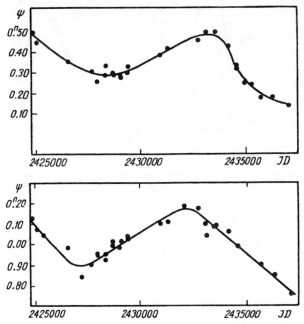

FIGURE 13. Variation of the Blazhko effect of AR Herculis (top — the phase ψ of the highest maximum, bottom — the phase of the midpoint of the rise branch).

Ultrashort-period stars also show the Blazhko effect. Broglia /207/ has established that the famous star RV Arietis, whose light curve is given in Figure 14, is characterized by maxima of variable "height" and minima of variable "depth", these variations being distinctly periodic. This effect differs from the classical Blazhko effect in that the period of the primary light variation is $P = 0^d.09312819$, whereas the Blazhko effect period is $\Pi_1 = 0^d.31633$, i.e., a factor of 4 longer. The $O-C$ residues and the maximum magnitude M are plotted in Figure 15. Note that the plot is greatly improved if we assume a third period $\Pi_2 = 0^d.27554$. Similar side periods were observed for the other stars as well.

The available data on the Blazhko effect are summarized in Table 2.

The tabulated data, which are completely reliable, show that the Blazhko effect is somewhat different for ultrashort-period and ordinary RR Lyrae stars. This difference lies in the ratio Π_1/P. Indeed, for ultrashort-period variables this ratio is between 3.36 and 4.06, whereas for ordinary RR Lyrae stars Π_1/P lies between 54 and 495. On the other hand,

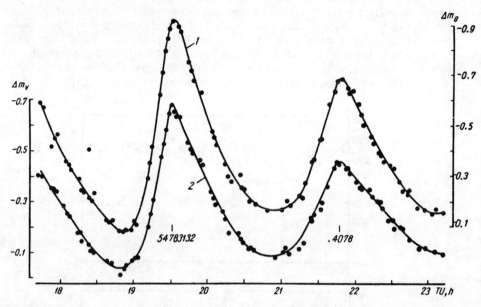

FIGURE 14.   Light variation curves of RV Arietis in blue (1) and yellow (2) light.

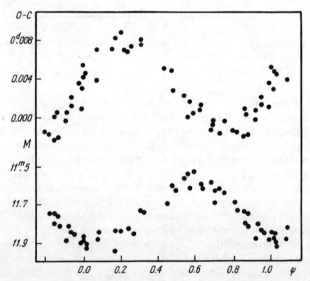

FIGURE 15.   The position and the height of the maximum of RV Arietis vs. the age ψ.

TABLE 2. Blazhko effect data

Star	P	Π_1	Π_2	Secular variation	Π_1/P
SX Phe	$0^d.0550$	$0^d.1928$	—	?	3.50
RV Ari	0.0931	0.3163	—	No	3.40
BP Peg	0.1095	0.3698	—	No?	3.36
AI Vel	0.1116	0.3792	—	?	3.38
VZ Cnc	0.1784	0.7163	—	No	4.02
RR Gem	0.3973	37	—	Yes	93
SW And	0.4423	36.83	—	Yes	83.2
RW Dra	0.4429	41.72	124	Yes	94.2
RV Cap	0.4478	221.86	—	Yes	495.4
XZ Cyg	$0^d.4665$	$57^d.24$	41.7?	Yes	122.7
AR Her	0.4700	31.55	—	Yes	67.1
XZ Dra	0.4765	78	—	No	164
RZ Lyr	0.5112	116.80	—	Yes	228.5
Y Lmi	0.5245	33.4	—	?	63.7
RW Cnc	0.5472	29.9	91.1	Yes	54.6
RR Lyr	0.5668	41.02	61.53	Yes	72.37
DL Her	0.5916	33.6	—	Yes	56.8
V365 Her	0.6131	40.64	—	Yes	66.3
TU Cas	2.1339	5.230	—	?	2.44

the general observed features have much in common for the two groups.
We have mentioned in the preceding that not only the shape and the amplitude of the light curve change, but also the position of the maximum
relative to the linear relation $M = M_0 + P \cdot E$. Given the phase ψ_1 of the
maximum deviation of the maximum and the phase ψ_2 of the maximum amplitude, we calculate the difference $\Delta = \psi_2 - \psi_1$. The numerical results are
as follows (Figures 16 — 19):

Star	Δ
BP Peg	0.30
RV Ari	0.40
RV Cap	0.08
AR Her	0.29
RZ Lyr	—0.25?
DL Her	0.15
V365 Her	0.25

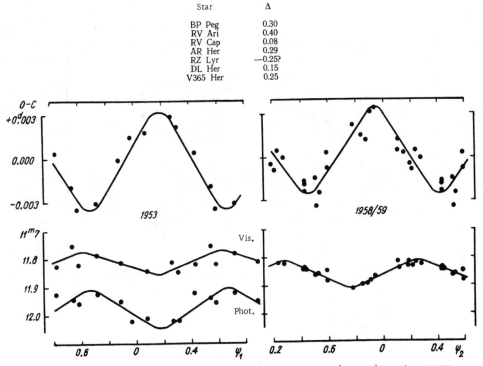

FIGURE 16. Position and height of the maximum of BP Pegasi vs. the age ψ for two observations seasons.

13

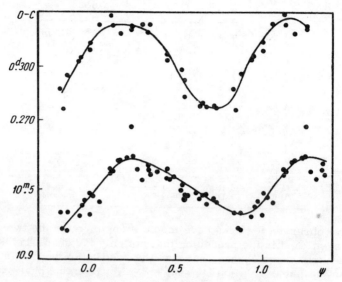

FIGURE 17. The position and the height of the maximum of RV Capricorni vs. the age ψ.

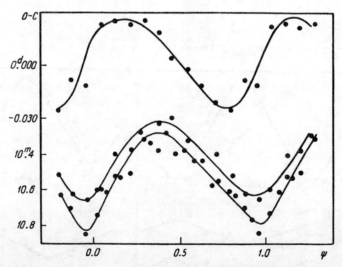

FIGURE 18. The position and the height of the maximum of AR Herculis vs. the age ψ.

14

For all objects, except the unique star RZ Lyrae, the values of Δ are comparable. In ultrashort-period stars no slow variations of the period have yet been discovered, whereas for the ordinary stars these secular variations are a general rule. SW Andromedae is a Population I star, whereas most other stars belong to Population II. The Blazhko effect is thus characteristic of both stellar populations, although it may show different features. The last point requires further study. It would seem, however, that the RR Lyrae stars in globular clusters should reveal a very pronounced Blazhko effect. This statement has been confirmed, but it requires further verification.

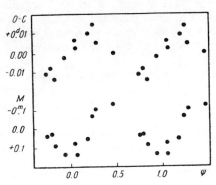

FIGURE 19. The position and the height of the maximum of DL Herculis vs. the age ψ.

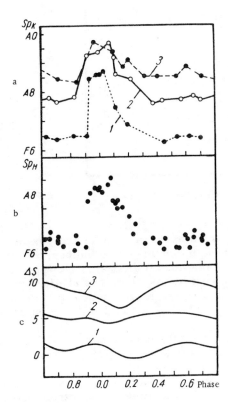

FIGURE 20. SpCaII, SpH, and ΔS as a function of phase:

a) stars of first group, b) second group, c) third group.

SPECTROPHOTOMETRIC OBSERVATIONS

In RR Lyrae stars, as in the classical cepheids, the spectral type changes simultaneously with light variation. One of the indirect indicators of the spectral type is the color index. It has been studied for numerous stars. A much more important task, however, is to study the changes in the spectrum.

Qualitative and even semiquantitative investigations of the spectra of RR Lyrae stars were carried out by Preston /341/ and Alaniya /1, 2/. Using a nebular spectrograph with the Lick Observatory reflector, Preston obtained a great number of spectrograms of RR Lyrae stars with 430 Å/mm dispersion in H_γ. Using these spectrograms he determined the spectral types of the stars from hydrogen lines, SpH, and from the lines of ionized calcium SpCaII. He could thus calculate the index $\Delta S = 10\,[SpH - SpCaII]$. A total of 129 stars were studied, six of which turned out to be stars of a different type. From the light-variation periods, Preston constructed the group average curves for the variation of the spectral type as a function of the light variation phase φ (Figure 20). Preston's main results amount to the following:

1) The SpH curves proved to be the same for all the stars investigated (Figure 20b).

2) The stars were divided into three groups depending on the variation of SpCaII. Stars of the first group showed normal quantity of calcium. In stars of the second group the calcium lines were weaker than what should have been expected from SpH. Finally, in stars of the third group the calcium lines were even fainter.

3) The curves of the spectral type as a function of the phase φ were different for the three groups (Figure 20a).

4) The ΔS curves of the three groups were also different (Figure 20c).

Preston studied the properties of various stars and tried to correlate them with the light-variation period, galactic concentration, and kinematic properties.

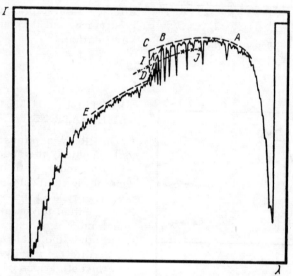

FIGURE 21. Energy distribution in the spectrum of a hot star and Chalonge's parameters.

We are mainly concerned with Preston's principal conclusion. In the following we will use SpCaII and ΔS at the minimum as the indices, remembering that SpH at the minimum is the same for all the stars, being F 5.5.

No detailed quantitative spectrophotometric investigations of RR Lyrae stars have been undertaken. The most comprehensive study in this direction was carried out by Fringant /235/, who studied in detail the brightest star of the group, RR Lyrae itself.

Chalonge and a group of co-workers, with the participation of Fringant, proposed a multidimensional spectral classification of stars proceeding from extensive spectrographic observations (the ordinary spectral classification based on line intensities does not take account of the energy distribution in the stellar spectrum). In this classification, the spectrum of a hot star is characterized by the following quantities: the absolute gradient in the blue-violet spectrum Φ_B, the absolute gradient in the ultraviolet spectrum Φ_{UV}, the Balmer jump D (Figure 21, segment CD); the Balmer jump wavelength λ_1. λ_1 is determined by drawing a continuous envelope of the microphotometer tracing (Figure 21). This gives two branches ABC and DE. In the region below the Balmer jump the envelope follows the line BKD. The segment CD is bisected to give the point I. A point K satisfying the equality $\log \frac{I}{i} = \frac{D}{2}$ is then found on the section BD of the envelope. The wavelength corresponding to the point K is λ_1.

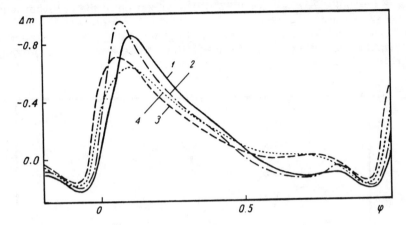

FIGURE 22. Light variation curves of RR Lyrae.

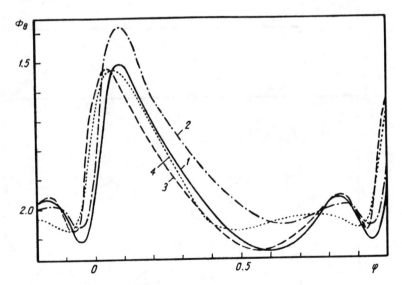

FIGURE 23. Curves of the gradient Φ_B.

17

Fringant studied 250 spectrograms of RR Lyrae and found all the necessary parameters. The observations used in her work were obtained with the 82-in. McDonald Observatory telescope (USA) and the 80-cm Haute-Provence Observatory telescope (France) in 1949—1957. This considerable length of time prevented normal reduction of measurements, since the two periods P and Π are variable. Correcting for their variation, Fringant studied the distribution of energy in the spectrum as a function of φ and ψ. Since relatively few observations were available, she had to plot averaged curves for four values of ψ (Figure 22, curve 1 $\psi = 0.19$, curve 2 $\psi = 0.42$, curve 3 $\psi = 0.80$, curve 4 $\psi = 0.92$). The variation of the relevant parameters is shown in Figures 22—26. The effects which occur in the spectrum of RR Lyrae thus show great variety. In general they can be attributed to the formation of a shock wave in the stellar atmosphere, which propagates toward the periphery.

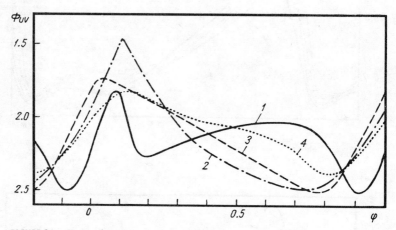

FIGURE 24. Curves of the gradient Φ_{UV}.

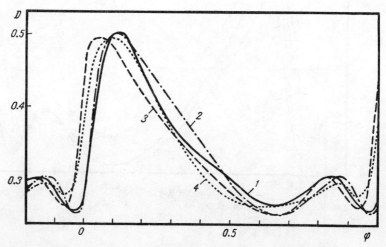

FIGURE 25. Curves of the Balmer jump D.

No conclusive interpretation of the effects studied by Fringant is available, and no mathematical theory has been developed for the propagation of shock waves in the extended shell of a pulsating star. There was, however, an attempt to construct a theoretical curve of RR Lyrae from observations.

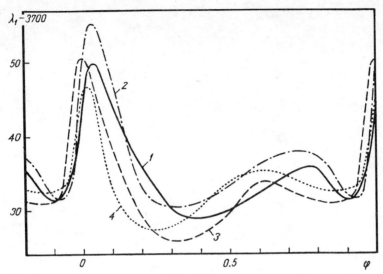

FIGURE 26. Curves of the wavelength λ_1-3700.

Following Bergé, Fringant assumed that the absolute stellar magnitude was a function of the parameters $\Phi_b, D,$ and $\lambda_1,$ so that

$$\Delta M = \frac{\partial M}{\partial \Phi_b} \Delta \Phi_b + \frac{\partial M}{\partial D} \Delta D + \frac{\partial M}{\partial \lambda_1} \Delta \lambda_1$$

The least squares method gave $\frac{\partial M}{\partial \Phi_b} = -5.2$; $\frac{\partial M}{\partial D} = -15.3$; $\frac{\partial M}{\partial \lambda_1} = -0.053$. Hence $M = C -5.2 \cdot \Phi_b - 15.3\, D - 0.053\,(\lambda_1 - 3700)$. Figure 27 shows the theoretical light curves and the actual results of observations. The fit between the

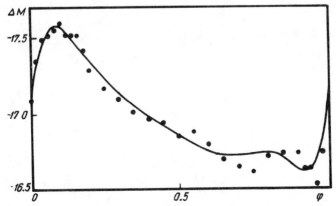

FIGURE 27. Theoretical and observed light curves of RR Lyrae.

19

theoretical curve and the observational data is purely formal however. Using Fringant's data for other inequality phases and retaining the same coefficients in the formula, we fail to obtain satisfactory fit between the "theoretical" curve and observation results. The solution of the problem is apparently much more complex.

Chapter 2

STABLE AND NON-STABLE OSCILLATIONS

It has been reliably established that the galactic population of RR Lyrae stars is not homogeneous. Some authors tried to distinguish between Population I and Population II stars among these variables.

In general, differentiation of homogeneous groups meets with a fundamental difficulty: most of RR Lyrae variables are high-magnitude stars, so that the spectral methods, which usually provide the most comprehensive information on stars, are inapplicable. The astronomers therefore have to resort to simpler criteria. One of these is the light-variation period P, another is the shape of the light curve. A third hitherto unused criterion can be introduced, namely the constancy of the period or some variation parameter if the period is variable. Since a short-period star goes through numerous pulsation cycles during a relatively short time, the length of the period and its possible variation are established fast and with high reliability.

NON-STABLE STARS OF THE GROUP $P \sim 0^d.36$

Among the RR Lyrae stars there is a group of objects with periods close to $0^d.36$.

IV Cygni

According to Mannino /314/, this RR Lyrae variable has a light curve with a large amplitude and a period of about $0^d.3$, which is very rare for variables of the subtype RRa. Mannino derived the following equation:

$$\text{Max hel JD} = 2433118.488 + 0.334345 \cdot E; \ P^{-1} = 2.990923. \qquad (6)$$

Tsesevich estimated the magnitude of this star from photographs taken with the seven-camera astrograph of the Odessa Observatory. Comparison stars used: $p = 0^s.7$, $q = 7^s.1$, $r = 13^s.5$ (Figure 28). Reduction to one period using (6) has shown that the star indeed has the above period and a RRa-type light curve. The average maximum

$$\text{Max hel JD} = 2436131.219; \ O - A = - 0^d.052 \text{ for } E = 9011.$$

In a later study Wenzel /365/ compared his estimates with the data of Mannino and Wachmann /363/ and showed that the period of this star was variable (Figure 29).

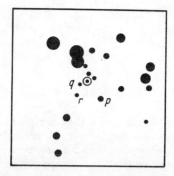

FIGURE 28. Comparison stars for IV Cygni.

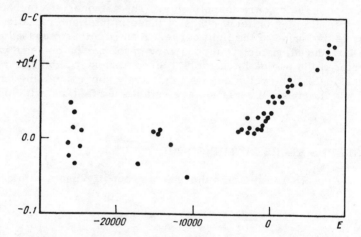

FIGURE 29. O−C plot for IV Cygni.

A complete list of the maxima is given in Table 3; the Odessa photographic observations are arranged in Table 4.

The O−B residues were calculated for the formula

$$\text{Max hel JD} = 2434122.519 + 0.33433895 \cdot E. \tag{7}$$

The period apparently changed abruptly and all the maxima can be fitted with the equation

$$\text{Max hel JD} = 2434122.482 + 0.33432871 \cdot E \tag{8}$$

for $-26324 < E < -3000$ and by (7) for $E > 0$. The period thus increased abruptly by $0^d.00001024$. The O−C residues were calculated from (8).

5443

22

TABLE 3. List of maxima

Source	Max hel JD	E	O − B	O − C
Wenzel /365/	2425321.660	−26324	$+0^d.281$	$+0^d.047$
	5326.622	−26309	.226	−.006
	5329.616	−26300	.211	−.021
	5470.406	−25879	.244	+.017
	5476.378	−25861	.198	−.029
	5495.499	−25804	.262	+.035
	5837.467	−24781	.201	−.015
	5851.532	−24739	.224	+.008
	8396.397	−17127	.101	−.037
	9086.495	−15063	.124	+.006
	9166.396	−14824	.118	+.003
	9230.254	−14633	.117	+.004
	9843.406	−12799	.091	−.003
	30541.432	−10711	.017	−.057
Wachmann /363/	2775.478	− 4029	.011	+.006
	2914.225	− 3614	.007	+.007
Mannino /314/	3118.52	− 3003	.02	+.027
	3119.50	− 3000	.00	+.004
	3126.53	− 2979	+.01	+.013
	3443.47	− 2031	−.01	+.010
	3467.56	− 1959	+.01	+.028
	3472.56	− 1944	.00	−.013
	3479.58	− 1923	−.01	+.012
	3562.50	− 1675	.00	+.019
	3601.28	− 1559	+.00	+.016
Wenzel /365/	3773.478	− 1044	+.009	+.035
	3896.497	− 676	−.009	+.021
Wachmann /363/	3926.259	− 587	−.003	+.028
Wenzel /365/	3947.343	− 524	+.018	+.049
Wachmann /363/	4122.518	0	−.001	+.036
Wenzel /365/	4132.569	+ 30	+.020	+.057
	4215.476	+ 278	+.011	+.051
	4538.445	+ 1244	+.008	+.058
	4658.458	+ 1603	−.006	+.047
	4811.600	+ 2061	+.008	+.067
	4812.598	+ 2064	+.003	+.062
	4990.475	+ 2596	+.012	+.076
Wachmann /363/	5044.295	+ 2757	+.003	+.069
	5398.351	+ 3816	+.001	+.071
Tsesevich	6131.219	+ 6008	−.008	+.090
Wenzel /365/	6611.671	+ 7445	−.001	+.112
	6659.48	+ 7588	.00	+.112
	6660.48	+ 7591	−.01	+.109
	6661.48	+ 7594	−.009	+.106
	6691.59	+ 7684	+.01	+.126
Wachmann /363/	6847.386	+ 8150	+.005	+.125

TABLE 4. Odessa observations (Tsesevich)

JD hel	s	JD hel	s	JD hel	s
243...		243...		243...	
6049.469	15.5	6105.267	[13.1	6406.459	9.7
6050.478	13.5	6128.215	2.0	6407.459	9.7
6051.459	[13.5*	6131.223	− 1.0	6423.404	6.1
6053.444	12.6	6138.212	12.1	6424.407	[7.1
6069.389	0.0	6371.486	[14.1	6426.411	13.5
6070.398	− 1.0	6372.478	17.5	6428.417	15.5
6071.393	− 2.0	6379.594	[15.1	6429.449	4.4
6075.376	− 1.0	6381.471	12.6	6453.367	[7.1
6076.378	− 1.0	6395.477	16.5	6454.382	[7.1
6079.358	18.5	6396.451	11.9	6461.360	14.5
6080.369	[13.5	6397.442	12.6	6483.262	14.5
6081.373	[13.5	6398.459	14.5	6485.260	[10.1
6082.368	15.5	6399.443	10.8	6487.281	6.2
6101.310	[12.1	6400.468	10.1	6489.290	1.0
6102.284	8.1	6404.456	7.1	6490.263	[7.1

* Stars fainter than the magnitude indicated are preceded by a bracket.

23

AQ Lyrae

A much studied variable. Considerable variation of its period was noted by Zverev and Tsesevich /39/, although the complete picture of the phenomenon could not be reconstructed. It is only recently that Lange and Migach have finally unravelled the most peculiar behavior of this star. Lange compiled a summary table of the maxima (Table 5), which showed that the star's period changed several times:

Years	E	Period
1928—1938	0—10200	$0^d.3571721$
1938—1941	10200—13600	0.3571285
1941—1951	13600—24100	0.3571611
1951—1956	24100—29100	0.3571861
1956—1963	29100—36200	0.3571427

Table 5 gives E and $O-A$ (Figure 30) calculated for the equation

$$\text{Max hel JD} = 2425301.623 + 0.3571611 \cdot E. \qquad (9)$$

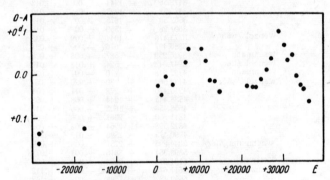

FIGURE 3. O—A plot for AQ Lyrae.

TABLE 5. List of maxima

Source	Max hel JD	E	O — A
Kurochkin	2415143.445	—28441	$-0^d.159$
	5290.266	—28030	— .131
	8923.320	—17858	— .122
Wenzel /366/	25301.610	0	— .013
	5331.636	+ 84	+ .011
	5380.528	+ 221	— .028
	5404.421	+ 228	— .064
Hoffmeister	5439.437	+ 386	— .050
/259, 260/	5443.415	+ 397	— .001
	5450.550	+ 417	— .009
	5453.426	+ 425	+ .010
	5454.458	+ 428	— .030
	5470.532	+ 473	— .028
	5478.369	+ 495	— .049
	5478.395	+ 495	- .023
	5483.382	+ 509	— .036
	5494.439	+ 540	— .051
	5498.371	+ 551	— .048
	5499.469	+ 554	— .021

TABLE 5 (continued)

Source	Max hel JD	E	O — A
	2425502.321	+ 562	—0^d.027
	5507.316	+ 576	— .032
	5512.329	+ 590	— .019
Wenzel /366/	5512.347	+ 590	— .001
Hoffmeister	5532.310	+ 646	— .039
/259, 260/	5536.229	+ 657	— .049
	5758.419	+ 1139	— .010
	5759.434	+ 1142	— .067
Wenzel /366/	5759.442	+ 1142	— .059
	6094.519	+ 2220	— .002
	6508.449	+ 3379	— .021
Zverev /39/	7661.422:	+ 6607	+ .036:
	7662.490:	+ 6610	+ .032:
Tsesevich /39/	7663.196	+ 6612	+ .024
Zverev /39/	7676.414	+ 6649	+ .027
	7691.415	+ 6691	+ .027
	7696.415	+ 6705	+ .027
	7907.510	+ 7296	+ .040
	7917.519	+ 7324	+ .048
	7945.41:	+ 7402	+ .081
	7950.41	+ 7416	+ .080
	7951.45	+ 7419	+ .049
	7955.379	+ 7430	+ .049
Wenzel /366/	9056.517	+10513	+ .059
	9080.443	+10580	+ .056
Kurochkin	9366.530	+11381	+ .049
Wenzel /366/	9376.538	+11409	+ .064
	9425.449	+11546	+ .044
	9436.467	+11577	— .010
	9514.350	+11795	+ .012
	9750.42	+12456	— .002
	9752.55!	+12462	— .015
	9759.33	+12481	— .021
	9851.492	+ 12739	— .006
	9872. 55!	+ 12798	— .019
	9876. 48!	+ 12809	— .020
	9877. 55	+ 12812	— .019
	30234.372	+ 13811	— .003
	0254.342	+ 13867	— .034
	0531.502	+ 14643	— .031
Kurochkin	0607.199	+ 14855	— .052
	0617.215	+ 14883	— .037
Wenzel /366/	2827.339	+ 21071	— .026
	3005.583	+ 21570	— .005
Filatov	3126.283	+ 21908	— .025
Fridel /139/	3137.338	+ 21939	— .042
Wenzel /366/	3153.431	+ 21984	— .022
Filatov	3175.198	+ 22045	— .041
Wenzel /366/	3444.494	+ 22799	— .045
Tsesevich /39/	3458.423	+ 22838	— .045
Fridel /139/	3484.501	+ 22911	— .040
Tsesevich /39/	3502.364	+ 22961	— .035
Migach	3517.350	+ 23003	— .050
	3531.323	+ 23042	— .006
Tsesevich /39/	3536.288	+ 23056	— .041
Tsesevich and Migach	3541.330	+ 23070	— .020
Tsesevich /39/	3555.234	+ 23109	— .025
	3837.394	+ 23899	— .022
Filatov	3842.365	+ 23913	— .050
Fridel /139/	3855.595	+ 23950	— .036
Filatov	3856.313	+ 23952	— .033
Tsesevich and Migach	3862.401:	+ 23969	— .016
	3863.449	+ 23972	— .040
	3871.331	+ 23994	— .015
	3920.254	+ 24131	— .024
Filatov	4181.352	+ 24862	— .010
Migach	4182.417	+ 24865	— .016
Fridel /139/	4212.770	+ 24950	— .022
Filatov	4269.213	+ 25108	— .010
Fridel /139/	4595.673	+ 26022	+ .004
Migach	4605.324	+ 26049	+ .012
	4606.393	+ 26052	+ .009
Kurochkin	4630.328	+ 26119	+ .014
Fridel /139/	5019.654	+ 27209	+ .035
	5566.513	+ 28740	+ .080
Filatov	5692.261	+ 29092	+ .103
Migach	5693.325	+ 29095	+ .100

25

TABLE 5 (continued)

Source	Max hel JD	E	O — A
	243 5722.282	+ 29176	+0^d.127
Filatov	5756.172	+ 29271	+ .086
Yudkina /166/	6010.454	+ 29983	+ .070
Filatov	6043.324	+ 30075	+ .081
	6047.238	+ 30086	+ .066
Tsesevich	6050.451	+ 30095	+ .065
Migach	6050.448	+ 30095	+ .062
Yudkina /166/	6065.450	+ 30137	+ .063
	6069.415:	+ 30148	+ .099:
Migach	6069.361	+ 30148	+ .045
Yudkina /166/	6074.384	+ 30162	+ .070
	6078.309	+ 30173	+ .064
	6079.371	+ 30176	+ .055
	6083.307	+ 30187	+ .062
	6102.232:	+ 30240	+ .057:
Fridel /139/	6114.021	+ 30273	+ .061
Filatov	6379.362	+ 31016	+ .027
Migach	6400.445	+ 31075	+ .041
Filatov	6432.243:	+ 31164	+ .048:
Migach	6463.287	+ 31251	+ .022
	6482.224	+ 31304	+ .030
Filatov	6782.232	+ 32144	+ .043
Migach	7193.298	+ 33295	— .004
	7519.363	+ 34208	— .027
Lange	7824.374	+ 35062	— .031
	7850.431	+ 35135	— .047
	7854.360	+ 35146	— .047
	7855.429	+ 35149	— .050
	7869.358	+ 35188	— .050
Migach	7869.383	+ 35188	— .025
	7881.529	+ 35222	— .024
Lange	7884.360	+ 35230	— .049
Migach	7886.518	+ 35236	— .034
	7900.452	+ 35275	— .029
Lange	7903.296	+ 35283	— .042
Migach	7903.304	+ 35283	— .036
	7904.375	+ 35286	— .035
Lange	7908.298:	+ 35297	— .040:
Migach	7909.388	+ 35300	— .022
	7910.441	+ 35303	— .040
Lange	7962.224	+ 35448	— .046:
	8210.425::	+ 36143	— .072:
	8229.372::	+ 36196	— .054:

Table 6 gives the averaged data of Table 5. It was prepared to simplify the construction of diagrams.

TABLE 6. Averaged data of Table 5

$\bar{E}$	$\overline{O—A}$	n	$\bar{E}$	$\overline{O—A}$	n	$\bar{E}$	$\overline{O—A}$	n
— 28441	—0^d.159	1	+ 11542	+0^d.032	5	+ 29075	+0^d.099	5
— 28030	— .131	1	+ 12651	— .015	7	+ 30141	+ .066	14
— 17858	— .122	1	+ 13839	— .018	2	+ 31162	+ .034	5
+ 447	— .027	22	+ 14794	— .040	3	+ 32144	+ .043	1
+ 1141	— .045	3	+ 21753	— .027	6	+ 33295	— .004	1
+ 2220	— .002	1	+ 22977	— .034	9	+ 34208	— .027	1
+ 3379	— .021	1	+ 23972	— .030	8	+ 35237	— .038	17
+ 6646	+ .029	6	+ 24946	— .014	4	+ 36170	— .063	2
+ 7381	+ .058	6	+ 26060	+ .010	4			
+ 10546	+ .058	2	+ 27209	+ .035	1			

V 759 Cygni

First studied by Tsesevich from Odessa photographs. A tentative equation was derived for the calculation of the O—A residues:

$$\text{Max hel } JD = 2\,433\,447.437 + 0.360\,021 \cdot E. \tag{10}$$

Visual observations confirmed the validity of this result and showed that the period started increasing. This led to a radical revision of the old Moscow, Simeise, and Harvard photographs. The present summary is thus based on all the material on hand.

Detailed visual observations reveal the Blazhko effect; this is evident from the following three records of visual maxima:

E	Max hel JD	O—A
5319	2435362.427	$+0^d.038$
5322	5363.502	$+ .034$
5413	5396.276	$+ .045$

The secular variation of the period was investigated using seasonal light curves constructed from the preliminary elements (10). These curves give the maxima listed in Table 7.

TABLE 7.   List of maxima

Source	Max hel JD	E	O — B
Tsesevich (Mos.)	2414430.37	— 58144	$+0^d.006$
	5163.36	— 56108	$+ .006$
	5257.32	— 55847	$+ .002$
	7554.20:	— 49467	— .012
Tsesevich (Har.)	25765.743	— 26659	— .045
	6653.549	— 24193	— .036
	7565.814	— 21659	— .048
	8011.874	— 20420	— .047
Tsesevich (Sim.)	8041.399	— 20338	— .043
Tsesevich (Mos.)	8427.36	— 19266	— .018
Tsesevich (Har.)	8736.590	— 18407	— .040
	9550.589	— 16146	— .035
	9847.584	— 15321	— .052
	30608.631	— 13207	— .076
	1323.672	— 11221	— .024
	1705.657	— 10160	— .015
	2832.485	— 7030	— .033
Tsesevich (Odes.)	3515.477	— 5133	$+ .011$
Tsesevich (vis.)	3754.518	— 4469	$+ .002$
Tsesevich (Har.)	3860.720	— 4174	$+ .000$
Tsesevich (Odes.)	3862.529	— 4169	$+ .009$
Tsesevich (Mos.)	4127.493	— 3433	$+ .002$
Tsesevich (Odes.)	4215.329	— 3189	— .006
	4567.427	— 2211	— .002
	4950.522	— 1147	$+ .037$
Tsesevich (Mos.)	4980.406	— 1064	$+ .040$
	5335.43	— 78	$+ .089$
Tsesevich (vis.)	5363.510	0	$+ .088$
Tsesevich (Mos.)	5724.28	$+ 1002$	$+ .123$
Tsesevich (vis.)	6461.322	$+ 3049$	$+ .215$
Tsesevich (Odes.)	7172.413	$+ 5024$	$+ .277$
	7522.398	$+ 5996$	$+ .327$
	7912.354	$+ 7079$	$+ .388$

The O — B residues were calculated from the equation

$$\text{Max hel JD} = 2435363.422 + 0.36001476 \cdot E. \tag{11}$$

Table 7 and the O — B diagram (Figure 31) lead to the following conclusions.

1. The period is subject to marked changes.

2. The changes were reliably traced for the last 30,000 light-variation cycles.

27

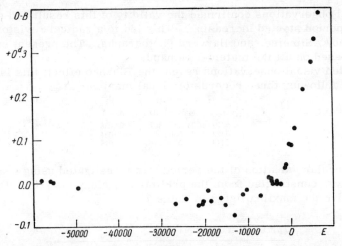

FIGURE 31. O—B diagram for V 759 Cygni.

Comparison stars (Figure 32): photographic scale, $u = -9^s.2$; $s = 0^s.0$; $k = +1^s.3$; $a = +6^s.7$; $b = +13^s.8$; $e = +15^s.6$; $c = +17^s.3$; $d = +17^s.9$; $f = 22^s.3$; visual scale $u = 0^s.0$; $a = 10^s.1$; $d = 16^s.9$.

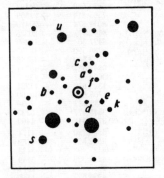

FIGURE 32. Comparison stars for V 759 Cygni.

Visual observations were used to derive the average light variation curve (Table 8). The phases were calculated from (10).

TABLE 8. Average visual light variation curve

φ	s	n	φ	s	n	φ	s	n
$0^p.018$	18.9	1	$0^p.180$	12.5	4	$0^p.585$	17.4	3
.062	14.0	3	.214	14.0	3	.653	17.6	3
.082	10.3	3	.292	14.6	3	.759	18.0	3
.103	9.3	4	.396	18.9	1	.856	18.7	4
.114	8.3	2	.503	17.3	4	.933	19.9	2
.127	8.7	3						

The photographic average light curve (Table 9) was obtained by an artificial technique. The phases were calculated from (10) or (11), according as the given observation fell in one of the two corresponding intervals.

The photographic and visual curves are shown in Figure 33 (1 and 2, respectively). Tsesevich's visual observations are summarized in Table 10, the photographic observations in Table 11—14.

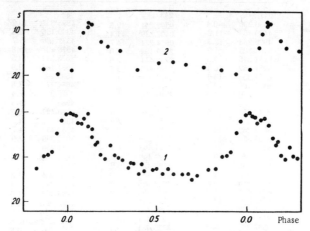

FIGURE 33. Light curves of V 759 Cygni.

TABLE 9. Average photographic light curve

φ	s	n	φ	s	n	φ	s	n
0ᵖ.008	0.2	5	0ᵖ.260	9.7	5	0ᵖ.589	13.5	5
.028	0.7	5	.278	10.0	5	.630	13.7	5
.042	0.9	5	.309	10.6	5	.674	13.5	5
.056	2.4	5	.335	12.1	5	.690	14.9	5
.072	2.9	5	.351	11.2	5	.722	14.0	5
.091	1.5	5	.367	11.4	5	.786	12.8	5
.117	3.1	5	.391	13.5	5	.825	12.5	5
.133	5.9	5	.409	11.6	5	.860	9.9	5
.152	7.4	5	.428	13.2	5	.888	9.8	4
.164	6.7	5	.472	12.8	5	.903	9.0	4
.182	9.5	5	.491	12.4	5	.933	4.7	5
.211	10.2	5	.527	13.5	5	.967	2.0	4
.233	7.5	5	.558	12.4	5	.990	0.4	4

TABLE 10. Visual observations (Tsesevich)

JD hel	s	JD hel	s	JD hel	s
243...		243...		243...	
3754.452	14.1	5363.422	18.9	5396.311	13.5
.474	14.1	.428	18.9	.327	13.5
.518	9.0	.441	19.9	6455.360	14.2
5362.417	12.8	.448	19.9	.375	13.8
.418	10.1	.487	14.6	.399	14.6
.423	7.4	.494	13.5	.422	14.6
.427	9.0	.499	7.9	6461.343	5.6
.433	8.7	.505	8.1	.352	5.6
.448	11.5	.509	8.3	.360	6.7
.462	14.2	.514	10.1	.373	7.8
.491	15.8	5364.261	13.9	.396	11.8
.251	18.9	.266	14.4	6462.283	15.0
.274	17.9	.305	14.6	.305	14,6
.301	16.9	5367.253	16.5	.322	14.2
.308	16.9	.287	18.4	.424	6.7
.323	16.9	.299	19.9	.442	8.5
.341	16.9	5396.237	18.9	.472	10.1
.351	16.1	.253	13.9	6466.285	13.9
.366	17.9	.269	12.7	.347	5.5
.386	19.9	.272	8.3	.394	8.7
.393	16.1	.278	7.4	6482.290	11.8
.401	18.9	.291	11.5		
5363.416	17.9	.298	13.2		

29

TABLE 11. Moscow observations (Tsesevich)

JD hel	s	JD hel	s	JD hel	s
241...		243...		243...	
4430.396	—0.4	0592.341	12.3	4295.337	10.2
4499.403	11.2	0607.339	11.4	4330.156	2.6
4926.283	12.3	0614.274	14.7	4331.166	15.7
5163.395	1.2	2169.122	8.0	4518.407	15.8
5251.346	9.5	3766.444	2.6	4678.211	13.5
5257.352	0.4	3775.480	3.8	4681.168	9.9
6351.333	13.1	3917.234	—0.4	4683.203	9.9
6462.166	12.3	3917.291	2.6	4684.178	7.8
7554.170	3.2	3949.316	1.2	4978.340	7.4
242...		3951.184	9.9	4980.354	14.8
8043.351	10.9	3951.228	12.3	4980.422	—0.7
8081.263	11.2	3952.250	3.8	4982.341	13.5
8082.252	12.3	3953.195	9.5	4982.452	14.7
8408.354	12.3	3953.219	3.0	5012.399	12.9
8426.367	11.2	4121.469	8.7	5041.258	0.4
8427.343	0.4	4127.492	—1.3	5074.169	10.5
8433.256	12.3	4128.427	10.2	5335.434	—1.3
8653.485	1.2	4131.463	—1.1	5337.398	11.3
8750.385	9.5	4223.356	7.2	5347.445	11.3
8759.457	11.2	4223.383	7.2	5361.281	11.3
8776.370	13.5	4224.345	0.0	5361.306	11.4
8781.369	12.3	4224.447	11.3	5363.276	11.2
243...		4250.403	10.8	5365.273	13.5
0587.292	18.2	4250.431	9.9	5724.281	0.8
0589.316	13.5	4281.301	8.0		

TABLE 12. Old Odessa observations (Tsesevich)

JD hel	s	JD hel	s	JD hel	s
243...		243...		243...	
3442.406	2.9	3568.340	14.8	3862.528	—1.7
3447.428	1.7	3570.312	14.8	3864.544	14.8
3457.346	12.2	3572.342	2.6	3867.493	11.8
3458.362	15.8	3598.221	5.4	3870.497	6.2
3468.347	2.9	3762.490	10.6	3895.525	12.9
3470.341	13.8	3763.537	2.7	3896.480	12.0
3482.362	—0.9	3767.504	0.0	3897.508	6.7
3484.526	1.0	3768.441	13.8:	3900.495	12.8
3499.320	5.6	3768.532	5.6	3912.391	9.1
3506.319	15.8	3769.470	13.8	3913.457	13.8
3508.320	7.7	3770.482	11.8	3918.385	9.1
3509.321	10.2	3771.530	11.7	3920.373	15.8
3510.313	15.8	3776.520	5.0	3947.380	15.8:
3511.440	5.7	3800.501	15.8	3949.325	1.9
3515.486	2.7	3801.499	12.2	3950.350	2.2
3516.507	13.8:	3802.476	9.7	3951.340	12.0
3517.478	15.8	3811.453	6.7	4158.474	2.9
3519.488	3.8	3812.471	2.2	4161.467	10.8
3530.388	13.8	3823.437	13.8	4162.456	5.2
3531.451	14.8	3824.417	6.7	4177.432	10.8
3532.400	2.9	3824.440	11.1	4178.404	15.8
3536.423	10.2	3829.413	1.9	4181.458	12.0
3538.441	14.8	3831.402	13.8:	4196.368	12.8
3539.414	12.6	3831.404	13.8:	4208.348	13.8
3540.406	11.4	3837.349	4.2	4209.346	4.5
3541.414	3.1	3839.368	14.8	4211.358	4.0
3542.430	10.8	3839.367	14.8	4214.341	10.8
3544.394	12.0	3842.386	4.5	4215.343	3.1
3545.410	10.8	3850.330	5.6	4217.506	2.9
3559.407	2.9	3850.510	12.8	4220.353	2.2
3564.354	15.8	3852.324	13.8	4241.564	6.7:
3565.373	14.8	3854.343	12.9	4243.443	4.0
3566.362	15.8	3859.401	10.5	4250.420	13.8

30

TABLE 12 (continued)

JD hel	s	JD hel	s	JD hel	s
243...		243...		243...	
4279.419	2.0	4566.498	11.0	4627.346	15.8
4513.486	3.8	4567.427	0.0	4638.422	9.7
4538.475	11.8	4570.341	2.2	4650.386	6.7:
4542.422	13.8	4570.480	13.8:	4654.370	11.4
4543.433	13.8	4576.468	4.5	4655.390	10.2
4548.440	9.1	4594.339	11.8	4950.522	—1.7
4565.392	11.0	4625.454	6.7		

TABLE 13. New Odessa observations (Tsesevich)

JD hel	s	JD hel	s	JD hel	s
243...		243...		243...	
6756.485	16.4	7170.391	13.8	7526.409	4.8
6757.499	15.0	7172.403	—2.2	7544.350	—3.4
6760.506	—1.8	7173.357	15.0	7545.348	16.8
6765.501	14.3	.409	15.0	7549.314	16.7
6766.510	14.4	7174.380	14.4	7555.294	14.2
6780.412	13.8	7175.378	9.2	7557.294	10.1
6789.422	11.3	7176.375	—3.9	7848.490	14.3
6790.512	12.2	.403	—1.8	7872.387	—1.5
6791.471	—1.2	7189.340	—2.2	7873.474	—2.6
6792.480	17.0	7192.361	15.8	7881.449	5.8
6806.345	10.9	7195.334	15.6	7882.425	14.8
6807.360	4.5	7196.310	14.4	7884.442	14.9
6809.380	15.3	7197.327	6.3	7886.409	3.1
6817.327	14.3	7198.333	—1.8	7900.384	18.8
6834.285	12.5	7204.340	16.8	7902.362	11.7
6837.327	14.5	7461.497	14.8	7903.374	0.0
6862.242	15.9	7472.505	14.5	7904.391	17.8
6863.236	14.8	7473.472	—1.7	7906.387	14.3
6868.244	6.7	7493.455	15.2	7908.355	10.6
7135.512	15.1	7496.449	3.5	7909.363	15.6
7136.495	9.4	7497.484	16.7	7910.346	14.6
7137.486	—1.2	7501.464	15.0	7911.340	5.5
7144.450	14.7	7518.394	14.6	7912.343	—1.3
7145.455	2.4	7519.422	16.4	7913.366	16.3
7161.397	14.7	7521.407	5.8	7939.295	16.8
7162.415	10.1	7522.380	—1.0	7959.217	4.4
7165.396	12.2	7523.409	16.4	7962.231	15.2
7166.384	10.5	7524.422	14.7	7963.251	14.4
7167.389	2.2	7525.409	14.8	7964.230	2.7
7169.420	16.8				

TABLE 14. Simeise observations (Tsesevich)

JD hel	s	JD hel	s	JD hel	s
242...		242...		242...	
7960.400	2.4	8025.455	16.3	8078.341	14.3
	—1.7		14.9	8092.231	10.6
7962.445	15.0	8037.296	17.6		7.9
	17.8		15.5	8123.221	9.1
8014.422	3.0	8039.415	13.8		9.5
	4.0		13.8	8318.474	16.2
8017.319	1.3	8041.422	—0.2		15.9
	4.9		1.3	8330.347	17.0
8019.310	15.9	8043.386	17.0		17.6
	16.9		15.9	8345.422	9.8
8021.309	5.3	8044.357	9.0		13.8
	9.1	8078.341	15.9	8360.350	4.4
					4.9

31

SS Tauri

First investigated by C. Hoffmeister in 1919. Very little observed. The observations were repeated between 1958 and 1962. The period showed substantial variation. To establish the nature of these variations, V. Satyvaldyev, at Tsesevich's request, estimated the magnitude of the star on the photographs of the Dushanbe collection.

In the compilation of the table of average maxima, several different times determined by various observers were averaged (Table 15).

TABLE 15. Individual maxima

Source	Max hel JD	Source	Max hel JD
Hoffmeister /254/	2420894.250	Hoffmeister /254/	2421276.410
	0895.320		1282.319
	0897.190	Kukarkin /48/	6631.332
	1116.595		6634.294
	1138.424		6650.210
	1222.390		6657.240
	1250.512		6687.232
	1251.260		6707.197
	1252.377	Lange	37943.398
	1256.415		7960.418
	1259.395		7961.523
	1269.374		7974.487

Tsesevich's visual observations gave two maxima from the average curves. Several maxima on seasonal curves were obtained from the observations of Satyvaldyev and Tsesevich's Harvard observations. Alaniya's maximum /1/ was added, and the complete list is given in Table 16.

TABLE 16. List of maxima

Source	Max hel JD	E	O — C
Tsesevich (Har.)	2416052.748	— 59223	+0^d.018
	6171.836:	— 58901	— .008:
	6803.709	— 57193	+ .044
	7257.595	— 55966	+ .040
	8120.671	— 53633	+ .096
	9801.551	— 49089	+ .066
	20843.612	— 46272	+ .066
Hoffmeister /254/	0895.340	— 46132	+ .006
Tsesevich (Har.)	1240.538	— 45199	+ .070
Hoffmeister /254/	1250.513	— 45172	+ .057
Tsesevich (Har.)	1522.757	— 44436	+ .041
Hoffmeister /254/	1567.190	— 44316	+ .084
Tsesevich (Har.)	2985.829	— 40481	+ .085
	4024.897	— 37672	+ .051
	4765.859	— 35669	+ .067
	5157.603	— 34610	+ .067
	5586.727	— 33450	+ .085
	5946.643	— 32477	+ .070
Tsesevich (vis.)	6305.47:	— 31507	+ .076:
Tsesevich (Har.)	6335.438	— 31426	+ .081
Kukarkin /48/	6669.080	— 30524	+ .056
Tsesevich (Har.)	7364.533	— 28644	+ .062
	7753.688	— 27592	+ .063
	8121.768	— 26597	+ .074
	8456.910	— 25691	+ .070
	.8843.856	— 24645	+ .081

TABLE 16 (continued)

Source	Max hel JD	E	O — C
Tsesevich (Har.)	2429245.603	— 23559	$+0^d.096$
	9550.405	— 22735	+ .085
	9918.492	— 21740	+ .103
	30280.639:	— 20761	+ .100
	0667.581	— 19715	+ .107
	1028.603	— 18739	+ .089
	1392.603	— 17755	+ .089
	1760.698	— 16760	+ .115
	2118.794	— 15796	+ .130
	2470.949	— 14840	+ .122
	2857.884	— 13794	+ .122
Satyvaldyev	2866.365	— 13771	+ .095
	3190.413	— 12895	+ .104
Tsesevich (Har.)	3212.625	— 12835	+ .111
Satyvaldyev	3604.371	— 11776	+ .113
	4285.380	— 9935	+ .102
	5055.528	— 7853	+ .080
Alaniya /1/	5070.342	— 7813	+ .097
Satyvaldyev	5480.568	— 6704	+ .083
	5810.525	— 5812	+ .073
Tsesevich (vis.)	6495.585	— 3960	+ .044
Satyvaldyev	6780.440:	— 3190	+ .061:
Tsesevich (vis.)	7198.452	— 2060	+ .065
Satyvaldyev	7313.503	— 1749	+ .072
Lange	7960.419	0	.000
	8294.477	+ 903	+ .022
	8652.568	+ 1871	+ .031

The O—C residues were calculated for

$$\text{Max hel JD} = 2437960.419 + 0.3699186 \cdot E \tag{12}$$

and are plotted in Figure 34. The period undergoes gradual, but irregular fluctuations. Deviations from the linear ephemeris reach one third of its value.

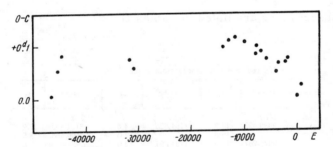

FIGURE 34. O—C diagram for SS Tauri.

Comparison stars (Figure 35):

	s_1	s_2	s_3
a	—3.2	0.0	0.0
k	0.0	—	—
b	5.8	6.9	10.1
c	17.1	12.4	15.8

The new visual observations (Tsesevich) were applied to calculate the average light curve (Table 17) from the equation

$$\text{Max hel JD} = 2436495.585 + 0.36993 \cdot E; \quad P^{-1} = 2.703214. \tag{13}$$

33

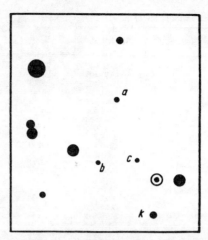

FIGURE 35. Comparison stars for SS Tauri.

TABLE 17. Average light curve

φ	s_1	n	φ	s_1	n	φ	s_1	n
$0^p.012$	5.3	3	$0^p.316$	14.9	3	$0^p.909$	14.6	4
.043	5.3	5	.462	15.0	4	.926	12.6	4
.072	7.4	5	.636	15.4	5	.942	9.7	4
.114	10.6	5	.711	15.9	5	.956	6.3	4
.206	14.3	5	.796	15.5	5	.976	4.6	4
.267	14.8	3	.871	15.5	5	.991	5.1	3

The observations are listed in Tables 18—21.

TABLE 18. Simeise observations (Tsesevich)

JD hel	s_3	JD hel	s_3	JD hel	s_3
241...		242...		242...	
8986.493	17.8	1194.325	4.6	8101.480	12.2
	15.8		2.0	8466.488	8.5
9021.287	16.8	1194.394	12.0		8.5
242...			11.5	243...	
0069.458	12.0	4447.520	16.8	3210.502	11.5
	12.4		16.8		11.2
0073.502	9.0	4449.459	18.8	3216.508	19.3
	9.1	5186.464	5.9	3238.350	18.8
0811.376	13.4	7751.513	7.1		18.3
	13.2		11.7		
		8101.480	12.7		

34

TABLE 19. Dushanbe observations (Satyvaldyev)

JD hel	s₂	JD hel	s₂	JD hel	s₂
242...		243...		243...	
9990.176	[12.4	4384.160	—2.0	5811.180	[12.4
243...		5024.445	0.0	5894.117	4.6
2858.257	5.0	5037.329	[12.4	6102.412	2.3
2866.380	3.9	5038.339	[12.4	6214.135	— 2.0
2881.219	6.9	5095.188	[14.4	6226.181	[12.4
2883.230	[12.4	5098.167	12.4	6244.131	[13.4
2941.158	5.2	5099.221	3.4	6248.180	12.4
2947.153	[12.4	5395.351	9.1	6249.137	[12.4
3157.444	11.3	5477.216	7.8	6255.122	[14.4
3179.436	[12.4	5477.240	1.2	6545.172	0.0
3181.442	6.9	5478.165	[12.4	6576.102	[12.4
3183.483	9.1	5478.187	[12.4	6580.181	8.3
3185.480	[12.4	5480.151	[12.4	6598.133	4.6
3190.387	3.4	5480.174	4.9	6983.212	2.0
3190.415	1.7	5484.151	[12.4	7194.377	6.9
3192.434	[12.4	5484.171	12.4	7194.395	3.0
3244.213	[14.4	5503.145	4.1	7206.446	9.3
3274.183	9.1	5503.168	5.5	7290.192	0.0
3275.244	10.6	5504.140	[14.4	7313.121	— 2.0
3328.118	8.0	5504.165	[12.4	7316.120	2.6
3595.208	8.7	5510.144	3.9	7320.150	3.9
3597.293	9.6	5510.167	2.3	7320.170	— 3.0
3600.215	[14.4	5514.143	[12.4	7340.106	10.0
3604.385	0.0	5514.165	12.4	7340.125	3.4
3646.099	5.8	5540.128	0.0	7340.130	— 1.0
3888.443	1.0	5542.137	8.7	7608.280	0.0
3891.424	—4.0	5755.365	4.6	7637.292	2.3
4271.415	8.7	5780.212	2.3	8027.563	[12.4
4281.412	4.9	5783.223	5.2	8046.261	0.0
4283.370	[12.4	5809.247	[12.4	8052.123	6.9
4283.420	12.4	5810.243	10.6	8053.246	[12.4
4285.446	6.1	5811.134	[9.9		

TABLE 20. Old visual observations (Tsesevich)

JD hel	s₁	JD hel	s₁	JD hel	s₁
242...		242...		242...	
5097.438	15.4	6305.490	4.6	6324.400	10.8
5099.420	15.4	6322.234	12.0	6325.193	12.9
.440	16.5	.276	13.8	.219	14.6
6011.332	9.0	.312	14.8	6325.252	14.6
6025.258	11.2	.373	15.3	.280	14.6
6298.233	13.8	.415	15.8	6326.314	13.3
.329	14.3	.441	14.8	6336.368	7.8:
6305.476	3.5				

TABLE 21. New visual observations (Tsesevich)

JD hel	s₁	JD hel	s₁	JD hel	s₁
243...		243...		243...	
6495.384	13.2	6495.564	5.2	7196.479	16.1
.397	13.7	.567	4.9	.493	16.6
.435	13.7	.571	4.4	.509	16.7
.453	15.3	.577	3.9	.522	16.3
.463	15.3	.581	3.9	.538	16.5
.470	15.3	.587	5.8	.568	15.8
.490	15.5	.592	4.0	.576	11.2
.498	15.5	.595	3.6	.581	9.4
.509	14.6	.606	4.1	.586	6.2
.516	14.8	.609	4.2	.592	3.4
.533	14.8	.615	6.9	.598	5.9
.547	13.3	.624	6.8	7197.409	14.8
.554	12.0	7196.464	15.6	.434	15.2
.556	10.9	.472	16.1	.461	15.4

35

TABLE 21 (continued)

JD hel	s_1	JD hel	s_1	JD hel	s_1
243...		243...		243...	
7197.497	16.3	7198.436	9.8	7198.521	14.5
.521	16.7	.441	5.9	.538	14.2
7198.396	16.7	.446	5.0	.550	14.2
.399	16.3	.450	5.5	.563	15.0
.412	16.3	.457	6.1	.573	14.3
.416	15.4	.464	5.5	7204.386	5.3
.419	15.0	.474	6.8	.392	7.8
.423	14.6	.481	9.9	.399	9.4
.426	13.7	.486	10.6	.407	10.3
.429	12.2	.498	13.0	.442	13.5
.432	11.9	.507	12.5	.459	14.3
				.477	15.0

The four stars considered in this section lead to the conclusion that RRa stars with periods between $0^d.334$ and $0^d.370$ are non-stable. The reasons for the sudden changes in their structure are not clear. It would be wrong to suggest, however, that all stars with periods between these limits are non-stable. We will now describe three more stars which have the same light curves and yet perfectly stable periods. These are V 559 Scorpii, AA Aquilae, and SS Cancri.

STABLE STARS OF THE GROUP $P \sim 0^d.36$

V 559 Scorpii

Tsesevich investigated this variable star using the Simeise collection and recent photographs taken with the 16-in. astrograph of the Shternberg Astronomical Institute. "Seasonal" light curves were constructed (although the "seasons" in the Simeise collection cover very prolonged intervals of time) using the equation

$$\text{Max hel JD} = 2437078.468 + 0.3506408 \cdot E. \qquad (14)$$

The resulting maxima are listed in Table 22.

TABLE 22. List of maxima

Time	Max hel JD	E	O — C
2419920—2421369	2420634.458	— 46897	—0^d.003
2423193—2429071	5388.460	— 33339	+ .009
2430137—2434131	33034.512	— 11533	— .017
2436702—2436751	6720.483	— 1021	+ .015
2437047—2437137	7078.468	0	— .004

The O — C deviations were calculated for the final equation

$$\text{Max hel JD} = 2\,437\,078.472 + 0.35\,064\,101 \cdot E; \ P^{-1} = 2.85191969. \qquad (15)$$

36

Although these deviations are small, the contribution from the Blazhko effect is nevertheless noticeable. Comparison stars (Figure 36): $a = 0^s.0$; $b = 7^s.3$; $c = 14^s.2$. Expression (15) was applied to derive the average light curves (Tables 23, 24). Tsesevich's observations are summarized in Tables 25, 26.

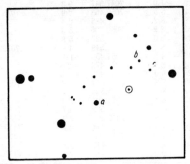

FIGURE 36. Comparison stars for V 559 Scorpii.

TABLE 23. Average light curve (Simeise)

φ	s	n	φ	s	n	φ	s	n
$0^P.008$	0.4	2	$0^P.444$	9.5	5	$0^P.880$	11.3	4
.121	3.4	4	.531	10.6	5	.845	5.3	4
.180	4.0	4	.664	10.0	4	.934	3.4	2
.325	7.2	5	.716	11.2	2	.994	—1.0	2

TABLE 24. Average light curve (Moscow)

φ	s	n	φ	s	n	φ	s	n
$0^P.066$	—1.9	4	$0^P.564$	+8.3	5	$0^P.845$	+8.4	4
.117	—0.1	4	.623	+7.8	5	.903	+5.2	5
.228	+2.6	5	.688	+8.6	5	.946	+0.6	3
.334	+5.6	5	.746	+7.7	5	.985	—0.8	3
.463	+7.1	5	.796	+8.5	5			

TABLE 25. Simeise observations (Tsesevich)

JD hel	s	JD hel	s	JD hel	s
241...		242...		242...	
9920.450	10.4	5382.442	5.5	7961.433	4.0
242...		5385.465	7.3	.483	4.9
0630.492	11.7	5388.489	—1.0	7963.449	10.1
0634.464	—0.3	5735.482	11.9	9071.384	8.7
0636.497	11.0	5744.497	5.3	243...	
0992.529	4.8	5745.471	5.2	0137.478	1.6
1008.429	11.2	5747.457	7.3	0161.379	1.2
1369.461	2.4	5764.404	3.4	2686.452	7.0
3193.376	6.8	5770.348	4.9	3031.494	6.8
3932.356	5.7	5771.349	—2.5	3034.505	2.9
4296.366	11.9	6486.488	11.8	3802.362	11.6
4312.376	4.9	6853.359	8.3	3803.364	6.4
4644.497	6.4	7572.44:	11.8	3823.343	11.6
4652.454	0.6	7931.490	10.7	3824.333	11.9
4668.349	10.1	7961.383	14.2	4131.496	9.6

TABLE 26. Moscow observations (Tsesevich)

JD hel	s	JD hel	s	JD hel	s
243...		243...		243...	
6702.496	8.1	6749.442	7.3	7077.379	—1.0
6703.497	8.2	6750.422	8.2	.401	—3.0
6720.444	5.3	6751.424	2.4	.468	—1.5
6720.489	—1.0	7047.547	9.3	.513	2.1
6721.432	5.5	7050.463	2.9	7078.371	8.3
.479	8.2	.485	2.4	.415	8.4
6722.382	8.2	7050.506	5.3	.459	1.2
.429	9.0	.527	5.8	.504	—3.0
.475	8.2	.549	4.9	7079.345	5.3
6724.419	4.2	7051.450	—4.0	.392	7.3
6728.381	8.5	.472	—2.0	.440	8.4
.428	9.0	.493	—1.5	.486	8.3
.475	7.3	.514	1.2	7080.347	7.3
6729.442	8.2	.536	—1.0	.402	8.3
.482	8.4	7053.491	8.3	.450	9.3
6733.439	—1.5	.522	8.7	7087.434	7.3
6734.432	8.7	7072.397	8.2	7088.444	8.2
.478	8.7	.444	8.7	7116.364	—2.5
6747.336	9.0	7074.393	7.3	7135.304	—2.5
.380	8.2	.439	5.8	7136.300	5.2
.418	9.0	.485	7.3	7137.303	7.3

AA Aquilae

Observed repeatedly by Tsesevich in 1951 and 1956. Same comparison stars as in /157/: $a = 0^s.0$; $x = 3^s.3$; $b = 10^s.6$; $d = 15^s.1$; $f = 22^s.5$. The light curves near the maximum were constructed. The phases were calculated from

$$\text{Max hel JD} = 2424347.3966 + 0.36178688 \cdot E; \quad P^{-1} = 2.7640596. \quad (16)$$

The curves are listed in Tables 27, 28.

TABLE 27. The light curve from 1951 observations

φ	s	n	φ	s	n	φ	s	n
0P.942	19.4	2	0P.008	5.2	2	0P.170	11.9	2
.972	12.0	1	.044	7.3	2	.285	12.5	5
.980	9.2	3	.085	8.9	3	$\varphi_{Max} = 0^P.007$		
.995	6.4	2	.134	10.1	3			

TABLE 28. The light curve from 1956 observations

φ	s	n	φ	s	n	φ	s	n
0P.915	20.2	3	0P.020	5.2	3	0P.141	9.2	2
.965	13.3	1	.034	5.1	2	.186	11.5	2
.975	13.2	2	.052	5.0	3	.278	13.6	1
.994	8.0	2	.080	6.8	5	$\varphi_{Max} = 0^P.028$		
.009	5.6	3	.112	8.5	4			

This leads to two maxima:

Max hel JD	E	O — A
2433884.459	26361	—^{d}.002
5721.257	31438	+ .004

38

The O−A residues were calculated from (16). The visual observations of Tsesevich are listed in Table 29.

TABLE 29. Visual observations (Tsesevich)

JD hel	s	JD hel	s	JD hel	s
243...		243...		243...	
3884.435	19.5	3889.306	16.6	5721.317	10.6
.446	12.0	.330	18.3	.348	13.6
.448	9.5	3890.293	9.2	.372	17.6
449	9.2	.303	11.5	.266	19.7
.450	8.9	.310	12.3	.297	19.7
.453	7.2	.344	13.0	.320	13.3
.456	5.7	3892.351	19.5	.323	13.2
.458	5.7	.359	19.3	.332	7.7
.460	4.6	3897.291	14.5	.339	5.5
.471	7.0	.292	17.1	.345	5.9
.473	7.7	.298	18.9	5722.351	5.9
.483	9.8	.311	17.6	.360	6.5
3885.283	13.2	.387	19.3	.375	8.1
.284	11.6	.407	19.1	.397	12.4
.285	12.1	.432	19.3	5726.315	6.1
.287	12.8	5713.295	5.3	5742.267	8.9
.295	12.7	.308	3.9	.284	9.1
.299	13.4	.320	7.4	5743.290	20.3
.315	13.9	.330	8.8	.308	13.3
.344	17.6	5719.283	19.3	.313	8.3
.398	17.3	.313	19.2	.318	5.6
.431	18.3	.351	20.3	.321	5.0
3886.297	9.2	.362	20.0	.325	4.8
.299	7.6	.408	20.7	.329	4.3
.311	9.2	5720.301	19.6	.334	5.3
.316	12.0	.321	19.4	.344	6.8
3887.301	19.3	.337	19.2	.349	6.3
.302	20.0	.370	19.2	.365	9.3
.308	19.5	5721.272	7.1	5748.278	20.3
.330	19.3	.289	8.1	5749.269	17.6

A comprehensive series of visual observations has been recently obtained by Lange. The maxima determined from his and previous observations are listed in Table 30.

TABLE 30. List of maxima (Lange)

Max hel JD	E	O−C	Max hel JD	E	O−C
2426923.336	7120	$+0^d.017$	2437884.3781	37417	$+0^d.0018$
6927.300	7131	+.001	7885.4610	37420	−.0006
6928.388	7134	+.004	7900.2911	37461	−.0038
6974.335	7261	+.004	7901.3794	37464	−.0009
6977.230	7270	+.005	7908.2536	37483	−.0008
37494.3769	36339	+.0069	7925.2543	37530	−.0039
7519.3371	36408	+.0028	7955.2830	37613	−.0035
7520.4215	36411	+.0028	7959.2668	37624	+.0006
7523.3155	36419	+.0025	7963.2436	37635	−.0022
7524.4010	36422	+.0027	8223.3717	38354	+.0011
7528.3824	36433	+.0044	8228.4346	38368	−.0010
7851.4560	37326	+.0023	8231.3344:	38376	+.0045:
7855.4365	37337	+.0022	8233.5009	38382	+.0003
7871.3528	37381	+.0008	8236.3957	38390	+.0008
7872.4371	37384	−.0002	8240.3769	38401	+.0023

Lange and Kanishchev used these and previously published lists to derive the summary data of Table 31.

39

The O−C residues calculated from (16) show that the period remained constant over 50,000 epochs !

TABLE 31. Normal and average maxima

Source	Max hel JD	E	O−C	Number of obs.	Number of max	Remarks
Photographic	2419625.352	−13052	−0d.002	—	—	—
Bolin /201, 202/	23739.245	− 1681	+.012	—	—	—
	3748.281	− 1656	+.004	—	—	—
Belyavskii /190, 191/	4028.6633	− 881	+.0010	—	7	Average
Bolin, Ivanov /278, 279/	4073.1632	− 758	+.0011	—	7	Average
Ivanov, Vorontsov-Vel'yaminov /362/	4350.6513	+ 9	−.0014	—	11	Ditto
Lange	4384.349	+ 102	−.0020	66	—	Normal
Tsesevich /157/	4384.657	+ 103	−.0040	63	--	Ditto
Ivanov, Blazhko, Vorontsov-Vel'-yaminov /278, 279/	4399.8548	+ 145	−.0009	—	11	Average
Tsesevich /157/	4752.600	+ 1120	+.0020	277	—	Normal
Lange	4764.179	+ 1152	+.0020	210	—	Ditto
Tsesevich /157/	5094.490	+ 2065	+.0030	313	—	Normal
Lange	6149.752	+ 4982	+.0010	40	—	Ditto
Tsesevich /157/	6567.320	+ 6136	−.0010	51	—	"
Lange	6942.686	+ 7173	+.011	86	—	"
Lange	2426951.9038	+ 7199	+.0035	—	4	Average
Tsesevich /157/	6964.202	+ 7233	+.0010	72	—	Normal
Florya /119/	6971.438	+ 7253	+.0010	200	—	Ditto
Lange	7315.500	+ 8204	+.0040	180	—	"
	7636.649	+ 9092	+.0020	39	—	"
	8011.811	+10129	.0000	161	—	"
Balazs /185/	8242.0305	+10765	−.0019	—	7	Average
Tsesevich	30623.314	+17347	.0000	45	—	Normal
Lange	0978.949	+18330	−.0010	102	—	Ditto
Tsesevich	3884.459	+26361	−.002	46	—	"
Born, Sofronievitsch /203/	3900.3782	+26405	−.0010	—	10	Average
Alaniya /1/, Born, Sofronievitsch /204/	4633.3594	+28431	+.0052	—	4	Ditto
Tsesevich	5721.257	+31438	+.004	45	—	Normal
Tsarevskii /141/	6100.4070	+32486	+.0018	105	—	Ditto
Lange	7519.3380	+36408	+.0037	90	6	Average
	7894.1439	+37444	−.0006	161	13	Ditto
	8232.0545	+38378	+.0010	68	6	"

SS Cancri

Variability established by Wolf /370/. Observed by Vogt /361/, Nijland /325/, Graff /242/, Baade /183/, Alaniya /1/, and Tsesevich. Tsesevich estimated the magnitude from Simeise photographs, Chumak from the photographs of the Odessa collection. A summary of all the available maxima is listed in Table 32. The O−A residues were calculated from

$$\text{Max hel JD} = 2\,423\,078.594 + 0.3\,673\,374 \cdot E. \qquad (17)$$

TABLE 32. List of maxima

Source	Max hel JD	E	O − A
Heidelberg	2415403.42:	−20894	−0^d.026:
	8307.64:	−12988	+ .024:
	8329.33:	−12929	+ .041:
	20516.42:	− 6975	+ .004:
	2363.38:	− 1947	− .008:
	2370.39:	− 1928	+ .023:
	2373.30:	− 1920	− .006:
Heidelberg	2422373.70:	− 1919	+^{d}.026
Vogt /361/	2374.412:	− 1917	+ .004
Heidelberg	2378.44:	− 1906	− .009:
Vogt /361/	2379.539	− 1903	− .012
Heidelberg	2380.283	− 1901	− .003
	2405.24:	− 1833	− .025:
	2406.34:	− 1830	− .027:
Vogt /361/	2407.460	− 1827	− .009
Heidelberg	2438.317	− 1743	− .008
	2454.477	− 1699	− .011
	2456.327	− 1694	+ .003
Nijland /325/	2778.476	− 817	− .003
	3034.874	− 119	− .007
Graff /242/	3078.594	0	.000
Baade /183/	3079.320	+ 2	− .009
	3079.692	+ 3	− .004
	3080.424	+ 5	− .007
Graff /242/	3081.532	+ 8	− .001
	3090.346	+ 32	− .003
	3113.487	+ 95	− .004
	3137.361	+ 160	− .007
Nijland /325/	3260.786	+ 496	− .007
	3416.545	+ 920	+ .001
	3449.968	+ 1011	− .004
	3511.310	+ 1178	− .007
	3731.345	+ 1777	− .008
	3804.082	+ 1975	− .003
	3846.324	+ 2090	− .005
	3885.631	+ 2197	− .003
	4144.237	+ 2901	− .003
Tsesevich	6322.546	+ 8831	− .005
	6325.481	+ 8839	− .007
	6336.500	+ 8869	− .009
	7425.295:	+11833	− .002:
	7443.285	+11882	− .012
	7458.347	+11923	− .011
	7460.183	+11928	− .012
Alaniya /1/	34393.322	+30802	+ .001
Tsesevich	6201.362	+35724	+ .007
Chumak	7363.263	+38887	+ .020

We see from Table 32 that the period changed most insignificantly over 60,000 cycles. Rejecting old, unreliable maxima, we averaged the accurate data to obtain Table 33, which shows that the period increased. The O − A values can apparently be fitted by a quadratic relation. However, the gap in observations makes the determination of this relation most uncertain. Two formulas are therefore proposed:

$$\text{Max hel JD} = 2423078.589 + 0.36733699 \cdot E; \quad E < 12000; \tag{18}$$
$$\text{Max hel JD} = 2434393.321 + 0.3673395(E - 30802);$$
$$30000 < E < 40000. \tag{19}$$

Table 33 also lists the differences $\delta = \overline{O - A} - (\overline{O - A})_{\text{calc}}$.

Comparison stars (Figure 37): Tsesevich's visual observations (s_1), $r = 0^s.0$; $s = 11^s.2$; $m = 20^s.4$; $p = 23^s.8$; $t = 29^s.0$; photographic observations (s_2), $a = 0^s.0$; $r = 8^s.2$; $s = 15^s.3$; $p = 26^s.7$. Chumak's photographic scale (s_3), $a = 0^s.0$; $r = 15^s.5$; $s = 26^s.0$; $p = 47^s.0$. Chumak's observations are given in Table 34, Tsesevich's in Tables 35, 36.

41

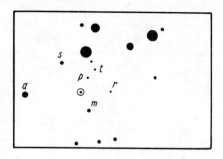

FIGURE 37. Comparison stars for SS Cancri.

TABLE 33. Averaged data of Table 32

$\bar{E}$	$\overline{O-A}$	n	$\overline{O-A}_{calc}$	δ	$\bar{E}$	$\overline{O-A}$	n	$\overline{O-A}_{calc}$	δ
−1882	−d.006	3	−d.004	−d.002	1885	−d.005	4	−d.006	+d.001
−1713	− .005	3	− .004	− .001	2901	− .003	1	− .006	+ .003
− 817	− .003	1	− .005	+ .002	8846	− .007	3	− .009	+ .002
− 119	− .007	1	− .005	− .002	11911	− .012	3	− .010	− .002
+ 114	− .004	8	− .005	+ .001	30802	+ .001	1	—	—
+ 496	− .007	1	− .005	+ .002	35724	+ .007	1	—	—
+1036	− .003	3	− .005	+ .002	38887	+ .020	1	—	—

TABLE 34. Observations of O. Chumak

JD hel	s_3	JD hel	s_3	JD hel	s_3
243...		243...		243...	
6933.480	− 2.0	6996.404	21.2	7350.420	1.0
6954.350	45.0	7015.316	35.0	7363.261	0.0
6959.441	30.0	7016.296	21.8	7377.335	24.0
6964.294	19.7	7017.302	19.3	7378.312	1.6
6993.372	6.5	7020.309	26.0	7719.314	30.0
6995.390	32.0	7317.409	26.0	8086.348	23.5

TABLE 35. Simeise observations (Tsesevich)

JD hel	s_2	JD hel	s_2	JD hel	s_2
242...		242...		242...	
4178.310	22.9	5998.346	16.4	8930.364	5.9
	20.7		17.1		11.0
4531.520	14.3	6004.326	23.8	243...	
4557.409	20.3	6011.339	23.2	0025.421	13.2
4558.270	5.1		22.3		13.5
4559.228	20.3	7102.428	2.7	2921.353	4.5
	17.3		1.2		4.1
4889.468	23.5	8169.606	11.9	3306.443	11.8
	21.4		11.5		12.6
5983.557	7.0	8171.581	21.5	3688.449	23.6
	6.9		21.5		25.6

42

TABLE 36. Visual observations (Tsesevich)

JD hel	s₁	JD hel	s₁	JD hel	s₁
242...		242...		242...	
5985.471	15.4	6336.503	8.1	7458.303	21.5
.482	14.5	.510	8.2	.321	19.8
.503	18.2	.517	8.9	.327	17.0
6322.469	19.6	.557	10.0	.331	14.7
.481	21.3	.630	12.8	.336	9.4
.500	20.7	6337.351	17.1	.342	7.2
.515	21.0	.450	14.4	.347	5.1
.523	19.6	7424.386	27.3	.352	6.1
.543	6.3	.394	28.1	.356	6.3
.545	4.9	.407	28.0	.362	7.8
.548	4.8	.418	27.3	.367	9.0
.550	4.9	.432	28.0	.374	8.8
.554	6.0	.446	26.9	.390	9.7
.559	7.1	.453	26.4	.395	10.2
.564	6.4	.463	26.9	.410	11.0
.568	7.0	.492	29.0	7459.358	19.2
.582	8.5	7425.253	25.8	.434	13.8
.589	9.5	.266	26.3	.440	8.2
.649	17.5	.292	5.6	.444	5.6
6324.367	9.7	.300	5.6	.457	5.2
.374	8.7	.306	6.1	.488	6.8
.379	8.9	.319	7.5	7460.166	15.3
.409	9.4	.328	9.5	.173	10.2
.446	10.5	.344	11.2	.181	5.6
.483	13.3	.362	13.2	.199	7.5
.514	17.0	.396	14.9	.208	8.1
6325.392	21.0	.452	16.9	.217	8.6
.398	19.9	.495	18.6	.227	10.4
.410	21.3	7443.146	17.3	.244	10.2
.421	21.6	.158	17.0	.270	11.2
.433	22.2	.169	18.2	.301	15.3
.444	21.6	.210	20.4	.352	17.3
.466	10.7	.234	22.1	.381	18.1
.471	8.2	.267	16.5	.419	18.8
474	7.1	.272	15.4	.460	21.9
479	4.8	.277	9.7	.463	22.3
487	5.1	.280	8.0	.483	21.9
.517	7.9	.287	6.4	.514	22.3
.537	9.8	.319	9.2	7461.301	7.1
.582	12.8	.327	9.8	.315	6.8
.621	16.8	.339	10.4	.324	8.5
6326.343	14.4	.354	12.7	.354	10.4
.354	15.4	7458.144	18.4	243...	
.375	15.4	.180	18.4	6201.349	14.9
6336.326	18.2	.257	19.0	.357	6.8
.347	18.2	.262	19.1	.359	5.3
.388	18.2	.269	18.1	.360	3.4
.449	21.8	.279	18.6	.362	3.1
.468	19.6	.282	19.2	.365	5.1
.482	10.6	.288	19.8	.368	5.6
.488	9.5	.294	20.4		
.494	8.3	.298	20.4		

STARS OF CONSTANT PERIOD

We have so far established the existence of stars whose light-variation
period remains constant over thousands of cycles. We will now proceed
with a discussion of objects whose periods are definitely known to be
constant. A fundamental reservation should be made, however. To prove
the constancy of a period is much more difficult than to detect its variation.
This requires an extensive series of highly reliable observations covering
an extremely long time which are suitably (uniformly) distributed over the
entire interval. For numerous objects the question therefore remains open
to this day.

43

The periods of various stars changed abruptly, in a jump-like fashion. Occasionally, these stars are known to have exhibited two or three jumps of this kind. Since both before and after the jump the period remains constant, such stars can be safely classified as constant-period variables, provided each has been observed for a sufficient length of time. Our classification is thus not to be interpreted as comprising stars with mathematically constant periods. We can only say that during a given range of observations the period remained constant.

TZ Aurigae

This variable, discovered by L. P. Tserasskaya, is remarkable in that its period remained constant over 50 years. It was observed mostly by Soviet astronomers. Unfortunately, Blazhko's old observations were not published and apparently not even processed properly. They are only briefly mentioned in /195/. Bottlinger and Graff /205/ also observed this star, but the reduction of the observation data was done most superficially. From these observations Tsesevich constructed a seasonal light curve and determined the exact times of maxima. All the maxima are listed in Table 37. The $O-A$ residues were calculated from

$$\text{Max hel JD} = 2419902.434 + 0.39167466 \cdot E. \qquad (20)$$

The least squares method was then applied to derive an improved formula, for which the $O-B$ residues were computed:

$$\text{Max hel JD} = 2419902.4324 + 0.391674615 \cdot E. \qquad (21)$$

A study of the $O-B$ residues shows that the light-variation period remained constant for 46,000 cycles.

TABLE 37. List of maxima

Source	Max hel JD	E	O — A	O — B
Bottlinger and Graff /205/	2419902.434	0	$0^d.000$	$+0^d.002$
Blazhko /195/	20070.458	+ 429	—.004	—.003
Tsesevich	6065.435	+ 15735	.000	+.003
Lange /56/	7343.465	+ 18998	—.004	+.002
Lange /58/	7790.756	+ 20140	—.006	—.003
Gur'ev /27/	8476.580	+ 21891	—.004	—.001
Batyrev /6/	33389.348	+ 34434	—.011	—.008
	3402.291	+ 34467	+.006	+.010
	3598.524	+ 34968	+.010	+.014
	3635.324	+ 35062	—.007	—.004
Batyrev /10/	3675.284	+ 35164	+.002	+.005
	3733.247	+ 35312	—.003	+.001
	3733.248	+ 35312	—.002	+.002
	3747.344	+ 35348	—.006	—.003
	4007.415	+ 36012	—.007	—.004
Alaniya /2/	5188.319	+ 39027	—.002	+.001
Lange	7946.4890	+ 46069	—.0049	—.001
	7959.4135	+ 46102	—.0057	—.002
	7968.4202	+ 46125	—.0075	—.004

Table 37 also lists the maximum determined from the old observations of Tsesevich (Table 38). Comparison stars (Figure 38): $p = 0^s.0$; $q = 5^s.2$; $r = 16^s.0$; $s = 19^s.7$.

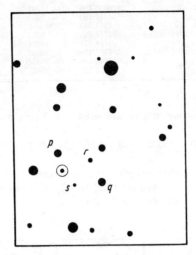

FIGURE 38. Comparison stars for TZ Aurigae.

TABLE 38. Visual observations (Tsesevich)

JD hel	s	JD hel	s	JD hel	s
242...		242...		242...	
5879.426	3.5	6065.412	15.0	6067.421	8.1
.443	8.3	.417	12.4	.446	11.3
6011.372	5.4	.422	8.3	.460	13.7
.379	4.0	.426	7.7	6068.252	16.9
.463	12.1	.433	6.2	.401	18.3
6039.451	18.5	.440	6.2	.442	18.8
.462	19.0	.449	7.3	6071.284	16.5
.481	18.6	.454	7.2	.290	14.5
6058.368	11.5	.464	9.1	.298	8.5
.412	7.8	.476	10.2	.302	6.3
.420	9.8	6066.370	14.9	.308	7.4
6062.395	14.9	.431	16.6	.313	6.8
.427	15.2	6067.301	17.8	.322	7.1
6064.356	15.0	.357	18.1	.329	9.2
.403	16.0	.379	13.1	.354	11.7
6065.340	18.0	.387	7.9	.363	12.8
.384	18.0	.396	6.2	.388	14.1
.405	16.9	.407	6.7	.431	15.5

V 445 Ophiuchi

Discovered by Hoffmeister /264/ and studied by Jensch /287/. Filin /118/ improved Jensch's equation. Tsesevich observed the star visually in 1944, but the results remained unreduced. The behavior of the star was studied by Tsesevich from Simeise and Odessa photographs. The maxima are listed in Table 39.

TABLE 39. List of maxima

Source	Max hel JD	E	O — C
Tsesevich (Sim.)	2420636.499	— 17397	— 0^d.028
	4644.490	— 7302	+ .012
Jensch /287/	7543.543	0	.000
Tsesevich (visual)	31257.295	+ 9354	— .005
Filin /118/	3488.571	+ 14974	.000
Tsesevich (Odes.)	7107.428	+ 24089	— .012
	7402.416	+ 24832	— .012
	7813.339	+ 25867	— .008

The O — C residues were calculated from Filin's equation

$$\text{Max hel JD} = 2427543.543 + 0.3970234 \cdot E; \; P^{-1} = 2.518743228. \qquad (22)$$

This equation, which apparently does not require any correction, has been used to construct the visual (Table 40) and photographic (Table 41) light curves from Odessa observations.

TABLE 40. Average visual light curve

φ	s_1	n	φ	s_1	n	φ	s_1	n
0^p.017	7.6	5	0^p.621	19.3	5	0^p.873	20.9	5
.089	12.2	5	.679	20.6	5	.925	16.1	5
.193	18.2	5	.740	21.4	5	.941	10.8	3
.318	19.3	5	.779	21.1	5	.972	7.9	3
.394	19.3	5	.821	21.3	5	.984	6.8	2
.532	20.5	5						

TABLE 41. Average photographic light curve

φ	s_2	n	φ	s_2	n	φ	s_2	n
0^p.028	0.0	2	0^p.350	7.6	5	0^p.766	10.6	5
.088	1.7	3	.404	8.7	5	.823	10.6	5
.115	1.3	3	.457	9.1	5	.842	8.2	4
.184	5.2	5	.518	10.0	5	.899	3.6	2
.251	7.6	5	.596	11.3	5	.950	—2.3	4
.314	6.8	5	.675	12.0	5	.967	—2.5	3

Tsesevich's observations are listed in Tables 42 — 44. Comparison stars (Figure 39):

	s_1	s_2	s_3
k	—	—7.3	—7.2
a	0.0	0.0	0.0
m	8.8	—	—
b	—	11.2	11.2
c	16.7	15.3	14.3
d	28.2	—	—
e	27.1	—	—

TABLE 42. Simeise observations (Tsesevich)

JD hel	s₃	JD hel	s₃	JD hel	s₃
241...		242...		242...	
9188.449	9.8	4296.366	4.6	6486.486	13.3
	10.2		3.4	6834.445	2.8
9556.324	9.2	4312.376	12.8		1.5
	8.6		10.1	6853.359	10.2
9560.331	8.6	4644.497	—1.0	7931.490	7.1
	8.3		—0.9		6.1
9920.460	11.2	4652.453	1.5	7961.432	9.2
	10.2		—1.0		7.4
242...		5382.441	9.5	7963.449	8.8
0253.498	7.1	5385.464	5.8	9071.384	6.5
	7.8		6.5		4.7
0630.492	6.1	5388.489	—1.0	243...	
.446	11.2		5.1	2686.452	13.0
.528	4.1	5739.495	2.0		13.5
0633.457	9.8		—1.0	3031.494	14.3
	9.7	5741.491	1.0		15.3
0634.463	6.1		0.0	3034.505	10.1
	6.7	5744.496	9.0		12.8
0636.497	6.1		7.8	3445.356	4.7
	1.1	5745.470	3.7	3766.420	7.8
0655.464	12.8		2.8		8.9
1006.423	9.3	5747.456	1.1	3771.499	7.2
	9.8		2.7		9.2
1008.429	0.2	5764.409	10.0	3802.362	5.1
	8.6		8.9		5.1
1369.461	5.6	5770.348	17.3	3803.364	7.5
	7.8	5771.349	9.3		10.2
3193.376	2.8		9.9	3806.451	7.8
	5.1	5774.384	3.1		9.0
3554.367	7.8		6.1	3807.410	3.1
	5.6	5799.346	9.7		6.2
3931.331	11.2		10.2	3823.343	4.6
.408	7.8	5802.359	10.1		4.6
.369	11.2	5820.337	16.3	3824.333	6.7
3937.477	8.4		12.2		5.1
	9.2	5826.355	9.2	4131.496	5.1
3942.462	7.8		9.2		5.6
	8.8	6486.486	10.6	4540.370	0.0
					1.5:

TABLE 43. Odessa observations (Tsesevich)

JD hel	s₂	JD hel	s₂	JD hel	s₂
243...		243...		243...	
6744.358	9.3	7107.370	10.3	7794.444	9.0
6749.405	3.1	.422	—2.7	7810.408	11.9
.474	2.4	7111.370	3.4	.454	9.8
6756.344	9.8	.396	—1.8	7812.363	7.5
.401	10.2	.424	1.5	7813.389	—1.3
6757.362:	13.2:	7128.352	15.3	.437	6.2
6758.363	13.5	7373.626	10.1	7818.443	7.5
.389	10.3	7378.596	—2.4	7821.369	7.0
6759.385	2.0	7402.538	6.7	.424	9.0
6761.379	7.5	7404.551	9.8	7839.382	13.8
6781.330	13.9	7405.546	7.1	7840.383	—1.8
7073.437	9.5	7406.554	9.8	7844.418	5.6
7075.464	6.8	7427.477	2.8	7847.394	11.2
7077.396	5.6	7454.398	3.7	7850.370	6.2
.451	9.3	7457.412	10.1	7881.364	8.4
7078.388	10.2	7458.394	—3.2	7884.390	—2.2
436	—1.0	7462.425	6.5	8113.559	8.7
7079.367	7.8	7464.423	1.4	8114.507	7.8
.418	13.5	7763.560	9.0	.597	9.0
.467	15.3	7789.427	9.0	8143.510	7.9
7080.451	—1.5	.480	7.5	8144.514	—1.3
7087.453	14.5	7790.403	5.9	8164.439	6.2
7099.361	10.6	.501	9.2	8165.435	11.2
7101.400	10.2	7793.425	11.2	8170.447	8.7
7102.407	6.1	.480	—3.2		
.456	10.2	.526	4.2		

47

TABLE 44. Visual observations (Tsesevich)

JD hel	s_1	JD hel	s_1	JD hel	s_1
243...		243...		243...	
1231.329	18.4	1243.310	19.9	1272.342	20.2
.349	18.4	1252.202	18.4	.353	18.2
.364	19.5	.287	20.4	.369	7.9
.410	18.6	1253.380	17.3	.379	6.2
.430	18.6	.408	18.5	.392	6.2
1232.227	21.2	1256.240	19.4	6722.415	14.1
.261	10.1	.377	20.8	.424	14.1
.320	9.8	1257.192	21.4	.454	14.6
.338	10.4	.206	22.2	6734.412	22.0
.373	14.0	.220	22.2	.423	22.9
.409	14.4	.232	21.9	.434	24.4
.452	15.7	.247	22.2	.456	24.4
1235.218	19.9	.258	21.2	.465	23.7
.262	20.7	.271	21.0	.470	22.9
.330	18.6	.282	17.4	.473	24.7
.342	19.1	.286	8.8	6736.337	23.8
.454	7.3	.290	7.5	.353	23.8
1236.225	22.6	.295	7.5	6748.358	22.7
.279	15.3	.303	7.5	.368	23.5
.333	20.2	.309	8.3	.378	24.4
.348	22.9	.321	8.3	.386	24.0
.415	24.2	1259.365	19.9	.418	24.0
1241.245	20.4	1270.220	20.4	6781.318	20.9
.331	20.9	.277	21.9	.334	21.9
.367	19.9	.295	21.6	6782.302	10.0
.392	8.8	.312	21.9	.318	12.8
.396	7.2	.392	21.6	.332	14.3
.420	7.2	1272.211	21.9	.341	14.5
.434	8.8	.244	21.2	6784.314	13.4
1242.352	19.6	.254	21.6	.327	14.5
.363	19.3	.273	21.9	.335	17.9
.372	17.5	.293	22.1	.353	17.9
.415	19.1	.303	21.9	6806.288	22.5
1243.241	19.9	.323	21.9		

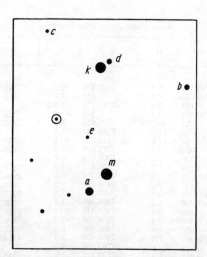

FIGURE 39. Comparison stars for V 445 Ophiuchi.

TW Herculis

Discovered by Cannon /213, 214/, observed by various authors. Lange compiled a summary table of the individual maxima (Table 45) and the table of average and normal maxima (Table 46). The O−A residues were calculated from

$$\text{Max hel JD} = 2421545.2267 + 0.39960057 \cdot E. \qquad (23)$$

Lange obtained the improved formula

$$\text{Max hel JD} = 2421545.2340 + 0.399600104 \cdot E, \qquad (24)$$

for which the O−C residues were calculated.

TABLE 45. List of maxima

Source	Max hel JD	E	O − A
Hoffmeister	2421545.253	+ 0	$+0^d$.026
/257, 258/	1642.732	+ 244	+ .003
	1665.523	+ 301	+ .016
	1669.510	+ 311	+ .008
	1721.453	+ 441	+ .002
	1731.448	+ 466	+ .007
	1749.438	+ 511	+ .015
	1820.560	+ 689	+ .008
	2191.382	+1617	+ .001
	2607.371	+2658	+ .006
	4376.403	+7085	+ .006
	4380.401	+7095	+ .008
	4386.394	+7110	+ .007
	4406.369	+7160	+ .002
	4408.365	+7165	+ .000
	4410.360	+7170	− .003
Tsesevich	4967.407	+8564	+ .001
	5067.304	+8814	.002
	5069.303	+8819	.001
	5071.300	+8824	− .002
	5077.300	+8839	+ .004
	5083.287	+8854	− .003
	5085.298	+8859	+ .010
	5087.288	+8864	+ .002
	5089.286	+8869	+ .002
	5091.281	+8874	− .001
	5093.281	+8879	+ .001
	5095.273	+8884	− .005
	5097.275	+8889	− .001
	5099.281	+8894	+ .007
	5103.270	+8904	.000
	5107.264	+8914	− .002
Lange /58/	7309.463	+ 14425	− .0019
Solov'ev /107, 108/	7599.176	+ 15150	+ .0007
	7969.202	+ 16076	− .0035
	8373.203	+ 17087	+ .0014
Gur'ev /27/	8394.784	+ 17141	+ .0039
Tsesevich	30616.1602	+ 22700	+ .0006
Batyrev /9/	3122.444	+ 28972	− .0104
	3191.184	+ 29144	− .0017
	3203.169	+ 29174	− .0047
	3211.166	+ 29194	+ .0003
	3217.161	+ 29209	+ .0013
Born /203/	3858.517	+ 30814	− .0017
	3872.504	+ 30849	− .0007
	3898.479	+ 30914	+ .0003
Sofronievitsch /203/	3898.480	+ 30914	− .0007
Tsarevskii /141/	6090.6842	+ 36400	− .0032
Lange	6756.4066	+ 38066	− .0154
	6758.4059	+ 38071	− .0141
	6760.4006	+ 38076	− .0174
	6762.4010	+ 38081	− .0150

TABLE 45 (continued)

Phase	Max hel JD	E	O — A
Lange	2436784.3767	+ 38136	—0d.0173
	7130.4288	+ 39002	— .0193
	7134.4272	+ 39012	— .0169
	7136.4263	+ 39017	— .0158
	7142.4178	+ 39032	— .0183
	7144.4160	+ 39037	— .0182
	7158.4066	+ 39072	— .0136
	7164.4019	+ 39087	— .0123
	7170.3930	+ 39102	— .0152
	7494.4695	+ 39913	— .0148
	7496.4670	+ 39918	— .0153
	7518.4453	+ 39973	— .0150
	7520.4396	+ 39978	— .0167
	7522.4417	+ 39983	— .0146
	7781.3834	+ 40631	— .0141
	7789.3763	+ 40651	— .0132
	7809.3651:	+ 40701	— .0044:
	7813.3563	+ 40711	— .0092
	7817.3551	+ 40721	— .0064
	7841.3332	+ 40781	— .0043
	7847.3290	+ 40796	— .0026
	7869.2959	+ 40851	— .0137
	7871.2963	+ 40856	— .0113
	7872.4926	+ 40859	— .0138
	7873.2939	+ 40861	— .0115
	7878.4869	+ 40874	— .0135
	7881.2885	+ 40881	— .0091
	7883.2845	+ 40886	— 0111
	7884.4816	+ 40889	— .0128
	7899.2704:	+ 40926	— .0092:
	7903.2650	+ 40936	— .0106
	7925.2374	+ 40991	— .0163
	7943.2207	+ 41036	— .0150
	7955.2040	+ 41066	— .0197
	7959.2030	+ 41076	— .0167
	7961.2018	+ 41081	— .0157
	7963.2006	+ 41086	— .0151
Sakharov	7869.2966	+ 40851	— .0130
	7871.3042:	+ 40856	— .0034:
	7873.2950	+ 40861	— .0106
	7878.4912	+ 40874	— .0092
	7884.2871	+ 40881	— .0105
	7899.2773:	+ 40926	— .0023:
	8179.3919	+ 41627	— .0077
	8181.3924	+ 41632	— .0052
	8209.3603	+ 41702	— .0094
	8231.3486	+ 41757	+ .0009
	8235.3405	+ 41767	— .0032
	8259.3045	+ 41827	— .0152
	8263.3177	+ 41837	+ .0020
	8267.3007	+ 41847	— .0110
	8273.2910	+ 41862	— .0148
Lange	8207.3591	+ 41697	— .0126
	8209.3657	+ 41702	— .0130
	8221.3489	+ 41732	— .0088
	8223.3438	+ 41737	— .0119
	8229.3374	+ 41752	— .0123
	8231.3328	+ 41757	— .0149
	8233.3324	+ 41762	— .0133
	8235.3325	+ 41767	— .0112
	8259.3031	+ 41827	— .0166
	8263.3016	+ 41837	— .0141
	8267.2983	+ 41847	— .0135
	8269.2943	+ 41852	— 0155
	8271.2932	+ 41857	— .0146
	8273.2905	+ 41862	— .0153
	8275.2901	+ 41867	— .0137
	8283.2800	+ 41887	— .0158
	8285.2765	+ 41892	— .0173
	8289.2789	+ 41902	— .0109

TABLE 45 (continued)

Source	Max hel JD	E	$O - A$
Lange	2438293.2810	+41912	—0^d.0048
	8295.2728	+41917	— .0110
	8297.2778	+41922	— .0040

TABLE 46. List of average maxima

Source	Max hel JD	E	$O - C$
Hoffmeister	2421686.2954	+ 353	+0^d.0046
	2092.2865	+ 1369	.0000
	4394.7820	+ 7131	— .0003
Tsesevich	5063.7108	+ 8805	— .0021
	5096.8767	+ 8888	— .0030
Lange	7309.463	+14425	— .0030
Solov'ev	7599.176	+15150	— .0004
	7969.202	+16076	— .0033
	8373.203	+17087	+ .0020
Gur'ev	8394.784	+17141	+· .0046
Tsesevich	30616.1602	+22700	+· .0038
Batyrev	3189.1847	+29139	+· .0033
Born	3882.0944	+30873	+· .0064
Tsarevskii	6090.6842	+36400	+· .0064
Lange	6764.3982	+38086	— .0054
	7136.0249	+39016	— .0068
	7159.2046	+39074	— .0039
	7510.4530	+39953	— .0040
	7820.9494	+40730	+ .0032
Sakharov	7875.6924	+40867	+ .0009
Lange	7880.4869	+40879	+· .0002
	7944.4189	+41039	— .0038
	8215.3521	+41717	+· .0006
	8232.5336	+41760	— .0007
Sakharov	8233.3386	+41762	+ .0051
Lange	8264.8992	+41842	— .0028
	8275.6886	+41868	— .0026
	8292.0774	+41909	+ .0026

Lange reports that numerous observations of TW Her made between 1959 and 1963 gave negative residues for the GCVS elements. The average period from 1918 to 1963 (41,000 E) apparently remained constant. Deviations of the maxima from the constant-period elements, however, cannot be attributed to observation errors. The maxima possibly fluctuate with a period of 292 days. This conclusion requires further verification. Small deviations superimposed on a generally stable period are observed for TW Her, TV Lib, and AA Aql.

AR Persei

Variability established by Schneller and studied by various authors. A summary list of maxima is given in Table 47.

TABLE 47. List of maxima

Source	Max hel JD	E	O — C
Kukarkin /44-47/	2415049.27:	— 27088	+0^d.093:
Tsesevich (Har.)	5236.878	— 26647	+ .034
Kukarkin /44-47/	5312.24:	— 26470	+ .074:
Tsesevich (Har.)	6303.726	— 24140	+ .030
	7006.729	— 22488	+ .025
	7850.589	— 20505	+ .021
	8778.706	— 18324	+ .015
	20511.541	— 14252	+ .013
	1670.737	— 11528	+ .012
	2500.986	— 9577	+ .014
Tsesevich (Har.)	3014.626	— 8370	+ .016
	3562.730	— 7082	+ .013
	4124.873	— 5761	+ .005
	4462.762	— 4967	+ .008
	5036.835	— 3618	+ .015
Guthnick /246/	5556.00	— 2398	+ .009
Beyer /193/	5939.426	— 1497	+ .015
Tsesevich (Har.)	6156.866	— 986	.000
Beyer /193/	6251.344	— 764	+ .006
Florya /132 - 135/	6576.458	0	.000
	6585.400	+ 21	+ .005
	6597.322	+ 49	+ .012
Tsesevich (Har.)	6600.295	+ 56	+ .006
Florya /132-135/	6603.268	+ 63	.000
	6605.404	+ 68	+ .009
	6628.371	+ 122	— .004
	6649.220	+ 171	— .007
	6761.144	+ 434	— .002
Beyer /193/	6979.464	+ 947	+ .011
	6985.425	+ 961	+ .014
Tsesevich (Har.)	7236.922	+ 1552	+ .011
Lange /58-60/	7353.519	+ 1826	+ .008
Tsesevich (vis.)	7460.323	+ 2077	— .001
Solov'ev /108/	7576.510	+ 2350	+ .011
Lange /58-60/	7782.470	+ 2834	+ .005
Tsesevich (vis.)	7858.209	+ 3012	— .004
Tsesevich (Har.)	7954.818	+ 3239	+ .006
Gur'ev /27, 28/	8087.164	+ 3550	+ .006
Solov'ev /108/	8397.389	+ 4279	+ .005
Kukarkin /44-47/	8632.292	+ 4831	+ .005
Tsesevich (Har.)	8702.932	+ 4997	+ .004
	9226.792	+ 6228	+ .013
	9613.623	+ 7137	+ .019
Payne-Gaposchkin /329/	9624.675	+ 7163	+ .007
Tsesevich (Har.)	9968.529	+ 7971	+ .017
	30337.891	+ 8839	+ .002
Tsesevich (Har.)	0705.987	+ 9704	— .002
	1075.802	+ 10573	+ .011
	1431.551	+ 11409	.000
Miczaika /320/	1793.270	+ 12259	+ .002
Tsesevich (Har.)	1962.642	+ 12657	+ .006
	2530.751	+ 13992	+ .006
	3030.763:	+ 15167	— .002:
	3588.667	+ 16478	+ .006
Batyrev /19/	3898.463	+ 17206	+ .003
	3916.338	+ 17248	+ .004
	3951.233	+ 17330	+ .004
	4339.342	+ 18242	+ .012
Alaniya /1/	4343.593	+ 18252	+ .008
Batyrev /19/	4385.301	+ 18350	+ .012
	4399.329	+ 18383	— .003
Tsarevskii /141/	6070.4666	+ 22310	+ .002
Geyer /239/	6231.3203	+ 22688	— .002
Ahnert /170, 171/	6605.394	+ 23567	+ .014
Geyer /239/	6607.5100	+ 23572	+ .002

52

Beyer /193/ determined one average maximum from his extensive observations covering a considerable length of time. Tsesevich revised these observations and obtained three maxima from the seasonal curves, which are also included in the list. The O−C residues (Figure 40) were calculated from the equation

$$\text{Max hel JD} = 2426576.458 + 0.42554937 \cdot E. \tag{25}$$

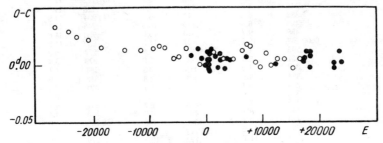

FIGURE 40. O−C residues for AR Persei (O Harvard observations).

These data point to constant light-variation period. However, the two maxima determined by Kukarkin from old Moscow photographs deviated from the elements (25) by two hours! It should be noted, however, that these maxima were based on 9 observations only. Tsesevich therefore decided to investigate the star on the basis of Harvard photographs (over two thousand). The observations gave 28 seasonal light curves, and the maximum was determined from each. These Harvard results are also included in the list. The light-variation period is indeed variable, though between very small limits. It actually became somewhat longer (see Figure 40). Tsesevich observed the star visually for three seasons using the following comparison stars (Figure 41): $a = \text{BD} + 46°858 = 0^s.0$; $b = \text{BD} + 47°961 = 3^s.7$; $c = \text{BD} + 47°970 = 7^s.5$; $e = \text{BD} + 46°864 = 8^s.8$; $d = \text{Anonyma} = 13^s.3$. The observations are listed in Table 48.

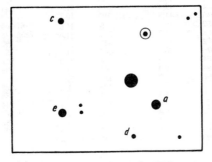

FIGURE 41. Comparison stars for AR Persei.

TABLE 48. Visual observations (Tsesevich)

JD hel	s	JD hel	s	JD hel	s
242...		242...		242...	
6596.283	10.6	6606.252	4.5	7425.441	1.3
.292	11.3	.261	3.2	.449	1.5
6597.232	10.3	.270	2.9	.464	3.4
.242	11.4	.277	4.4	.475	4.0
.267	11.0	.288	6.3	.484	4.6
.275	8.3	.293	6.2	.516	6.2
.338	4.6	.306	5.7	7437.112	11.2
.348	4.9	.321	7.4	.123	11.0
.382	6.5	.347	6.7	7443.234	11.9
.429	6.8	.380	10.3	.245	11.6
.458	9.7	.410	10.6	.250	11.8
6600.267	10.7	.427	11.5	.260	11.2
.293	4.8	7414.390	1.8	.277	6.2
.301	3.4	.396	2.0	.279	5.4
.317	4.0	.415	2.3	.284	4.1
.324	4.8	.421	2.8	.318	1.9
.354	6.4	.436	5.1	.324	3.5
.365	7.4	.456	5.8	.334	2.5
.386	7.3	.475	6.7	.347	4.1
.407	8.3	.494	6.0	.353	4.4
.440	10.8	.546	8.0	7458.256	4.8
6604.296	11.2	7415.112	11.1	.264	6.3
.308	11.7	.173	12.3	.279	6.0
.315	11.8	.182	11.0	.367	9.2
.323	12.7	.202	8.0	7460.172	11.4
.332	11.8	.207	5.2	.215	11.5
.338	11.5	.214	4.1	.250	12.0
.346	11.8	.222	3.8	.265	12.3
.356	12.1	.228	3.9	.277	12.1
.363	11.6	.235	3.8	.287	9.7
.371	12.0	.241	4.5	.289	9.4
.378	11.5	.253	4.8	.297	7.0
.382	10.6	.264	5.8	.303	4.4
.388	11.5	.300	5.2	.314	3.3
.397	11.5	.314	6.2	.323	2.9
.402	10.6	.333	6.8	.336	2.2
.411	10.6	.342	7.5	.346	2.9
.417	10.8	.367	8.3	.352	4.3
.427	11.5	.391	9.8	.377	4.3
.432	11.5	.398	10.4	.392	4.8
.439	11.8	.442	9.8	.405	6.0
.446	10.8	.455	10.6	7461.116	11.9
.457	11.3	.472	11.0	.147	5.6
.463	11.8	·490	11.4	.152	4.8
.480	11.8	.502	12.1	.187	1.9
6605.163	11.8	7422.119	9.2	.195	2.9
.208	14.3	7424.213	5.8	.232	4.8
.226	14.3	.221	6.2	.249	5.6
.238	13.8	.258	7.3	.257	6.0
.255	12.0	.376	11.6	.279	6.4
.263	13.3	.395	11.6	7484.111	6.7
.273	11.5	.408	11.4	.115	6.7
.286	11.5	.430	11.4	.125	3.6
.303	13.8	.468	11.9	.129	3.8
.324	12.4	.514	11.4	.141	2.1
.340	11.5	7425.197	10.7	.148	2.6
.352	10.6	.305	11.6	7861.303	7.0
.358	11.3	.314	11.6	7858.178	6.5
.372	10.6	.337	11.2	.185	3.2
.376	8.1	.342	11.6	.193	4.4
.384	5.5	.354	11.0	.202	3.0
.394	2.0	.362	11.4	.209	3.4
.398	3.2	.389	11.4	.215	3.0
.413	2.4	.403	6.2	.228	3.2
.419	4.2	.409	4.7	.236	4.0
.436	4.6	.413	4.1	.244	4.0
.462	6.8	.418	2.9	.277	5.6
.480	5.1	.421	1.9	.282	6.7
.504	8.0	.428	1.0	.321	7.2
6606.220	7.9	.435	0.8		

CZ Lacertae

Very little observed. Discovered twice, first by A. G. Belyavskaya and then by A. I. Parenago. Observed by Florya /136/, Dzigvashvili /32/, and Batyrev /11/. This was long ago, so that at Tsesevich's request B. A. Dragomiretskaya prepared magnitude estimates from the Odessa photographs. The results emerging from the reduction of these data point to a suspicion of a Blazhko effect, which is also evident from the table of the individual maxima compiled by Batyrev (Table 49). The average maxima are listed in Table 50.

TABLE 49. Individual maxima

Max hel JD	E	O − C	Max hel JD	E	O − C
2433894.532	+ 2352	−0^d.009	2433926.510	+ 2426	−0^d.007
3898.423	+ 2361	− .007	3934.290	+ 2444	− .005
3912.256	+ 2393	− .002	3950.273	+ 2481	− .010
3915.285	+ 2400	+ .003			

TABLE 50. List of average maxima

Source	Max hel JD	E	O − C
Florya /136/	2428744.296	− 9567	−0^d.004
	8771.102	− 9505	+ .011
	9146.155	− 8637	− .001
	9462.447	− 7905	− .009
Dzigvashvili /32/	32878.241	0	+ .007
Batyrev /11/	3926.513	+ 2426	− .004
Dragomiretskaya	7943.351	+11722	− .000

The O − C residues were calculated from Batyrev's equation:

$$\text{Max hel JD} = 2432878.234 + 0.43210346 \cdot E. \tag{26}$$

If the possible Blazhko effect is ignored, the period is seen to have remained constant over 21,000 cycles. Dragomiretskaya used the following comparison stars in her observations (Table 51, Figure 42): $a = 0^s.0$; $q = 7^s.0$; $c = 12^s.3$; $d = 17^s.5$.

TABLE 51. Observations of B. A. Dragomiretskaya

JD hel	s	JD hel	s	JD hel	s
243...		243...		243...	
7521.483	11.2	7901.456	6.1	8285.367	11.4
7523.491	[12.3	7903.449	14.9	8287.415	9.6
7526.492	13.3	7906.461	17.5	8288.416	15.4
7546.401	11.4	7931.304	10.5	8289.425	11.2
7555.368	10.5:	7942.370	[17.5	8291.390	12.3
7578.308	11.0	7943.339	4.2	8292.353	14.4
7583.303	15.3:	7962.308	15.8	8294.374	12.3
7606.262	10.0	7963.322	7.9	8295.378	14.6
7629.208	5.0	7964.300	13.3	8296.353	9.1
7881.535	15.3	7968.316	17.5	8298.375	15.8
7882.495	7.0	8262.448	14.0	8309.260	12.3
7884.521	16.0	8269.418	12.3	8313.277	6.3
7886.482	14.9	8270.430	9.3	8319.301	4.2
7887.467	15.8	8283.357	14.9	8320.318	10.2

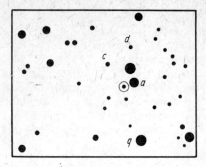

FIGURE 42. Comparison stars for CZ Lacertae.

U Trianguli

In his extensive study of this variable, Martynov /73/ called attention to the exceptional constancy of its period. The star has passed through another 20,000 cycles since the publication of Martynov's paper. The complete lists of all the maxima available to me is given in Table 52.

TABLE 52. List of maxima

Source	Max hel JD	E	O — A	O — B
Blazhko /197/	2418490.173	— 1364	$0^d.000$	$-0^d.001$
	9199.516	+ 222	.000	— .002
Tsesevich /143/	24774.974	+12688	—.001	.000
Kukarkin /49/	6679.377	+16946	—.001	.000
Martynov /73/	6975.461	+17608	+.001	+.003
	7339.523	+18422	—.001	+.001
Tsesevich (visual)	7663.332	+19146	—.003	—.001
Martynov /73/	8773.420	+21628	+.002	+.005
Alaniya /1/	33972.283	+33252	—.006	—.001
Tsesevich (Odessa)	6463.486	+38822	—.003	+.003
Lange	7612.4738	+41391	—.0089	—.0026
	7944.7823	+42134	—.0095	—.0031

The O — A residues were calculated from Martynov's equation

$$\text{Max hel JD} = 2419100.2259 + 0.44725319 \cdot E. \qquad (27)$$

It was improved by the least squares method:

$$\text{Max hel JD} = 2419100.2275 + 0.44725300 \cdot E. \qquad (28)$$

The O — B residues were computed from the improved relation. Table 52 includes one maximum obtained by Tsesevich from magnitude estimates based on Odessa photographs (Table 53). Comparison stars (Figure 43): $u = 0^s.0;$ $a = 3^s.8;$ $b = 14^s.1.$

TABLE 53. Odessa observations (Tsesevich)

JD hel	s	JD hel	s	JD hel	s
243...		243...		243...	
6072.549	12.2	6165.398	11.0	6487.460	11.5
6075.557	11.3	6184.315	13.6	6488.465	14.1
6076.542	11.5	6187.319	11.6	6489.443	—1.5
6079.519	12.2	6189.309	−0.0	6498.506	5.0
6081.563	0.0	6194.310	2.5	6518.397	12.4
6082.519	4.8	6213.258	11.5	6540.381	7.9
6083.514	12.2	6221.239	13.1	6541.293	1.9
6084.523	13.1	6461.506	12.0	6542.316	8.4
6085.546	—2.0	6462.509	13.5	6543.386	12.4
6100.514	12.4	6463.498	—1.5	6544.374	12.7
6102.492	13.6	6465.521	12.2	6545.402	3.0
6111.435	13.1	6466.481	12.0	6547.388	11.8
6112.514	10.4	6468.516	5.9	6549.305	13.0
6129.480	5.4	6469.557	12.6	6574.284	11.5
6138.403	3.8	6483.440	12.0	6575.304	1.7
6141.430	9.4	6484.466	12.0	6576.268	6.6
6163.362	1.3	6485.463	1.6	6581.252	11.3
6164.353	3.8				

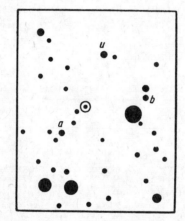

FIGURE 43. Comparison stars for U Trianguli.

ST Ophiuchi

Discovered by Cannon /211, 212/. Observed by various authors. Nachapkin /79/ reduced all the old data by Hertzsprung's technique and prepared a summary table of the maxima (Table 54). It was supplemented with new data by Tsesevich. The O−A residues were calculated from

$$\text{Max hel JD} = 2418159.662 + 0.45035643 \cdot E. \qquad (29)$$

Tsesevich observed the star visually (Table 55) and on Odessa photographs (Table 56). Comparison stars (Figure 44): visual scale $A = 0^s.0$; $a = 9^s.9$; $b = 22^s.9$; photographic scale $A = -6^s.8$; $k = 0^s.0$; $a = 9^s.6$; $b = 22^s.4$.

57

TABLE 54. List of maxima

Source	Max hel JD	E	O — A
Guthnick /245/	2418098.866	— 135	+0^d.002
Shapley /349/	8550.576	+ 868	+ .005
Florya /234/	26146.292	+ 17734	+ .009
Dombrovskii /35/	7310.920	+ 20320	+ .016
Selivanov /97/	7624.353	+ 21016	.000
Solov'ev /108/	7987.791	+ 21823	+ .001
	8426.891	+ 22798	+ .003
Alaniya /1/	34568.397	+ 36435	— .002
Tsesevich	6079.347	+ 39790	+ .003
	7077.350	+ 42006	+ .016
Lange	7848.343	+ 43718	— .001
	7852.399	+ 43727	+ .001

TABLE 55. Visual observations (Tsesevich)

JD hel	s	JD hel	s	JD hel	s
243...		243...		243...	
6079.267	10.8	6079.322	6.8	6079.361	—2.2
.276	11.2	.326	4.0	.370	—2.0
.287	11.2	.330	1.3	.382	1.1
.295	11.2	.335	—1.8	.393	2.6
.308	10.1	.340	—2.4	.402	3.2
.313	10.1	.347	—2.6		
.319	7.6	.354	—2.6		

TABLE 56. Odessa observations (Tsesevich)

JD hel	s	JD hel	s	JD hel	s
243...		243...		243...	
6755.453	10.6	6781.360	20.3	7087.488	16.6
6756.374	8.5	6791.381	—3.1	7101.424	17.1
.433	15.1	6805.303	5.2	7107.397	8.7
6757.391	17.1	6807.286	13.5	7135.431	—1.9
.417	13.3	6814.300	6.7	7169.331	9.6
.444	15.1	7075.503	14.9	7170.308	12.8
6758.419	18.5	7077.423	4.8	7172.304	15.4
.446	12.4	.491	12.8	7173.302	—2.5
6761.420	16.5	7078.412	11.7	7488.387	13.6
6778.418	7.5	7079.392	17.6	7493.345	15.6
6779.379	15.5	.441	14.7	7494.342	4.8
6780.351	17.1	7080.425	17.1	7498.350	16.9

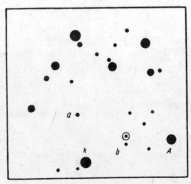

FIGURE 44. Comparison stars for ST Ophiuchi.

DX Delphini

Discovered by Belyavskii [Beljawsky] /21/ from Simeise photographs. Elements determined by Tsesevich /146/ from visual observations. Hoffmeister /271/ using 126 photographic observations improved the equation. Batyrev /12/ obtained a still better fit:

$$\text{Max hel } JD = 2425807.494 + 0.47261498 \cdot E; \ P^{-1} = 2.11588723. \tag{30}$$

Later Alaniya /1/ carried out photographic observations. Quite recently the star was observed by Kanishcheva and Sakharov. Tsesevich compiled a summary table of maxima (Table 57). The last observations showed positive variance, and it seems that the light-variation period somewhat increased.

Hoffmeister reported the times of brightness enhancement, and not the actual maxima. Tsesevich was therefore reduced to measurements of the old Simeise photographs (greatly overexposed, unfortunately). The maximum was then determined from the seasonal light curve; the deviation from (30) was $-0^d.031$ for $E = 4045$. This indicates that the star retained a constant period.

TABLE 57. List of maxima

Source	Max hel JD	E	O — A
Hoffmeister /271/	2425807.50	0	$+0^d.006$
	5826.42	+ 40	+.021
	6206.38	+ 844	—.001
	6214.37	+ 861	—.045
	6920.49	+ 2355	—.012
	7327.42	+ 3216	—.004
	7655.42	+ 3910	+.001
Tsesevich	7719.191	+ 4045	—.031
Hoffmeister /271/	9845.50	+ 8544	—.016
	9846.47	+ 8546	+.008
	9847.42	+ 8548	+.013
	30606.43	+ 10154	+.003
	0608.32	+ 10158	+.003
	0612.58	+ 10167	+.009
	0613.52	+ 10169	+.004
	0614.47	+ 10171	+.009
Tsesevich /146/	0950.490	+ 10882	.000
Batyrev /12/	3869.367	+ 17058	+.007
	3886.384	+ 17094	+.010
	3895.353	+ 17113	—.001
	3896.294	+ 17115	—.005
	3897.238	+ 17117	—.007
	3903.377	+ 17130	—.012
	3912.366	+ 17149	—.002
Alaniya /1/	4621.307	+ 18649	+.016
Alaniya /2/	5697.454	+ 20926	+.019
Sakharov	7871.488	+ 25526	+.024
Kanishcheva	7871.486	+ 25526	+.022

The O — A residues were calculated for equation (30). The visual observations and the maxima determined by Alaniya give the table of average maxima (Table 58). These figures lead to the following equation for the ephemeris:

$$\text{Max hel } JD = 2430950.488 + 0.47261673 \cdot E. \tag{31}$$

TABLE 58. Average maxima

Source	Max hel JD	E_1	O — A	O — B
Tsesevich	2430950.490	0	$0^d.000$	$+0^d.002$
Batyrev	3895.353	+ 6231	—.001	—.010
Alaniya	4621.307	+ 7767	+.016	+.005
	5697.454	+10044	+.019	+.004
Sakharov	7871.488	+14644	+.024	+.001
Kanishcheva	7871.486	+14644	+.022	—.001

In photographic observations (Table 59), Tsesevich used the following comparison stars (Figure 45): $a = 0^s.0$; $b = 4^s.5$; $c = 13^s.3$.

TABLE 59. Simeise observations (Tsesevich)

JD hel	s	JD hel	s	JD hel	s
242...		242...		242...	
2795.556	— 1.0	7364.292	— 2.5	7752.211	2.2
5451.465	5.4		— 2.0	7754.191	— 3.5
6181.487	2.2	7393.195	1.1	7774.179	2.7
	2.2		— 2.0		2.7
6184.384	1.1	7713.300	4.5	7785.173	4.5
7307.444	5.5		3.4		3.4
	4.5	7715.278	5.5	8394.394	— 1.0
7309.270	3.1		3.4		2.2
	2.2	7717.265	1.5	8402.337	1.7
7328.317	— 3.0		— 2.0		— 1.0
	— 1.5	7719.251	— 1.0	243...	
.373	— 2.0		— 3.0	3502.317	— 1.0
	— 1.0	7721.278	1.7		0.9
7338.344	— 1.0		2.7	3860.394	— 3.0
	— 2.0	7750.203	3.6		— 3.0
7357.268	— 0.5		5.8	4223.384	— 2.0
	— 1.5	7752.211	2.2		— 1.0

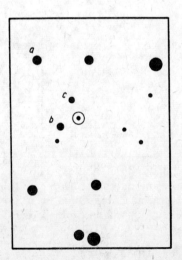

FIGURE 45. Comparison stars for DX Delphini.

AV Serpentis

Discovered by Hoffmeister /267/ and investigated by Solov'ev /103/. Observed visually by Batyrev /8/ and Tsarevskii.

Tsesevich studied the star from Simeise and Odessa photographs. Table 60 lists the maxima. Solov'ev published the average maximum and 18 individual maxima. Tsesevich divided them into 4 groups (according to time) and the averages are included in the table.

TABLE 60. List of maxima

Source	Max hel JD	E	O − A	O − B
Tsesevich (Simeise)	2418884.236:	−19401	−0^d.016	−0^d.005
	25774.404:	− 5269	+.005:	+.005:
Solov'ev /103/	8361.864	+ 38	+.003	.000
	8686.088	+ 703	+.002	−.002
	8699.741	+ 731	+.003	.000
	9087.351	+ 1526	+.006	+.002
	30545.632	+ 4517	+.006	−.001
Batyrev /8/	3461.224	+10497	+.010	.000
Tsesevich (Simeise)	3766.438	+11123	+.014	+.003
Tsesevich (Odessa)	6749.304	+17241	+.010	−.005
	7080.364	+17920	+.019	+.003
Tsarevskii	7135.455	+18033	+.016	.000
Tsesevich (Odessa)	7406.535	+18589	+.015	−.001
Tsarevskii	7454.320	+18687	+.020	+.003
Tsesevich (Odessa)	7761.475	+19317	+.014	−.003
	8198.326	+20213	+.014	−.003

The O − A residues were calculated from

$$\text{Max hel JD} = 2428343.334 + 0.4875564 \cdot E, \tag{32}$$

which was then improved by the least squares method:

$$\text{Max hel JD} = 2428343.337 + 0.48755712 \cdot E;$$

$$P^{-1} = 2.05104173. \tag{33}$$

The O − B residues were calculated from (33) and the average light curve was constructed from the Odessa photographs (Table 61). Comparison stars for the observations in Tables 62 and 63 (Figure 46): $n = 0^s.0$; $m = 9^s.0$; $p = 15^s.1$; $q = 21^s.7$. The period retained the last value over 40,000 epochs!

TABLE 61. Average light curve

φ	s	n	φ	s	n	φ	s	n
0^P.028	1.2	5	0^P.411	12.8	5	0^P.735	16.5	5
.074	2.7	5	.471	13.5	5	.790	17.8	5
.113	3.6	5	.514	15.5	5	.840	17.5	4
.168	4.0	5	.542	14.3	5	.885	16.0	5
.200	6.0	5	.572	14.2	5	.924	5.8	3
.250	7.3	5	.619	17.0	5	.953	4.1	4
.291	9.4	5	.661	16.0	5	.967	2.7	3
.325	12.4	5	.703	16.0	5	.982	0.5	3
.352	10.4	5						

61

TABLE 62. Simeise observations (Tsesevich)

JD hel	s	JD hel	s	JD hel	s
241...		242...		242...	
8878.307	12.5	5027.341	19.5	7607.357	7.8
	12.5		18.4		13.9
8880.333	2.5	5388.489	9.0	7716.219	22.7
	2.0		9.0		20.4
8884.333	6.0	5764.404	9.0	243...	
	6.3	5770.348	12.7	3034.505	9.0
9556.324	18.4		18.4		11.0
	13.1	5771.349	12.7	3412.426	10.0
9888.504	11.4		15.1		11.0
	12.3	5774.384	3.0	3445.356	13.6
9920.460	12.7	5799.346	7.0	3447.353	13.6
	10.2		6.0		22.7
242...		5802.359	19.1	3766.420	3.7
0253.498	19.1	5820.337	7.6		4.1
	11.0		8.0	3803.364	10.2
0633.457	12.7	5826.355	17.7		9.0
	18.4		19.1	3806.452	4.9
0634.464	13.6:	6486.486	12.3		4.2
	18.4		12.7	3807.410	2.0
3554.367	20.4	6860.355	8.1		0.8
	19.0		13.1	4131.496	13.6
3942.462	13.6	7573.417	19.5		19.5
	19.1	7595.384	10.2	4183.358	4.9
4312.376	7.2		7.2		4.9
	9.0	7597.358	4.2		

TABLE 63. Odessa observations (Tsesevich)

JD hel	s	JD hel	s	JD hel	s
243...		243...		243...	
6702.460	6.2	7111.370	16.8	7811.462	16.9
6703.469	2.5	.396	19.1	7812.363	8.2
6726.385	0.0	.424	17.3	7813.389	13.9
6728.409	6.3	7373.606	14.4	.437	12.7
6729.407	5.2	.626	15.1	7814.421	14.1
6730.384	6.7	7378.596	15.8	7817.386	16.4
6732.387	11.4	7400.556	13.9	7818.443	18.7
6744.358	21.7	7402.509	17.3	7821.369	17.7
6749.405	5.0	.538	17.3	.424	1.0
.474	6.4	7404.526	18.9	7839.356	16.9
6756.344	19.5	.551	7.2	7840.383	12.0
.401	19.5	7405.546	0.8	7844.418	3.6
6757.362	18.8	7406.529	−1.0	7847.394	5.7
6758.363	17.9	.554	1.0	7849.378	10.2
.389	20.1	7427.452	17.7	7850.370	13.6
6759.385	11.8	.477	5.4	7881.364	17.1
6761.379	18.4	7453.381	1.6	7884.390	1.3
6781.330	17.3	7454.398	4.1	8090.621	1.5
7016.565	3.2	7457.384	9.0	8113.559	5.3
7019.551	7.4	.412	17.3	8114.507	5.4
7020.563	9.0	7458.368	9.0	.597	6.3
7028.553	19.1	.394	11.6	8143.483	14.1
7046.503	15.1	7462.425	18.9	.510	12.7
7052.465	17.3	7464.397	15.1	8144.492	13.6
7073.437	18.9	.423	11.7	.514	12.5
7075.464	6.4	7758.565	0.0	8164.409	11.4
7078.388	3.6	7761.562	3.0	.439	8.0
.436	4.1	7763.536	7.4	8165.411	12.7
7079.367	5.5	.560	7.8	.435	8.1
.418	3.0	7764.529	7.7	8170.396	17.7
.467	4.5	7789.427	13.3	.447	16.9
7080.372	0.0	.480	12.4	8172.426	13.4
.400	3.4	7790.403	11.6	8173.431	4.1
.451	5.4	.501	12.4	.465	0.8
7087.453	16.2	7793.425	12.5	8195.349	13.1
7099.361	1.8	.480	17.7	8198.360	0.0
7101.400	11.4	.526	15.1	8200.395	7.4
7102.407	7.0	7794.444	16.7	8203.368	11.4
.456	12.0	7810.408	7.8		
7107.370	12.5	.454	18.7		

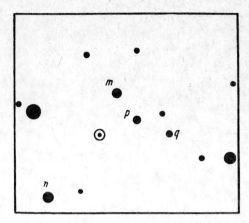

FIGURE 46. Comparison stars for AV Serpentis.

ET Pegasi

Discovered by Kaho /291/ who classified it as an RR Lyrae star.
Kaho's determination of the light-variation period, however, was inaccurate.
Visual observations of Tsesevich supplemented by Mandel's photographic
observations gave the following equation of the period:

$$\text{Max hel JD} = 2437872.494 + 0.489834 \cdot E. \tag{34}$$

The average light curves give two maxima:
Mandel, photographic Max hel JD = 2436869.315;
Tsesevich, visual Max hel JD = 2437872.494.
Comparison stars in visual observations (Figure 47): $k = -12^s.0$; $u = 0^s.0$;
$a = 11^s.2$; $b = 21^s.2$; $c = 21^s.7$. The theoretical light curve (Table 64) from
visual observations is based on the tentative equation

$$\text{Max hel JD} = 2437872.498 + 0.489834 \cdot E;$$

$$P^{-1} = 2.04150794. \tag{35}$$

The rise branch shows considerable scatter, which probably points
to the presence of Blazhko effect. Tsesevich's visual observations are
given in Table 65.

TABLE 64. Average visual light curve

φ	s	n	φ	s	n	φ	s	n
$0^P.009$	5.7	5	$0^P.696$	19.6	5	$0^P.949$	13.0	5
.026	6.4	5	.763	19.8	5	.958	13.5	5
.053	6.1	5	.824	19.7	5	.965	10.8	5
.086	7.2	5	.863	21.0	5	.975	7.0	5
.117	9.2	5	.886	19.9	5	.985	5.8	4
.148	11.7	5	.911	19.2	5	.995	5.0	4
.210	15.4	5	.926	17.9	5	$\dfrac{\text{Max} - \text{Min}}{P} = 0.12.$		
.610	19.5	5	.938	17.1	5			

63

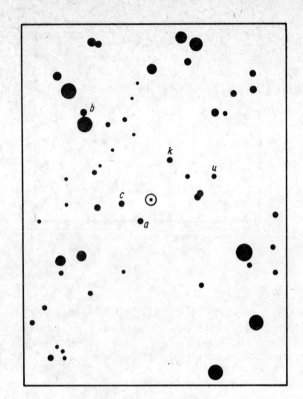

FIGURE 47. Comparison stars of ET Pegasi.

TABLE 65. Visual observations (Tsesevich)

JD hel	s	JD hel	s	JD hel	s
243...		243...		243...	
7582.379	19.4	7873.402	19.8	7901.369	15.7
.395	19.6	.416	20.7	.374	12.9
.416	19.6	.432	20.0	.378	11.2
.437	20.2	.444	17.9	.382	8.7
.448	20.2	.452	16.4	.388	5.3
7583.256	19.4	.457	14.4	.393	4.7
.300	19.4	.458	13.8	.396	4.3
.310	19.7	.461	10.1	.404	4.7
.324	19.7	.463	8.6	.412	4.9
.336	19.6	.469	7.0	.418	5.3
.343	19.6	.471	6.2	.425	5.6
.349	19.4	.476	5.9	.432	6.0
.388	19.6	.480	6.2	.437	5.8
7872.432	23.7	7900.337	19.6	.443	6.1
.470	17.3	.354	19.8	.449	8.4
.478	15.6	.361	19.2	.456	10.2
.482	14.4	.376	19.8	7902.306	19.4
.486	6.0	.386	17.3	.324	19.8
.496	5.7	.394	14.7	.333	18.6
.501	6.6	.396	12.2	.336	17.8
.506	8.2	7901.282	19.6	.341	17.7
.511	8.2	.334	21.7	.345	17.9
7873.316	19.4	.341	20.0	.352	12.4
.330	20.0	.348	20.0	.354	8.6
.348	19.8	.354	19.6	.360	7.0
.373	20.2	.358	18.8	.367	6.0
.383	19.1	.363	17.5	.384	6.5

TABLE 65 (continued)

JD hel	s	JD hel	s	JD hel	s
243...		243...		243...	
7902.391	6.0	7902.476	15.1	7903.390	6.8
.400	6.0	7903.297	20.7	.411	8.2
.406	6.7	.322	17.5	.416	9.6
.420	9.9	.329	17.3	.423	9.9
.433	9.2	.338	14.0	.429	14.0
.440	8.7	.345	9.2	.437	12.6
.447	11.4	.349	5.2	.468	15.9
.455	10.6	.353	4.1	.487	15.9
.462	14.0	.362	4.6		
.470	15.9	.373	4.7		

The tentative equation of the elements was used to derive the photographic curve from Kaho's observations (Figure 48 and Table 66). The average maximum time Max hel JD = 2 435 366.992.

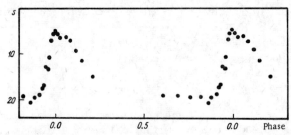

FIGURE 48. Light curve of ET Pegasi.

TABLE 66. Average photographic curve from Kaho's observations

φ	m	n	φ	m	n	φ	m	n
$0^P.060$	12.95	5	0.475^P	13.78	5	0.818^P	13.96	5
.099	12.88	5	.495	14.01	5	.851	14.16	3
.145	13.14	5	.524	13.98	5	.881	13.99	3
.216	13.48	5	.568	14.06	5	.938	13.08	3
.285	13.45	5	.621	13.98	5	.959	12.83	4
.328	13.64	5	.661	14.04	5	.988	12.73	4
.377	13.87	5	.724	14.05	5			
.440	13.85	5	.772	14.05	5	$\varphi_{Max} = 0^P.990.$		

Tsesevich determined the magnitude of the variable from old Moscow and Simeise photographs. Unfortunately, it proved impossible to improve the tentative equation of the period. It seems, however, that the period is actually longer. These observations are given below (the number of photographs is indicated in parentheses):

Moscow		Simeise		
JD	m	JD	m	
2418564.407	13.29	2423669.320	13.48	(2)
8566.383	13.92	4759.404	14.02	(2)
8595.222	13.34	4828.278	13.29	(2)
8598.225	13.77	7694.450	13.86	(2)
8599.209	13.77	8757.488	13.87	(2)
8919.387	13.14	9134.496	13.72	(1)
8925.301	13.67			

Y Lyrae

Discovered by Williams /367, 369/; very few observations available. The most extensive series of observations is that of Martynov /74/. A summary of all the available average maxima is given in Table 67.

TABLE 67. List of average maxima

Source	Max hel JD	E	O — A	O — B
Williams /367, 369/	2415020.274	0	$\overset{d}{0.000}$	$\overset{d}{+0.020}$
Hartwig /250/	5689.345	+ 1331	—.017	+.003
Tsesevich	26065.483	+21972	—.015	—.005
Martynov /74/	6954.755	+23741	—.012	—.002
	7344.344	+24516	—.012	—.002
	8748.373	+27307	—.006	+.003
Payne-Gaposchkin /329/	30938.614	+31666	—.014	—.008
Alaniya /2/	3868.327	+37494	—.010	—.006
Strelkova /115/	6268.206	+42268	+.001	+.003

The O — A residues were calculated from

$$\text{Max hel JD} = 2415020.274 + 0.50269545 \cdot E. \qquad (36)$$

The variation of the O — A values shows that the period probably remained constant over 40,000 epochs, and that Williams's first maximum should be ignored. The improved equation is thus

$$\text{Max hel JD} = 2415020.2535 + 0.50269589 \cdot E. \qquad (37)$$

Comparison stars used in Tsesevich's visual observations (Table 68): $c = -0^s.8$; $n = 7^s.6$; $x = 12^s.6$; $y = 17^s.0$. The chart of these stars was unfortunately lost.

TABLE 68. Visual observations (Tsesevich)

JD hel	s	JD hel	s	JD hel	s
242...		242...		242...	
6062.445	11.3	6065.465	11.4	6066.559	10.1
6064.359	17.0	.474	9.2	6067.410	20.1
.409	17.9	.478	7.7	.458	14.5
.436	18.1	.490	6.9	.462	14.8
.445	18.1	.502	7.1	.480	8.8
.455	15.9	.536	8.0	.489	6.8
.470	11.4	.550	8.8	.506	7.8
6065.408	19.2	6066.482	6.7	.514	8.9
.451	15.9	.531	8.9		

AO Pegasi

Discovered by Haas /249/ and investigated by Dubyago /224/. Also observed by Lange and Tsesevich. The maxima are listed in Table 69.

TABLE 69. List of maxima

Source	Max hel JD	E	O — C
Haas /249/	2424408.363:	— 2021	$-0^d.007$:
Tsesevich	5475.494	— 71	+.001
Dubyago /224/	5514.348	0	+.001
Lange /58/	7339.398	+ 3335	—.003
Tsesevich (Sim)	34300.319:	+16055	—.007:
Lange /58/	7196.329	+21347	—.004
	7218.219	+21387	—.004
	7868.349	+22575	+.002
	7903.357:	+22639	—.013:
	7908.298	+22648	+.002
	7909.392	+22650	+.002

The O — C residues were calculated from

$$\text{Max hel } JD = 2425514.347 + 0.54724252 \cdot E. \tag{38}$$

The period most probably remained constant. Comparison stars used in visual observations (Table 70, Figure 49): $s_1 - a = -9^s.7$; $c = 0^s.0$; $d = 3^s.3$; $s_2 - a = -10^s.5$; $c = 0^s.0$; $d = 7^s.9$; in photographic observations (Table 71, Figure 49): $s_3 - a' = 0^s.0$; $c = 8^s.0$; $d = 18^s.6$.

TABLE 70. Visual observations (Tsesevich)

JD hel	s_1	JD hel	s_1	JD hel	s_2
242...		242...		243...	
5469.472	4.3	5475.429	4.3	6104.335	5.6
5474.383	— 0.3	.449	3.0	.347	6.3
.398	— 1.0	.454	2.1	.386	8.9
.408	— 1.0	.468	1.1	.406	6.6
.439	— 1.0	.479	— 1.0	.421	8.9
.458	+ 0.7	.487	— 1.5	6105.262	6.5
.507	1.3	.506	— 1.0	.290	5.9
5475.296	4.3	.523	0.5	.309	5.3
.304	2.6	.549	0.9	.317	5.3
.313	3.8	5478.310	0.9	.366	6.5
.323	3.8	.321	1.1	.450	8.9
.334	3.8	.336	1.4	6106.260	1.8
.341	3.8	.374	2.1	.283	2.3
.349	3.8	5483.435	4.3	6112.235	— 1.5
.355	2.8	.451	4.3	6128.349	6.1
.361	2.8	.489	2.9	.370	5.6
.367	3.8	.501	3.8	6129.258	3.7
.376	4.8	.562	3.2	268	3.5
.385	4.0	5485.453	0.8	.346	4.6
.396	4.3	.526	2.9		
.410	4.3				

TABLE 71. Simeise observations (Tsesevich)

JD hel	s_3	JD hel	s_3	JD hel	s_3
242...		243...		243...	
9514.294	4.6	3148.292	14.7	4294.229	14.2
	4.6		16.3	4300.319	3.6
		4294.229	13.3		4.4

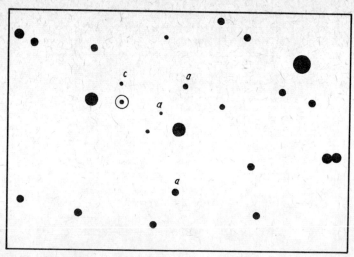

FIGURE 49. Comparison stars of AO Pegasi.

W Canum Venaticorum

Discovered by L. P. Tserasskaya /217/. Observed by various authors, especially by Prof. S. N. Blazhko. Blazhko's results were carefully reduced by Bugoslawski /209/. However, Bugoslawski determined not the maxima, but the times of passage through a certain magnitude on the rise branch. To obtain the maximum times, her figures should be incremented by $0^d.050$ (Table 72). Nachapkin /80/ reduced old observations by Hertz-sprung's method. These data were included in the table, as they seem to be highly reliable. The O$-$A residues were calculated from

$$\text{Max hel JD} = 2421077.9856 + 0.5517593 \cdot E. \qquad (39)$$

The variation of the O$-$A residues shows that the star retained a constant period, but equation (39) should be improved. The least squares method gives

$$\text{Max hel JD} = 2421077.9868 + 0.551758775 \cdot E. \qquad (40)$$

TABLE 72. List of maxima

Source	Max hel JD	E	O $-$ A	O $-$ B
Payne-Gaposchkin				
/330/	2415804.828	$-$ 9557	$+0^d.006$	$0^d.000$
Blazhko /209/	8002.469	$-$ 5574	$-.010$	$-.014$
Zinner /376/	20226.610	$-$ 1543	$-.011$	$-.013$
	0234.358:	$-$ 1529	$+.012:$	$+.010:$
	0238.243:	$-$ 1522	$+.035:$	$+.033:$
	0239.307:	$-$ 1520	$-.004:$	$-.006:$
	0240.413	$-$ 1518	$-.002$	$-.004$
	0245.374	$-$ 1509	$-.007$	$-.009$

TABLE 72 (continued)

Source	Max hel JD	E	O — A	O — B
Zinner /376/	2420250.358:	— 1500	+0^d.011:	+0^d.009:
	0251.450	— 1498	.000	—.002
	0255.354:	— 1491	+.042:	+.040
	0256.408:	— 1489	—.008:	—.010:
Robinson /344, 345/	1077.984	0	—.002	—.003
Blazhko /209/	1397.457	+ 579	+.003	+.002
	1402.423	+ 588	+.003	+.002
	1407.387	+ 597	+.001	.000
	1408.490	+ 599	+.001	.000
	1471.394	+ 713	+.004	+.003
Jordan /290/	2764.713	+ 3057	—.001	.000
Blazhko /209/	3304.336	+ 4035	+.002	+.003
	3309.298	+ 4044	—.002	—.001
	3544.355	+ 4470	+.005	+.006
	3545.457	+ 4472	+.004	+.005
	3945.486	+ 5197	+.007	+.009
	4253.367	+ 5755	+.007	+.008
	4264.394	+ 5775	—.002	.000
	4269.359	+ 5784	—.002	—.001
	4275.430	+ 5795	—.001	+.001
	4280.397	+ 5804	.000	+.002
	4285.361	+ 5813	—.001	.000
	4291.436	+ 5824	+.004	+.006
Parenago /85, 328/	4621.933	+ 6423	—.003	.000
Tsesevich (visual)	6072.503	+ 9052	—.008	—.004
Detre /221/	6516.123	+ 9856	—.002	+.002
	7048.020	+10820	—.001	+.003
Radlova /92/	7311.200	+11297	—.010	—.006
Solov'ev /108/	7628.468	+11872	—.004	+.001
Kleissen /297/	7926.435	+12412	+.013	+.018
Solov'ev /108/	7955.663	+12465	—.002	+.003
Nachapkin /80/	9360.431	+15011	—.013	—.007
Alaniya /1/	34130.377	+23656	—.027	—.015
Geyer /239/	5601.394	+26322	.000	+.013
	6232.5853	+27466	—.021	—.008
Strelkova /115/	6409.157	+ 27786	—.013	+.001
Satanova /94/	6877.601	+ 28635	—.012	+.002
Ahnert /176/	7025.458	+ 28903	—.027	—.013

Tsesevich's visual observations are given in Table 73. Comparison stars (Figure 50): $r = 0^s.0$; $m = 8^s.6$; $t = 14^s.0$.

TABLE 73. Visual observations (Tsesevich)

JD hel	s	JD hel	s	JD hel	s
242...		242...		242...	
6066.402	5.8	6071.518	6.3	6074.510	10.8
.415	5.8	.563	8.0	.556	10.8
.444	5.1	6072.404	12.2	6089.446	10.1
.456	1.6	.424	11.8	.465	10.9
.461	2.5	.443	11.7	.479	11.0
.479	5.4	.467	7.1	.519	11.6
.492	5.8	.476	4.9	6090.432	6.9
.572	7.0	.485	4.8	.445	9.7
6067.340	10.2	.495	4.3	.471	9.5
.443	11.1	.503	5.3	.495	10.9
.474	12.7	.517	3.7	.508	10.6
.505	7.0	.531	4.8	.520	10.9
.538	1.8	.541	4.6	6091.510	7.3
.546	2.0	.556	5.7	6092.404	4.3
.556	3.0	6073.334	11.6	.445	5.7
.577	4.8	.401	10.1	.466	6.5
6068.318	10.1	.412	10.0	.476	7.2
.388	10.3	.423	10.0	.496	7.6
.457	11.0	6074.337	9.3	6093.414	11.0
.513	10.6	.402	10.1	.446	6.4
.571	11.0	.461	10.5	.451	6.0
6071.319	11.7	.476	10.3	.460	3.8
.419	4.8	.491	9.4		

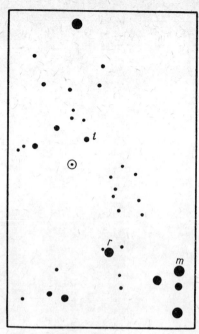

FIGURE 50. Comparison stars for W Canum
Venaticorum.

UY Cygni

Discovered by Williams, repeatedly observed by Luizet, Hartwig, Jordan
and others. Remarkable in that its period did not change over 35,000 epochs.

Mustel /77/ assembled the old observations and reduced them by Hertzsprung's technique. Mustel's summary of maxima (Table 74) was supplemented by later determinations.

TABLE 74. List of maxima

Source	Max hel JD	E	O — A
Williams /368, 369/	2415708.080	—11995	+0^d.007
Hartwig /250/	6911.894	— 9848	—.012
Luizet /309/	7472.040	— 8849	—.010
Jordan /290/	21074.029	— 2425	+.011
Tsesevich /77/	6152.322	+ 6632	+.001
Lause /302/	6406.876	+ 7086	—.005
Florya /77/	6428.745	+ 7125	—.004
Mustel /77/	7246.259	+ 8583	+.003
Selivanov /95/	7696.488	+ 9386	—.014
Gur'ev /27/	8064.890	+10043	+.005
Kleissen /297/	8424.289	+10684	—.008
Payne-Gaposchkin /329/	9886.615	+13292	.000
Solov'ev /27/	33516.632:	+19766	+.014:
Alaniya /1/	4626.255	+21745	+.003
Tsarevskii /142/	4862.319	+22166	+.010
Tsesevich (visual)	6041.470	+24269	—.001

Tsesevich carried out visual observations in two seasons, last in 1957. Tsesevich's old observations were reduced and published in part by Mustel.

The O—A residues were calculated from Mustel's equation

$$\text{Max hel JD} = 2422433.727 + 0.56070478 \cdot E; \quad P^{-1} = 1.78346973. \qquad (41)$$

We see from Table 74 that equation (41) does not require any improvement, although the star has gone through 16,000 cycles since Mustel's observations.

Comparison stars in visual observations (Figure 51): $x = 0^s.0$; $y = 7^s.1$; $a = 15^s.0$; $b = 19^s.8$. The observations were reduced to one period using (41). Two seasonal curves were constructed (Tables 75 and 76). It is significant that in both curves the rise branch shows a distinct stop. The visual observations of Tsesevich are shown in Table 77.

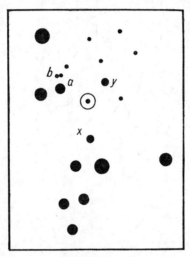

FIGURE 51. Comparison stars for UY Cygni.

TABLE 75. Average light curve for 1930

φ	s	n	φ	s	n	φ	s	n
0P.008	6.2	3	0P.289	11.8	3	0P.900	11.8	4
.029	7.6	3	.330	13.6	2	.911	12.2	4
.044	7.0	4	.453	14.4	4	.928	12.0	4
.056	9.3	2	.532	15.0	5	.943	9.6	3
.116	10.5	3	.626	16.1	2	.952	9.6	4
.145	10.4	2	.738	14.9	3	.965	6.0	3
.187	11.2	5	.821	16.2	5	.974	7.1	3
.223	11.1	3	.868	16.3	4	.990	6.1	5
.260	12.6	4						

TABLE 76. Average light curve for 1957

φ	s	n	φ	s	n	φ	s	n
0P.002	4.9	3	0P.258	13.0	2	0P.912	16.1	5
.014	6.9	3	.346	14.6	5	.930	11.4	5
.029	7.6	5	.401	15.4	5	.942	11.2	4
.044	8.3	4	483	14.6	4	.956	11.2	4
.072	9.8	5	.568	17.3	3	.973	7.2	4
.125	11.6	5	.721	17.0	3	.987	5.8	4
.180	11.7	3	.881	17.2	5			

TABLE 77. Visual observations (Tsesevich)

JD hel	s	JD hel	s	JD hel	s
242...		242...		243...	
6065.546	11.0	6216.359	8.6	6040.422	11.0
.578	8.3	.391	11.1	.438	11.3
6066.504	8.1	.396	12.9	.454	12.1
.531	5.5	6234.295	13.0	6041.327	17.1
.558	6.3	.350	14.2	.393	17.6
6067.504	13.5	.483	16.9	.411	17.9
.559	16.2	6239.328	9.5	.421	16.9
6068.480	14.3	.342	10.0	.428	13.7
.514	14.3	.376	12.0	.431	9.7
6071.525	10.5	.414	12.8	.434	10.7
6072.557	16.9	.479	14.6	.438	11.6
6074.528	14.7	.530	14.3	.444	11.3
6090.454	14.3	6250.374	14.3	.452	8.4
.498	14.3	.393	12.8	.455	7.1
6091.476	14.3	.404	12.0	.460	4.1
.494	14.3	.420	9.0	.464	4.1
6092.407	9.6	.428	8.6	.471	4.4
.421	12.2	.440	5.2	.479	6.1
6093.387	12.9	.451	5.2	.489	4.2
.397	13.0	.460	3.6	.494	5.0
.406	13.8	.468	7.1	.506	8.4
.418	11.4	.472	6.4	6043.356	16.9
.427	7.2	.509	10.0	.385	17.9
.433	5.9	.526	11.3	.440	17.9
.439	5.6	.554	11.4	.457	18.2
6185.347	10.5	6254.388	10.9	6050.376	13.9
.370	8.6	.430	12.3	.387	14.4
6189.398	9.1	6261.328	14.3	.403	12.0
6190.335	14.6	6298.228	13.6	.414	11.0
359	16.6	.291	14.3	.428	9.2
.383	16.7	243...		.433	6.5
.395	13.3	6027.384	18.8	.449	4.7
.400	12.0	.396	17.9	.456	8.4
.409	9.5	.400	17.9	.462	8.4
.417	8.2	.411	17.7	.470	9.4
.429	4.1	.413	13.2	.488	10.6
.439	6.8	.415	11.4	.501	12.7
.454	5.0	.418	11.2	.520	12.4
6191.315	15.0	.421	10.9	6052.308	16.7
.340	17.9	.426	11.0	.322	13.6
6194.324	10.9	.429	11.3	.329	15.0
.405	10.0	.434	11.3	.344	16.2
.516	12.8	.446	6.4	.352	15.0
6195.401	18.2	.450	6.2	.368	13.0
6209.407	14.6	.456	6.1	.382	13.0
.420	17.4	.462	9.9	.402	13.9
.444	17.7	.467	10.8	.431	16.2
.474	12.1	.474	10.5	.471	17.4
.481	12.0	.486	10.8	.510	17.1
.487	10.0	.493	9.7	6053.338	11.0
.494	6.8	6028.426	16.9	.353	12.0
.499	6.8	6040.351	4.2	6057.299	13.0
.509	7.2	.366	7.1	.332	13.0
.515	8.3	.377	8.2	.357	13.0
.527	8.2	.399	9.7	.373	12.9
.536	8.2	.405	10.7	.433	13.8
.545	8.6				

SW Draconis

Discovered by Leavitt /331, 332/. First studied by Sperra /353/, and then by Martin and Plummer /319/. Subsequently observed by Nijland /323/, Graff /243/, Jordan /290/, Tsesevich /148/, Lause and Robinson /346/. Florya /137, 138/ reduced all the old observations (except those of Tsesevich) using Hertzsprung's technique and obtained the maxima listed in Table 78. Tsesevich added new data.

TABLE 78. List of maxima

Source	Max hel JD	E	O−A	O−B
Sperra /353/	2418423.5192	− 13694	−0^d.0057	−0^d.0084
Martin and				
Plummer /319/	9598.1859	− 11632	+.0010	−.0011
Nijland /323/	20120.5758	− 10715	+.0033	+.0015
Jordan /290/	0912.4192	− 9325	+.0051	+.0037
	3886.6578	− 4104	−.0045	−.0044
Florya /137, 138/	6394.3497	+ 298	−.0008	+.0005
Lause /303/	6403.4648	+ 314	−.0004	+.0008
Florya /137, 138/	6929.8421	+ 1238	+.0016	+.0031
Solov'ev /108/	7571.286	+ 2364	−.003	−. 001
Payne-Gaposchkin				
/329/	7865.807	+ 2881	−.002	. 000
Solov'ev /108/	7976.892	+ 3076	−.002	. 000
Gur'ev /27/	8438.892	+ 3887	−.005	−. 003
Dziewulski /226/	8591.5872	+ 4155	+.0187	+.0211
Azarnova	33058.350	+ 11996	−.003	+. 002
	3156.340	+ 12168	+.004	+. 009
	3160.330	+ 12175	+.006	+. 011
	3218.432	+ 12277	+.002	+. 007
Alaniya /1/	4392.527	+ 14338	+.007	+. 012
Batyrev /20/	4526.380	+ 14573	−.013	−. 007
	5238.473	+ 15823	−.008	−. 002
	5241.320	+ 15828	−.009	−. 003
	5274.373	+ 15886	+.003	+. 009
	5279.486	+ 15895	−.011	−. 005
Lange	7878.317	+ 20457	−.015	−. 008
	7882.306	+ 20464	−.014	−. 007
Sakharov	7886.305	+ 20471	−.003	+. 004
Lange	7887.429	+ 20473	−.018	−. 011

The O−A residues were calculated from

$$\text{Max hel JD} = 2426224.5888 + 0.56967021 \cdot E. \qquad (42)$$

The period is constant, but equation (42) should be slightly modified. Tsesevich found a new equation, after omitting Dziewulski's maximum:

$$\text{Max hel JD} = 2426224.5876 + 0.56966993 \cdot E. \qquad (43)$$

RX Eridani

Observed relatively little. Investigated by Zinner /377/ who gave a correct determination of its period. Later observed by Tsesevich /157/. The observations of Gur'ev /29/, Alaniya /1/, and Kinman /294/ were also published; the Harvard photographs were reduced by Payne-Gaposchkin /329/. The maxima are summarized in Table 79.

TABLE 79. List of maxima

Source	Max hel JD	E	O−A	O−B
Zinner /377/	2419704.611:	− 3385	−0^d.044:	−0^d.040:
Payne-Gaposchkin				
/329/	21692.479	0	.000	.000
Tsesevich /157/	6592.454	+ 8344	+ .003	− .007
	7415.190	+ 9745	+ .008	− .003
Gur'ev /29/	8102.268	+ 10915	− .010	− .003
	8442.297	+ 11494	+ .024	+ .010
	9535.157	+ 13355	+ .021	+ .005
Alaniya /1/	34336.449:	+ 21531	− .002:	− .028:
Kinman /294/	6964.401	+ 26006	+ .029	− .003

The O—A residues were calculated from

$$\text{Max hel JD} = 2421692.479 + 0.587245 \cdot E, \tag{44}$$

and the O—B residues from the improved formula

$$\text{Max hel JD} = 2421692.479 + 0.58724622 \cdot E. \tag{45}$$

There is no reason to believe that the period of this star is variable.

ST Comae Berenices

Discovered by Guthnick and Prager from Berlin—Babelsberg photographs /247, 248/. Their formula

$$\text{Max hel JD} = 2424998.480 + 0.37460 \cdot E$$

was wrong, however. Tsesevich established the error /153/ and Solov'ev /110/ derived an improved formula on the basis of his and Tsesevich's observations:

$$\text{Max hel JD} = 2427862.583 + 0.5989312 \cdot E. \tag{46}$$

The star was observed by Kukarkin /50/, Alaniya /1/, Payne-Gaposchkin /329/, Jacchia /282/, Batyrev. Tsesevich measured the Simeise photographs and Gvozdik the Odessa photographs. The maxima are summarized in Table 80.

TABLE 80. List of maxima

Source	Max hel JD	E	O — A	O — B
Tsesevich (Sim.)	2420957.532	— 11529	$+0^d.027$	$+0^d.005$
Jacchia /282/	5760.396:	— 3510	+.062:	+.054:
Kukarkin /50/	6883.929	— 1634	0	—.005
Tsesevich (visual)	7459.504	— 673	+.002	—.001
	7853.598	— 15	—.001	—.003
Solov'ev /110/	7914.699	+ 87	+.009	+.007
	8305.184	+ 739	—.009	—.010
	8708.262	+ 1412	—.012	—.011
	9337.149	+ 2462	—.003	.000
Payne-Gaposchkin /329/	31210.610	+ 5590	+.002	+.009
Alaniya /1/	4478.387:	+ 11046	+.010:	+.027:
Gvozdik	6613.543	+ 14611	—.024	—.001
Tsesevich (Odessa)	6715.371	+ 14781	—.014	+.009
Gvozdik	7781.452	+ 16561	—.031	—.004

The star shows two peculiar features. First, the range of the light curve is relatively small, and the maxima are therefore pinpointed with considerable uncertainty. Second, the considerable spread of the points in seasonal light curves points to the presence of Blazhko'effect. Therefore, without attaching much importance to the scatter of the O—A values, we prefer to derive an improved formula based on the average maxima (except the old Simeise maximum, which is also taken into consideration), ignoring the slow variations in the period. The improved equation is

$$\text{Max hel JD} = 2427862.585 + 0.5989295 \cdot E;$$
$$P^{-1} = 1.66964559. \tag{47}$$

Comparison stars (Figure 52): Simeise observations (s_1) $k = 0^s.0$; $a = 10^s.3$; $c = 19^s.5$; Odessa photographic observations of Tsesevich (s_2) $k = 0^s.0$; $a = 10^s.8$; $b = 20^s.4$; Odessa photographic observations of Gvozdik (s_3) $k = 0^s.0$; $a = 8^s.1$; $c = 13^s.9$.

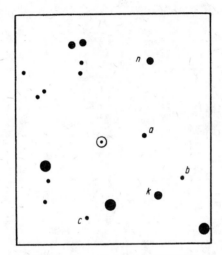

FIGURE 52. Comparison stars for ST Comae Berenices.

The comparison stars in Tsesevich's visual observations (s_4) $k = -11^s.7$; $n = -7^s.6$; $a = 0^s.0$; $c = 10^s.3$; the average light curve (Table 81) was obtained from equation (47).

The observations are summarized in Tables 82 —85.

TABLE 81. Average light curve

φ	s_4	n	φ	s_4	n	φ	s_4	n
$0^p.007$	— 1.6	5	$0^p.292$	5.2	10	$0^p.720$	9.4	10
.027	— 0.8	5	.379	7.0	10	.778	8.5	10
.053	+ 2.5	10	.434	7.2	10	.847	9.0	10
.091	2.4	10	.478	6.8	10	.889	7.4	10
.126	4.0	10	.510	8.2	10	.911	5.2	10
.164	3.7	10	.576	8.4	10	.939	2.9	10
.226	4.1	10	.650	8.7	10	.974	0.7	5

TABLE 82. Simeise observations (Tsesevich)

JD hel	s_1	JD hel	s_1	JD hel	s_1
242...		242...		242...	
0957.532	7.5	5716.295	15.8	6861.357	10.8
	5.9	5728.273	13.0	243...	
0967.568	15.4	5738.284	8.8	2996.307	13.7
	17.9	6061.370	17.9		16.0
0985.519	15.8	.561	9.8	3002.287	16.4
	16.7	.561	9.3		15.4
4963.353	13.1	6070.378	15.6	4127.334	9.2
	11.4		19.0		8.8
5709.293	17.4	6424.361	14.5		
	17.3		15.8		

TABLE 83. Odessa observations (Tsesevich)

JD hel	s_2	JD hel	s_2	JD hel	s_2
243...		243...		243...	
6288.452	8.7	6660.488	10.8	6997.601	12.0
6304.398	16.3	6661.466	6.3	7015.492	6.8
6313.447	7.2	6663.484	14.3	7016.455	16.6
6335.379	16.1	6668.442	14.6	7017.461	17.2
6340.364	14.0	6699.372	10.8	7019.474	16.9
6344.357	12.7	6702.396	10.8	7020.502	16.6
6345.349	8.8	6714.345	11.9	7052.407	15.6
6607.572	7.2	6715.353	5.8	7077.362	15.6
6608.596	16.1	6716.374	16.6	7078.359	9.4
6612.636	13.4	6722.339	15.6	7079.336	17.2
6613.589	6.0	6971.592	16.6		

TABLE 84. Photographic observations (Gvozdik)

JD hel	s_3	JD hel	s_3	JD hel	s_3
243...		243...		243...	
6288.452	4.0	7079.336	11.0	7780.426	8.1
6313.447	3.5	7100.368	10.4	7781.412	4.1
6340.364	5.1	7101.354	10.0	.437	3.6
6344.357	9.3	7105.345	5.1	7783.403	11.0
6345.349	3.5	7111.340	5.1	7786.391	6.3
6607.572	2.7	7373.557	6.3	7789.400	10.1
6608.596	5.8	7377.538	7.1	7807.350	9.3
6613.589	2.7	.565	9.1	7808.341	10.0
6660.488	5.1	7398.457	11.0	.365	5.4
6663.484	5.8	.484	11.0	7810.338	6.3
6667.417	2.7	7400.457	8.1	7812.336	10.6
6668.442	7.0	.481	5.1	8085.563	11.0
6699.372	4.0	7405.426	11.0	8090.484	3.6
6702.396	5.1	.449	8.1	.509	2.7
6714.345	6.1	7406.442	10.0	.575	5.4
6715.353	3.0	7425.378	11.0	8106.484	10.0
6716.374	10.4	.406	11.0	8138.422	3.0
6722.339	8.1	7426.391	5.1	.449	4.5
6971.592	6.1	.417	6.3	8141.455	4.8
6997.601	4.0	7734.518	11.6	8143.358	6.3
7015.492	3.6	.541	10 0	.382	10.4
7016.455	7.1	7758.468	13.9	.410	8.1
7017.461	7.1	.493	10.0	8144.405	4.1
7019.474	9.1	7759.448	10.4	8162.348	4.5
7020.502	6.1	7761.467	11.6	8163.355	10.0
7046.444	10.0	.491	11.0	.379	10.0
7052.407	10.0	7764.471	11.0	8165.324	5.1
7077.362	9.1	7780.400	6 1	.348	4.1
7078.359	3.0				

TABLE 85. Visual observations (Tsesevich)

JD hel	s_4	JD hel	s_4	JD hel	s_4
242...		242...		242...	
6733.485	3.5	7485.331	8.1	7600.226	1.0
6765.417	2.4	.335	6.2	.324	1.5
.437	1.7	.342	5.6	7601.222	7.4
.462	1.7	.351	4.2	.264	8.4
.487	3.3	.357	4.3	7602.198	4.9
.495	2.1	.363	4.9	.232	4.3
.510	2.7	.371	4.6	.244	5.2
6770.385	3.4	.395	5.9	.347	6.7
.428	6.1	.403	5.4	7843.267	9.3
6771.383	1.6	.430	5.6	.282	8.9
.397	0.7	.453	6.9	.328	9.8
.421	1.0	.508	8.7	.346	9.2
.453	2.4	7490.339	7.1	.350	8.0
.481	3.3	.353	8.2	.363	5.4
.546	4.3	.402	8.6	.369	6.5
.594	5.4	.414	8.0	.380	5.6
6809.354	7.1	.464	9.7	.392	3.4
6812.312	5.4	.484	9.4	.436	—1.2
.329	5.0	7492.486	1.8	.438	—2.0
.350	5.6	.496	2.5	7857.200	—1.4
.369	6.8	.509	2.1	.222	—1.3
7424.454	6.5	.521	2.7	.236	—1.5
.468	8.5	.532	2.7	.278	4.3
.492	8.0	7508.219	9.5	.305	5.7
.529	8.7	.234	9.6	7858.195	9.8
.549	9.0	.255	8.7	.218	9.3
7425.428	5.2	.280	8.3	.228	9.6
.451	5.6	.288	6.3	.243	9.8
.475	5.6	.293	6.0	.257	9.3
.529	6.7	.298	6.0	.264	9.8
7459.364	8.5	.314	6.1	.284	8.9
.381	9.0	.335	7.9	.316	9.2
.396	8.8	.353	7.4	.326	8.4
.431	8.2	7509.329	6.3	.335	6.5
.444	7.8	.364	5.6	.346	4.2
.495	1.5	.394	7.5	.362	1.8
.509	— 2.1	.420	7.5	.369	—1.0
.513	— 1.7	7510.185	8.4	.385	—1.5
.522	— 0.2	.198	9.2	.394	—1.0
.530	1.0	.229	9.1	7861.206	9.0
.536	0.0	.282	8.9	.240	8.5
.540	1.1	.298	9.1	.304	9.0
.549	1.9	.317	9.4	.327	7.9
.558	2.8	.344	6.0	.335	6.7
7460.326	8.8	.348	5.6	.349	3.6
.354	8.4	.357	3.2	.363	1.9
.378	8.8	.358	3.1	.396	0.9
.386	8.3	.364	0.9	.412	0.0
397	9.2	.373	1.4	7862.220	7.9
.417	8.6	.378	1.7	.284	8.7
.422	7.8	.401	2.5	.325	9.2
.442	8.8	.416	—1.6	.370	8.2
.461	8.8	.433	—1.4	.391	9.4
.472	8.2	.447	—1.0	.435	9.2
.489	8.2	.465	1.7	.482	9.4
.508	8.7	.475	—0.5	.491	9.2
.535	9.2	.482	1.5	.504	8.5
7461.329	6.4	7510.487	1.7	.518	7.3
.337	6.2	.514	2.7	.530	8.0
.358	6.3	7599.206	1.7	.535	5.7
7484.303	8.6	.214	2.1	.540	4.6
.326	8.0	.259	2.2	.546	4.1
.359	8.7	.312	6.2	7864.235	9.4
.372	9.2	7600.180	7.6	.325	8.8
.377	8.6	.215	2.1	7865.232	9.8

77

AT Andromedae

Discovered by Morgenroth /321/. Lange determined its period /62/. Parenago /89/ obtained from the Moscow photographs

$$\text{Max hel JD} = 2429146.374 + 0.6169129 \cdot E; P^{-1} = 1.62097437. \qquad (48)$$

Subsequently observed by Alaniya /2/, Kippenhan /296/, Lange /69/, and Tsesevich. At Tsesevich's request, R.I. Chuprin revised Parenago observations and again determined the maxima (Table 86).

TABLE 86. List of maxima

Source	Max hel JD	E	O — A
Parenago /89/	2417469.447	—18928	$0^d.000$:
Lange /62/	28022.37:	— 1822	+.011:
Parenago /89/	8745.378	— 650	—.003
	9162.423	+ 26	+.009
Tsesevich (visual)	31700.402	+ 4140	+.009
Alaniya /2/	4961.450	+ 9426	+.055
Lange /69/	6427.179	+11802	—.001

Comparison stars in the visual observations of Table 87 (Figure 53): $m = 0^s.0$; $k = 12^s.2$; $a = 18^s.9$; $b = 25^s.4$.

TABLE 87. Visual observations (Tsesevich)

JD hel	s	JD hel	s	JD hel	s
243...		243...		243...	
1700.318	23.4	1700.408	7.8	1701.281	16.2
.363	17.4	.420	11.0	.288	16.3
.379	17.1	.427	14.0	.297	17.1
.395	7.1	.431	15.6	.322	16.3
.402	8.4	1701.272	16.0	.337	19.8

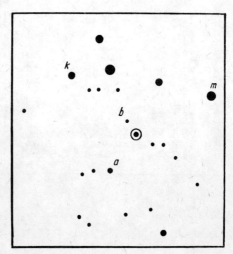

FIGURE 53. Comparison stars for AT Andromedae.

AV Virginis

Discovered by Hoffmeister /267/. Observed by Jensch /288/, Solov'ev /108/, and Payne-Gaposchkin /329/. Tsesevich estimated its magnitude from Simeise and Odessa photographs and plotted seasonal light curves using the equation

$$\text{Max hel JD} = 2427871.611 + 0.656913 \cdot E;$$
$$P^{-1} = 1.5222716.$$

(49)

The maxima are listed in Table 88.

TABLE 88. List of maxima

Source	Max hel JD	E	O — A	O — B
Tsesevich (Sim.)	2419922.373	—12101	$+0^d.067$	$+0^d.003$
Jensch /288/	27871.611	0	.000	—.003
Solov'ev /108/	8306.477	+ 662	—.010	—.010
	8308.445	+ 665	—.013	—.013
Tsesevich (Sim.)	9350.335	+ 2251	+.013	+.021
Payne—Gaposchkin /329/	30444.720	+ 3917	—.019	—.003
Tsesevich (Odessa)	7016.431	+13921	—.066	+.001

The O — B residues were calculated from the improved equation

$$\text{Max hel JD} = 2427871.614 + 0.6569080 \cdot E.$$

(50)

There is no indication of variable period at this stage. The observations listed in Tables 89 and 90 used the following comparison stars (Figure 54): $a = 0^s.0; \ b = 4^s.3; \ c = 15^s.1.$

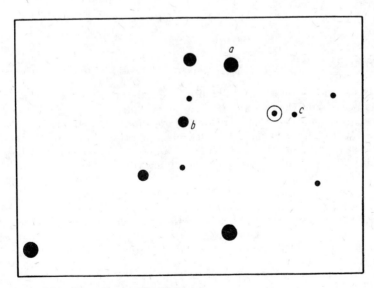

FIGURE 54. Comparison stars for AV Virginis.

TABLE 89. Simeise observations (Tsesevich)

JD hel	s	JD hel	s	JD hel	s
241...		242...		242...	
9886.321	2.6	6070.507	4.0	9364.386	11.2
	2.6		6.4	243...	
9892.325	6.4	6420.582	2.4	0108.447	1.1
	2.4		0.6		−0.4
9922.331	1.8	6423.563	9.2	0112.354	−1.2
	1.3	6806.555	11.9		−2.2
242...			11.2	2998.433	9.2
0579.517	10.4	8258.518	3.2		8.0
	10.2		2.7	3736.425	6.5
0580.337	11.3	8278.340	6.7		7.5
	9.8		5.7	4120.393	6.0
0955.504	11.7	9349.400	10.8		1.1
	13.1		10.0	4122.357	7.5
3523.306	12.0	9350.501	4.3		11.5
	11.5		5.4		

TABLE 90. Odessa observations (Tsesevich)

JD hel	s	JD hel	s	JD hel	s
243...		243...		243...	
6288.452	12.1	6667.418	10.2	7017.461	15.1
6304.399	2.2	6668.443	8.8	7019.474	13.1
6313.447	8.2	6699.373	10.5	7020.502	4.3
6344.329	3.3	6702.398	2.3	7028.496	13.4
6345.349	10.2	6714.347	8.8	7044.379	13.1
6347.348	11.9	6715.357	12.1	7046.444	11.8
6608.664	3.4	6716.376	7.1	7052.408	13.1
6612.635	10.6	6722.340	5.6	7077.363	12.9
6613.588	11.5	6971.591	12.9	7078.360	9.2
6660.489	4.3	6997.601	11.2	7079.337	12.1
6661.494	10.2	7015.492	11.9		
6663.485	11.0	7016.455	0.3		

SU Draconis

An extensively observed variable. Old observations examined by Florya /130/ and Dziewulski /227/. A considerable amount of new observations have become available since then. The list of maxima (Table 91) includes all the available data. Only one determination (by Dziewulski) has been omitted, since it apparently contains an error (possibly a misprint). The O−A residues were calculated from the equation

$$\text{Max hel JD} = 2420605.7569 + 0.66041926 \cdot E. \tag{51}$$

Although the latest observations deviate from (51) approximately by 10 min, the last equation can be used at this stage. The period remained constant for the duration of 32,000 cycles. The individual maxima are listed in Table 92.

TABLE 91. List of maxima

Source	Max hel JD	E	O — A
Payne-Gaposchkin /330/	2416556.738	— 6131	+0^d.012
Enebo /228/	8251.3591	— 3565	—.0031
Sperra /353/	8394.6736	— 3348	+.0004
Ginori /241/	9451.3499	— 1748	+.0059
Martin and Plummer /316, 318/	9724.7589	— 1334	+.0013
Jordan /290/	20688.3088	+ 125	—.0005
Strashnyi /114/	6258.9548	+ 8560	—.0010
Florya /130/	6322.3536	+ 8656	+.0076
Kukarkin /51/	6540.2710	+ 8986	—.0134
Florya /130/	6583.8783	+ 9052	+.0063
	6929.9310	+ 9576	—.0007
Kowalczewski, Kepinski /299/	7151.1668	+ 9911	—.0054
Solov'ev /112, 113/	7486.665	+10119	.000
Solov'ev /108/	7572.528	+10549	+.008
Dziewulski /227/	7882.2645	+11018	+.0082
Solov'ev /108/	7976.686	+11161	—.010
Opalski /227/	8036.1386	+11251	+.0046
Alaniya /2/	34807.415	+21504	+.002
Geyer /239/	5892.4618	+23147	—.0197
Ahnert /175/	6610.357	+24234	.000
Koshuba /43/	6902.263	+24676	.000
Spinrad /354/	7044.9131	+24892	.000
Lange /65/	7137.360	+25032	—.012
Sakharov	7881.6564	+26159	—.0079

TABLE 92. Individual maxima

Source	Max hel JD	E	O — A
Ahnert /175/	2436604.398	24225	—0^p.015
	6608.368	24231	—.008
	6610.364	24234	+.007
	6612.350	24237	+.011
	6614.323	24240	+.003
	6637.443	24275	+.009
	6660.551	24310	+.002
Sakharov	7868.447	26139	—.009
	7872.409	26145	—.009
	7878.351	26154	—.011
	7882.320	26160	—.005
	7886.277:	26166	—.010:
	7903.454	26192	—.004

STARS WITH PERIODS VARYING IN PROPORTION TO TIME

If the star's period varies in proportion to time, the maxima are predicted by the equation

$$Max = M_0 + PE + aE^2. \tag{52}$$

The General Catalogue of Variable Stars gives four equations:

SW And Max hel JD = $2418132.7913 + 0.4422792156 \cdot E - 1.229 \cdot 10^{-10} \cdot E^2$;

S Ara Max hel JD = $2420047.2382 + 0.451888176 \cdot E - 0.95 \cdot 10^{-10} \cdot E^2$;

RR Leo Max hel JD = $2430440.361 + 0.45238142 \cdot E + 1.80 \cdot 10^{-10} \cdot E^2$;

TU UMa Max hel JD = $2425760.441 + 0.557665 \cdot E - 4.03 \cdot 10^{-10} \cdot E^2$.

Equation (52) can be considered as part of a more complex interpolation formula. It therefore often happens that the established equation does not fit new observations. The requirements imposed by (52) are very rigid. The higher the index E, the greater is the inaccuracy involved, since the formula contains E^2. It will be shown in what follows that the formula for TU UMa fell through. The period of this star is subject to irregular fluctuations. Recent observations do not fit the equation for RR Leo either. It is nevertheless important to continue the search for additional stars with periods described by relation (52).

AV Pegasi

Discovered by Hoffmeister /263/ in 1931, and two years later independently by Shapley and Hughes at the Harvard Observatory /350/. Examination of the Harvard plates by C. Payne-Gaposchkin and S. Gaposchkin revealed variations in its period. The maxima (unfortunately unpublished) were fitted by two linear equations.

Recent observation series show that the star markedly deviates from the linear ephemeris. Tsesevich examined all the data available, and the list of maxima is given in Table 93. The O—A residues were calculated from

$$\text{Max hel JD} = 2426582.302 + 0.39037523 \cdot E. \qquad (53)$$

This formula is based on the maximum of Florya and the period derived from the second relation of Payne-Gaposchkin.

TABLE 93. List of maxima

Source	Max hel JD	E	O — A
Payne-Gaposchkin /329/	2416347.713	— 26218	+0^d.269
Tsesevich (Moscow)	8950.26	— 19551	+.184
Florya /120-124/	26582.302	0	.000
Payne-Gaposchkin /329/	6652.561	+ 180	—.009
Florya /120-124/	6970.329	+ 994	—.006
	7312.293	+ 1870	—.011
Dombrovskii /35/	7616.773	+ 2650	—.023
Solov'ev /108/	7631.228	+ 2687	—.012
Balascz /185/	7653.4746	+ 2744	—.0170
	7655.4263	+ 2749	—.0172
	7660.5002	+ 2762	—.0182
Solov'ev /108/	7981.388	+ 3584	—.019
Gur'ev /27/	8098.495	+ 3884	—.024
	8866.355	+ 5851	—.032
Batyrev /5/	33191.265	+ 16930	—.090
	3203.367	+ 16961	—.089
	3239.282	+ 17053	—.089
	3527.377	+ 17791	—.091
	3529.324	+ 17796	—.096
	3532.457	+ 17804	—.086
	3575.391	+ 17914	—.093
	3642.144	+ 18085	—.094
Sofronievitsch /203/	3888.478	+ 18716	—.087
Born /204/	3898.624	+ 18742	—.091
	3898.630	+ 18742	—.085
	3910.338	+ 18772	—.088
Sofronievitsch /203/	3910.340	+ 18772	—.086
Born /204/	3912.283	+ 18777	—.095
	3917.367	+ 18790	—.086

TABLE 93 (continued)

Source	Max hel JD	E	O — A
Sofronievitsch /203/	2433917.362	+ 18790	—0^d.091
Born /204/	3926.338	+ 18813	—.093
Sofronievitsch /203/	3926.342	+ 18813	—.089
	3928.291	+ 18818	—.092
Alaniya /1/	4244.495	+ 19628	—.092
	4252.304	+ 19648	—.091
Tsarevskii /141/	6146.3994	+ 24500	—.096
Lange /65/	7136.366	+ 27036	—.121
Lange /68/	7497.468	+ 27961	—.116
	7501.372	+ 27971	—.116
	7519.328	+ 28017	—.117
	7528.312:	+ 28040	—.111
	7868.317	+ 28911	—.123
	7871.441	+ 28919	—.122
Sakharov	7871.448	+ 28919	—.115
	7873.391	+ 28924	—.124
Lange	7873.398	+ 28924	—.117
Sakharov	7878.477	+ 28937	—.113
	7882.396	+ 28947	—.098
	7885.506	+ 28955	—.111
	7887.460	+ 28960	—.109
	7898.377:	+ 28988	—.122:
Lange	7898.393	+ 28988	—.106
	7900.342	+ 28993	—.109
Sakharov	7900.340	+ 28993	—.111
	7902.293	+ 28998	—.110
Lange	7902.292	+ 28998	—.111
Sakharov	7903.457	+ 29001	—.117
	7911.268	+ 29021	—.114
	7912.444	+ 29024	—.109
	7916.347	+ 29034	—.109
Lange	7916.343	+ 29034	—.113
	7918.288	+ 29039	—.120
	7923.377	+ 29052	—.106
	7925.322	+ 29057	—.113
Sakharov	7925.325	+ 29057	—.110
Kanishcheva	7925.326	+ 29057	—.109
	7932.350:	+ 29075	—.112
Sakharov	7932.351	+ 29075	—.111
Lange	7932.347	+ 29075	—.115
	7939.374:	+ 29093	—.115:
Kanishcheva	7939.372	+ 29093	—.117
Lange	7943.271	+ 29103	—.121

The variation of the O —A residues shows that the period changes progressively, and not abruptly. Since the reduction of all the data by the least squares method should take too much time, the figures of Table 93 were averaged for further treatment (Table 94).

TABLE 94. Average O—A values

E	O — A	n	$(O — A)_{calc}$	δ
— 26218	+0^d.269	1	+0^d.267	+0^d.002
— 19551	+ .184	1	+ .185	—.001
+ 180	— .009	1	.000	—.009
+ 955	— .006	3	— .005	—.001
+ 2668	— .018	2	— .017	—.001
+ 2752	— .018	3	— .018	.000
+ 3734	— .022	2	— .024	+.002
+ 5851	— .032	1	— .037	+.005
+ 16981	— .089	3	— .088	—.001
+ 17878	— .092	5	— .091	—.001
+ 18777	— .089	11	— .094	+.005
+ 19638	— .092	2	— .096	+.004
+ 24500	— .096	1	— .107	+.011
+ 27036	— .121	1	— .111	—.010
+ 27997	— .115	4	— .112	—.003
+ 29008	— .113	31	— .112	—.001

83

The O—A residues can be described by the equation

$$O-A = +0.0015 - 0.0718 \, (E/10000) + 0.0112 \, (E/10000)^2.$$

The last column of Table 94 gives the differences $\delta = (O-A)-(O-A)_{calc}$. They clearly fall within the margin of observation errors. The final equation is

$$\text{Max hel JD} = 2436582.3035 + 0.39036805 \cdot E + 0.0112 \cdot 10^{-8} \cdot E^2. \tag{54}$$

VX Herculis

Again observed by Tsesevich in 1957. Comparison stars from /158/: $a = \text{BD} + 18°3190 = 12^s.5$; $b = \text{BD} + 18°3189 = 20^s.3$; $n = 16^h 24^m 0^s + 18°40' = 32^s.0$; $g = 16^h 24^m 0^s + 18°50' = 47^s.6$. The equation of the period

$$\text{Max hel JD} = 2421750.483 + 0.4553719 \cdot E; \; P^{-1} = 2.19600726 \tag{55}$$

was used to construct the light curve; the resulting maximum

$$\text{Max hel JD} = 2436020.432,$$

deviates by $O-A = -0^d.040$ for $E = 31337$ from the elements of (55). Table 95 lists all the currently available maxima.

TABLE 95. List of maxima

Source	Max hel JD	E	O — B
Payne-Gaposchkin /329/	2420950.845	— 1756	—0^d.009
Esch /230, 231/	1750.483	0	—.003
Leiner /306, 307/	2475.438	+ 1592	+.001
Tsesevich /158/	5069.231	+ 7288	—.003
	6564.219	+ 10571	+.001
	6946.275	+ 11410	.000
Florya /125/	7267.314	+ 12115	+.002
Solov'ev /108/	7611.582	+ 12871	+.009
	7978.597	+ 13677	—.005
Kleissen /297/	7979.521	+ 13679	+.008
Solov'ev /108/	8379.330	+ 14557	+.001
Tsesevich /158/	30900.261	+ 20093	—.005
Azarnova	3139.311	+ 25010	—.017
Alaniya /1/	4593.331	+ 28203	+.002
Geyer /239/	5694.380	+ 30621	—.037
	5984.4614	+ 31258	—.0276
Batyrev	6016.342	+ 31328	—.023
Tsesevich	6020.432	+ 31337	—.031
Batyrev	6032.285	+ 31363	—.018
Lange	7842.349:	+ 35338	—.056
	7847.360	+ 35349	—.054
	7852.366	+ 35360	—.057:
	7868.308	+ 35395	—.053
	7872.406	+ 35404	—.053
	7873.324	+ 35406	—.046
	7878.329	+ 35417	—.050
	7883.333	+ 35428	—.055
Batyrev	8220.320	+ 36168	—.043
	8235.333	+ 36201	—.057
	8251.277	+ 36236	—.051
	8261.295	+ 36258	—.052
	8292.255	+ 36326	—.057

The O−B residues were calculated from

$$\text{Max hel JD} = 2421750.486 + 0.45537152 \cdot E. \tag{56}$$

The plot shows that the period varies in proportion to time. To simplify the calculations, the average O−B values were calculated (Table 96).

TABLE 96. Average O−B values

$\overline{E}$	$\overline{O-B}$	n	$\overline{(O-B)}_{calc}$	δ
− 1756	−.009	1	−.006	−.003
0	−.003	1	−.003	.000
+ 1592	+.001	1	−.001	+.002
+ 7288	−.003	1	−.002	−.001
+10990	.000	2	+.002	−.002
+12493	+.006	2	+.002	+.004
+13678	+.002	2	+.001	+.001
+14557	+.001	1	.000	+.001
+20093	−.005	1	−.007	+.002
+25010	−.017	1	−.017	.000
+31072	−.032	3	−.034	+.002
+35413	−.053	8	−.050	−.003
+36238	−.052	5	−.050	−.002

Table 96 gave the formula

$$O-B = -0.0033 + 0.01314\,(E/10000) - 0.00745\,(E/10000)^2.$$

The differences $\delta = \overline{O-B} - \overline{(O-B)}_{calc}$ are listed in the last column of the table. Finally,

$$\text{Max hel JD} = 2421750.4827 + 0.45537282 \cdot E - 0.00745\,(E/10000)^2. \tag{57}$$

Tsesevich's new observations are listed in Table 97.

TABLE 97. New observations (Tsesevich)

JD hel	s	JD hel	s	JD hel	s
243...		243...		243...	
6007.456	39.3	6009.406	41.7	6020.430	14.2
6008.389	41.6	.414	41.7	.436	10.5
.409	41.0	.421	41.3	.442	14.5
.442	41.0	.428	42.0	.449	16.8
.456	42.4	.435	41.1	.459	20.9
.464	40.9	.444	42.0	.464	17.2
.467	40.4	.458	41.3	.471	20.3
.471	40.4	6013.389	41.2	.478	22.1
.477	41.7	6016.346	12.5	.484	23.1
.485	41.3	.354	14.5	.491	23.8
6009.340	42.0	.362	16.7	6021.354	16.4
.356	41.6	6020.338	39.3	6027.350	25.4
.364	42.8	.383	39.8	.351	26.1
.372	42.4	.400	39.8	.356	26.0
.380	41.2	.406	39.2	.361	26.1
.384	41.7	.419	23.5	.378	27.0
.389	41.6	.420	22.1	.391	29.2
.394	40.4	.423	16.0		
.400	40.8	.428	14.2		

AN Serpentis

Discovered by Hoffmeister /267/. Relatively few observations available, by Solov'ev /100/, Van Schewick /360/, Batyrev /18/, and Alaniya /1/. S. Gaposchkin /238/ and C. Payne-Gaposchkin /329/ studied its behavior from old Harvard plates. Batyrev summarized the maxima of this star (Table 98). His summary was supplemented by the following data: the normal maximum according to Payne-Gaposchkin's elements, new maxima obtained from Odessa photographs, and a maximum determined from Tsesevich's visual observations.

TABLE 98. List of maxima

Source	Max hel JD	E	O — A
Payne-Gaposchkin /329/	2414708.950	— 25478	+0.013^d
Solov'ev /100/	26122.464	— 3616	— .031
Van Schewick /360/	7462.673	— 1049	+0.017^d
	7463.686	— 1047	—.014
	7521.635	— 936	—.015
	7543.546	— 894	—.031
	7599.461	— 787	+.022
	7657.384	— 676	—.005
	7874.620	— 260	+.049
	7931.487	— 151	+.010
	7953.453	— 109	+.049
	7955.488	— 105	—.004
Solov'ev /100/	8007.193	— 6	+.015
	8010.310	0	.000
	8011.346	+ 2	—.008
	8021.266	+ 21	—.008
	8033.263	+ 44	—.018
	8043.208	+ 63	—.007
Van Schewick /360/	8045.337	+ 67	+.048
	8046.331	+ 69	—.002
	8219.688	+ 401	+.027
Solov'ev /100/	8245.265	+ 450	+.022
	8256.227	+ 471	+.021
	8280.219	+ 517	—.003
Van Schewick /360/	8308.430	+ 571	+.016
	8332.444	+ 617	+.015
	8333.472	+ 619	—.001
	8369.513	+ 688	+.017
	8391.433	+ 730	+.010
	8403.423	+ 753	—.008
	8424.326	+ 793	+.012
	8425.360	+ 795	+.002
Tsesevich (visual)	31235.149	+ 6177	—.006
Batyrev /18/	3876.303	+ 11236	—.019
	3886.233	+ 11255	—.008
	3887.271	+ 11257	—.014
	3898.235	+ 11278	—.014
	3899.283	+ 11280	—.010
	3909.202	+ 11299	—.010
	3910.238	+ 11301	—.018
	4128.466	+ 11719	—.016
	4149.344	+ 11759	—.021
	4174.410	+ 11807	—.015
	4184.331	+ 11826	—.013
	4242.265	+ 11937	—.030
	4265.232	+ 11981	—.034
	4515.341	+ 12460	+.002
Alaniya /1/	4540.403	+ 12508	+.005
Batyrev	4562.305	+ 12550	—.020
	4621.288	+ 12663	—.032
	4632.256	+ 12684	—.027
	4986.234	+ 13362	—.015
	5248.312	+ 13864	—.017
	5274.420	+ 13914	—.013
Tsesevich (photogr.)	6660.497	+ 16569	—0^d.039
	7817.400	+ 18785	—.050
Batyrev	8205.308	+ 19528	—.042
	8229.314	+ 19574	—.051
	8252.281	+ 19618	—.056
	8264.292	+ 19641	—.052

The O−A residues were calculated from the equation

$$\text{Max hel JD} = 2428010.3105 + 0.5220729 \cdot E. \qquad (58)$$

The star has a variable period. The data listed in the first column are not homogeneous: some maxima were determined with higher precision than others. Batyrev furthermore detected the Blazhko effect for this star. The data of Table 98 were therefore averaged and presented in Table 99.

TABLE 99. Averaged data from Table 98

$\bar{E}$	$\overline{O-A}$	n	$\bar{E}$	$\overline{O-A}$	n
− 25478	$+0^d.013$	1	+ 6177	$-0^d.006$	1
− 3616	− .031:	1	+ 11272	− .015	7
− 898	− .004	6	+ 11838	− .022	6
− 156	+ .026	4	+ 12573	− .012	5
+ 21	− .004	6	+ 13713	− .015	3
+ 179	+ .028	3	+ 16569	− .039	1
+ 480	+ .013	3	+ 18785	− .050	1
+ 696	+ .008	8	+ 19590	− .050	4

The graph of the O−A residues can be fitted with an expression containing a term with E^2. However, the spread of observation points around $E \sim 0$ is so large that introduction of a formula of this type is premature. At present the formula derived from the last two maxima is applicable:

$$\text{Max hel JD} = 2437817.400 + 0.522068 \cdot E. \qquad (59)$$

The period thus decreased by $0^d.0000049$. Comparison star used in the observations of Tables 100−102 (Figure 55): photographic scale $k = 0^s.0$; $a = 9^s.8$; $b = 20^s.4$; visual scale $a = 0^s.0$; $c = 6^s.2$; $d = 17^s.7$.

TABLE 100. Visual observations (Tsesevich)

JD hel	s	JD hel	s	JD hel	s
243...		243...		243...	
1216.447	9.7	1230.392	16.2	1236.361	14.2
.468	12.3	.423	13.9	.433	15.0
1221.271	13.6	.430	8.4	1241.325	16.6
.303	15.2	.434	6.2	.377	12.5
.353	15.4	.439	4.9	.382	11.7
1224.206	4.7	1231.179	15.6	.389	10.0
.235	5.3	.245	16.1	.396	5.3
.254	7.3	.272	15.8	.403	5.3
.296	12.2	.313	16.2	.420	4.1
.328	12.4	.337	16.4	1242.316	16.2
.362	15.2	.400	16.5	.348	16.8
.375	14.3	.426	16.5	.369	17.2
.399	15.1	1232.231	13.9	.409	16.3
.411	16.1	.318	16.0	.422	13.2
1225.308	8.4	.383	16.3	.427	12.0
.330	9.7	1235.195	9.3	.432	5.2
.360	12.0	.236	11.4	.438	5.7
.390	15.0	.383	14.6	.448	5.0
1230.271	15.6	1236.252	8.1	1253.318	15.4
.360	16.4	.327	12.0	1254.198	14.5

87

TABLE 101. Odessa observations (Tsesevich)

JD hel	s	JD hel	s	JD hel	s
243...		243...		243...	
6286.530	15.1	6661.520	5.2	7405.522	3.6
6288.543	16.6	6662.529	15.3	7406.529	1.1
6292.566	11.0	6667.511	15.3	7432.389	16.3
6304.461	2.2	6668.535	15.6	7457.383	12.2
6313.504	13.7	6690.518	12.3:	7458.367	11.7
6314.534	13.0	6702.460	13.0	7462.399	1.1
6322.520	15.4	6703.469	12.3	7464.397	16.3
6335.484	13.0	6715.412	7.8	7733.622	13.3
6336.505	13.0	6722.427	16.6	7758.565	9.8
6338.441	5.2	6726.385	6.5	7761.562	4.2
6339.437	0.0	6728.409	6.2	7763.536	16.3
6340.423	11.5	6729.407	2.8	7764.529	15.3
6344.409	13.5	6730.384	15.1	7811.462	15.6
6345.408	13.7	6732.388	16.3	7814.420	7.8
6347.413	12.3	6773.326	9.8	7817.386	2.2
6362.389	2.6	7015.579	6.5	7839.355	3.3
6364.374	15.6	7016.564	4.4	7849.377	5.2
6371.371	11.0	7019.550	15.1	8090.621	12.3
6373.367	6.9	7020.562	15.6	8143.483	15.3
6376.374	15.9	7028.553	4.4	8144.492	15.6
6391.358	14.2	7029.568	7.6	8164.409	15.6
6393.330	9.8	7044.461	14.4	8165.411	15.6
6395.351	7.2	7046.503	13.0	8170.421	5.2
6396.335	5.2	7052.465	14.6	8172.426	1.5
6397.337	11.5	7071.456	9.8	8173.431	6.2
6398.348	17.2	7080.372	9.8	8195.349	8.3
6399.342	17.6	7373.605	7.2	8198.360	15.6
6400.326	15.6	7378.572	14.6	8200.394	14.4
6405.340	12.3	7400.556	15.6	8203.368	9.8
6647.605	9.8	7402.508	13.2	8194.364	0.0
6660.556	5.8	7404.526	5.1		

TABLE 102. Simeise observations (Tsesevich)

JD hel	s	JD hel	s	JD hel	s
241...		242...		243...	
8901.280	12.7	5066.341	12.2	2707.374	9.8
	8.8	6122.463	12.4	3395.516	6.2
8911.280	16.9		9.8		7.8
	16.3	243...			
8912.287	12.7	2707.374	9.8		

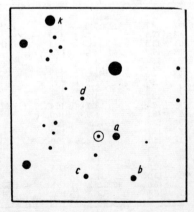

FIGURE 55. Comparison stars for AN Serpentis.

V 341 Aquilae

Discovered by Hoffmeister /265/ and studied by Selivanov /98/, Batyrev /14/, and Alaniya /2/. Batyrev derived the equation

$$\text{Max hel JD} = 2434244.395 + 0.5780168 \cdot E; P^{-1} = 1.730053521. \quad (60)$$

To establish the past behavior of the star, Tsesevich measured its magnitude from Simeise planetary plates. Comparison stars (Figure 56): $k = 0^s.0$; $a = 8^s.5$; $b = 16^s.5$. The seasonal average light curves gave 4 maxima. The summary of maxima (Table 103) shows that the period remained constant for almost 30,000 cycles. Tsesevich's observations are given in Table 104.

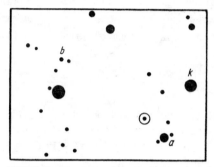

FIGURE 56.   Comparison stars for V 341 Aquilae.

TABLE 103.   List of maxima

Source	Max hel JD	E	O − A
Tsesevich (Sim.)	2419626.350	− 25290	0.000 d
	24357.406	− 17105	−.012
Selivanov /98/	7624.379	− 11453	+0d.010
	7628.427	− 11446	+.012
	7634.205	− 11436	+.010
	7635.348	− 11434	−.003
	7638.246	− 11429	+.005
	7639.385	− 11427	−.012
	7687.363	− 11344	−.009
	7690.260	− 11339	−.003
	7693.156	− 11334	+.003
	7694.303	− 11332	−.006
Tsesevich (Sim.)	9483.271	− 8237	.000
	33510.325	− 1270	+.011
Batyrev /14/	4240.353	− 7	+.004
	4244.395	0	.000
	4247.283	+ 5	−.002
	4251.329	+ 12	−.002
	4607.381	+ 628	−.009
	4618.362	+ 647	−.010
	4651.327	+ 704	+.008
	4625.309	+ 659	+.001
	4973.270	+ 1261	−.004
	5003.335	+ 1313	+.004
	5003.332	+ 1313	+.001
	5006.235	+ 1318	+.014
	5010.274	+ 1325	+.007
Alaniya /2/	5274.422	+ 1782	+.001
Karamysh	7145.484	+ 5019	+.023
	7882.467	+ 6294	+.034
	8234.479	+ 6903	+.034

TABLE 104. Simeise observations (Tsesevich)

JD hel	s	JD hel	s	JD hel	s
241...		242...		242...	
9623.329	9.5	4357.474	2.1	9483.353	5.7
9625.352	8.5		4.5	9485.348	11.9
	8.5	4382.408	9.9		12.1
9625.428	12.5		6.6	9495.349	9.6
	10.3	4697.359	1.3		10.3
9626.272	8.5	5805.380	1.4	243...	
.336	4.7		3.3	0207.406	13.8
9626.272	16.5	6187.503	4.2		12.7
.336	5.5	6209.324	5.7	0208.417	10.9
9628.327	10.5		10.8	2742.456	13.1
	10.8	6219.356	10.3		6.8
9630.305	12.9		10.8	2764.370	9.9
	13.5	6233.266	5.2		12.5
242...		6234.259	3.3	2768.352	11.9
1094.296	10.6	8369.420	3.8		13.1
	14.0		2.8	2772.400	11.9
1453.381	13.1	8398.339	1.7	3118.391	10.8
	13.8		5.0		9.9
2525.388	10.0	8845.231	6.3	3122.379	14.2
	10.1		6.9		13.5
3620.412	3.8	9103.455	11.2	3154.330	3.4
	4.0		10.8		5.1
3964.480	6.5	9468.469	10.3	3510.295	4.9
4351.419	10.4		10.5		3.8
	13.2	9483.353	4.5	3838.456	13.1

After the completion of this work, the magnitude estimates of
V. F. Karamysh from the plates of the Odessa collection became available
The seasonal light curves based on these data gave three new maxima:

Max hel JD	E	$O-A$
		d
2437145.484	5019	+0.023
7882.467	6294	+ .034
8234.479	6903	+ .034

To simplify further calculations, all the data were averaged (Table 105

TABLE 105. Averaged data

$\overline{E}$	$\overline{O-A}$	n	$\overline{(O-A)}_{calc}$	δ
	d		d	d
— 25290	0.000	1	+0.008	—0.008
— 17105	—.012	1	—.005	—.007
— 11397	+.001	10	—.007	+.008
— 8237	.000	1	—.006	+.006
— 1270	+.011	1	+.005	+.006
+ 2	.000	4	+.008	—.008
+ 660	—.002	4	+.010	—.012
+ 1385	+.004	6	+.012	—.008
+ 5019	+.023	1	+.023	.000
+ 6294	+.034	1	+.027	+.007
+ 6903	+.034	1	+.030	+.004

The residues $O-A$ can be described by the equation

$$O-A = +0.0082 + 0.02446 \ (E/10000) + 0.00864 \ (E/10000)^2,$$

whence the differences $\delta = \overline{O-A} - \overline{(O-A)}_{calc}$ are obtained and the improved
formula is established:

$$\text{Max hel JD} = 2434244.403 + 0.57801925 \cdot E + 0.00864 \ (E/10000)^2. \qquad (61)$$

SV Eridani

Discovered by Hoffmeister /261/. Studied by Lause /300/ and Florya /127/, but their period is wrong. Tsesevich derived the correct equation

$$\text{Max hel JD} = 2426590.319 + 0.7137306 \cdot E. \tag{62}$$

It was confirmed by S. Gaposchkin and C. Payne-Gaposchkin, who also established the variation of the period and obtained three linear equations:

$$
\begin{array}{lll}
(2414000-2421000) & \text{Max hel JD} = 2416755.766 + 0.71365272 \cdot E; & \\
(2421000-2426000) & \text{Max hel JD} = 2421159.757 + 0.71370172 \cdot E; & \quad(63) \\
(2426000-2432000) & \text{Max hel JD} = 2428153.409 + 0.71374288 \cdot E. &
\end{array}
$$

These formulas enabled Tsesevich to establish the correct count of the epochs. Gur'ev also observed this star and then, at Tsesevich's request, Tsarevskii observed it 8000 days later. The deviation from the ephemeris is so large that the few Odessa photographs had also to be measured (by B. A. Dragomiretskaya). These measurements gave three new, though highly uncertain, maxima. The known maxima are summarized in Table 106.

TABLE 106. List of maxima

Source	Max hel JD	E	O−B	O−C
Payne-Gaposchkin /329/	2416755.766	− 6171	+0.262 [d]	—
	21159.757	0	.000	—
Tsesevich /147/	6606.028	+ 7631	+.002	—
Florya /127/	6705.935	+ 7771	+.002	—
Tsesevich /147/	7424.671	+ 8778	+.040	—
Gur'ev /27/	8089.139	+ 9709	+.052	—
Payne-Gaposchkin /329/	8153.409	+ 9799	+.089	—
Gur'ev /27/	8398.202	+10142	+.082	+0.002 [d]
Tsarevskii	36607.137	+21643	+.734	−.005
	6612.141	+21650	+.742	+.002
Dragomiretskaya	7636.335	+23085	+.774	−.048
	7960.433	+23539	+.851	+.004
	8294.502	+24007	+.908	+.033

The curve of O−B residues calculated from the second equation in (63) can be fitted by an expression with a quadratic term. However, since most of the maxima were determined with insufficient accuracy, the last six maxima give a formula which is quite adequate for the time being for ephemeris calculations:

$$
\begin{aligned}
\text{Max hel JD} &= 2428398.200 + 0.7137590 \, (E-10142); \\
P^{-1} &= 1.4010331.
\end{aligned}
\tag{64}
$$

The O−C residues are given in the last column of Table 106. Dragomiretskaya (Table 107) used the following comparison stars (Figure 57): $a = 0^s.0$; $b = 8^s.7$; $c = 12^s.0$; $d = 15^s.1$.

Reliable formulas are thus available for two stars: AV Pegasi, (54), and V 341 Aquilae, (61). The periods of AN Serpentis and SV Eridani possibly vary in proportion to time.

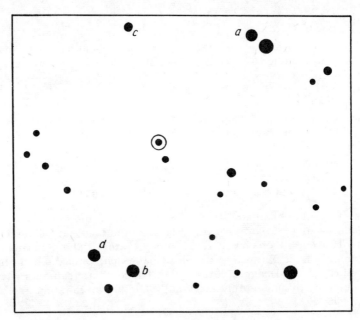

FIGURE 57.   Comparison stars for SV Eridani.

TABLE 107.   Observations of B.A. Dragomiretskaya

JD hel	s	JD hel	s	JD hel	s
243...		243...		243...	
7555.518	13.8	7911.566	13.8	8290.557	13.0
7560.512	13.8	7912.559	8.7	8291.543	12.0
.535	13.8	7942.505	10.9	8292.506	7.9
7561.541	6.8	7943.507	10.6	8294.522	6.1
7582.470	12.5	7946.477	13.2	8295.525	13.2
7583.519	5.6	.550	13.0	8296.498	13.2
7591.490	6.1	7960.415	6.5	8297.541	10.8
7605.431	13.0	7961.425	11.9	8298.540	13.4
7606.437	6.1	7963.462	11.2	8300.535	13.6
7636.345	5.5	7969.465	13.2	8319.456	6.1
7643.360	12.5	7973.500	10.8	8325.481	12.5
7645.329	13.4	7974.468	13.2		
7690.259	13.2	8288.567	13.4		

OTHER STARS WITH CONSTANT PERIODS

Table 108 lists other stars of constant period, giving the observation time ΔE, the range Δ JD, and the galactic coordinates. Asterisk (*) marks those stars whose period is possibly slightly variable.

To this list we should add several stars studied elsewhere. Constant periods (apart from Blazhko effect) are characteristic of the following stars: TV Lib, W Crt (Lange), KX Lyr (Batyrev), VY Ser (Odynskaya, Ustinov), BH Peg (Yudkina), S Com, XZ Dra, AE Dra, CG Peg (Tsesevich). The period of SW And decreases in proportion to time.

TABLE 108. Stars of constant period

Star	Max hel JD	P	ΔE	ΔJD	l	b

Galactic field stars

Star	Max hel JD	P	ΔE	ΔJD	l	b
XY And*	2435036.410	0.39872544	41500	2419689 — 2436075	100°	— 28°
ZZ And	2424828.379	0.5545326	33000	2418956 — 2437198	91	— 36
AA Aqr	2420748.582	0.60888937	27000	2421108 — 2437545	24	— 55
CH Aql*	2436050.3347	0.38918702	31300	2423964 — 2436129	8	— 27
SY Ari	2436105.346	0.5666815	31000	2420446 — 2437912	117	— 36
XX Boo	2429366.646	0.5814016	16000	2424609 — 2436105	11	+ 62
546. 1936 Cep	2437119.486	0.46786353	40500	2418180 — 2437119	69	+ 13
ST Cet	2425197.334	0.5265382	23500	2420805 — 2433163	92	— 60
SU Cet	2425197.349	0.5426428	23300	2420837 — 2433507	92	— 60
SVS 381 Com	2433005.444	0.5416158	31000	2419886 — 2436612	300	+ 85
V 833 Cyg	2436408.460	0.5381728	34200	2416731 — 2436134	50	— 6
835 Cyg	2436347.373	0.39375764	47000	2417840 — 2436347	45	— 12
SVS 396 Equ	2437582.384	0.5573455	23500	2424436 — 2437582	27	— 25
HV 10131 Equ	2426182.435	0.4671973	23500	2426182 — 2437177	23	— 34
CM Her	2429050.452	0.5627311	15693	2429050 — 2437881	20	+ 12
CW Her	2436721.358	0.6238405	31000	2418528 — 2437902	25	+ 38
CY Hya	2430052.326	0.57693446	30500	2419471 — 2436996	193	+ 35
WW Leo	2431150.188	0.6028456	21700	2420571 — 2433686	194	+ 40
AN Leo	2419858.410	0.5720244	30000	2419858 — 2437025	223	+ 62
AZ Lib	2427984.418	0.6513749	19000	2420245 — 2432674	311	+ 28
BR Lib	2428021.278	0.4931024	16800	2425798 — 2434120	314	+ 27
TT Lyn	2436255.883	0.5974379	31000	2418005 — 2436663	143	+ 43
FN Lyr	2433829.402	0.52739716	35000	2417793 — 2436081	41	+ 14
NR Lyr	2436079.374	0.6820293	30000	2415614 — 2436079	37	+ 13
NQ Lyr	2417852.251	0.5877888	31000	2417852 — 2436078	41	+ 14
CM Ori*	2425298.451	0.65592205	35000	2415100 — 2438084	168	— 5
ES Peg	2437901.2784	0.53867186	37123	2417904 — 2437901	59	— 24
SY Psc	2420398.3995	0.67354111	25500	2420398 — 2437578	98	— 57
V 756 Sgr	2429434.550	0.5239687	22200	2425801 — 2437497	334	+ 2
AI Tau	2424801.822	0.5685527	25000	2420461 — 2434688	138	— 25
WW Vir	2436722.344	0.6515827	26300	2420957 — 2438143	289	+ 56
AF Vir	2436717.406	0.4838254	18100	2429014 — 2437810	324	+ 58
AM Vir	2426859.275	0.61508784	29000	2420246 — 2438090	283	+ 45
SVS 122 Vir	2420636.380	0.61272478	28600	2420636 — 2438144	317	+ 47
SVS 123 Vir	2436716.419	0.5169041	36000	2419530 — 2438141	317°	+ 47°
BC Vul	2430258.453	0.4370163	23500	2426858 — 2437176	36	— 11

Stars from selected areas — "clouds"

Star	Max hel JD	P	ΔE	ΔJD	l	b
WX Aqr	2420387.308	0.5508409	33174	2420373 — 2437545	9	— 53
AI Aqr	2437910.364	0.6185331	28372	2420394 — 2437943	1	— 54
AK Aqr	2425424.765	0.6073594	23426	2420397 — 2434625	4	— 54
BE Aqr	2425439.731	0.4863896	29199	2420394 — 2434596	11	— 59
BG Aqr	2425424.60	0.505491	28092	2420396 — 2434596	10	— 59
BY Aqr	2425447.506	0.6578057	21116	2420715 — 2434605	5	— 35
CF Aqr	2427664.410	0.6323816	24335	2419216 — 2434605	2	— 35
CG Aqr	2425482.376	0.4533280	31397	2420372 — 2434605	5	— 34
CH Aqr	2425449.551	0.4970528	27923	2420726 — 2434605	2	— 36
CL Aqr	2425413.593	0.5954252	23309	2420726 — 2434605	3	— 38
CO Aqr	2424737.444	0.5610604	24737	2420726 — 2434605	3	— 39
CS Aqr	2420741.340	0.5698440	24323	2420741 — 2434601	8	— 37

TABLE 108 (continued)

Star	Max hel JD	P	ΔE	ΔJD	l	b
V 370 Oph	2436778.350	0.5694273	27073	2421749 — 2437165	349	+ 23
V 722 Oph	2437089.456	0.6190385	28925	2419556 — 2437462	342	+ 28
V 723 Oph	2437378.576	0.7146405	23596	2420630 — 2437493	335	+ 24
V 724 Oph	2437077.501	0.44233	927	2437052 — 2437462	338	+ 26
V 731 Oph	2437079.468	0.5286466	31887	2420636 — 2437493	336	+ 24
V 765 Oph	2436778.431	0.379873	1851	2436755 — 2437458	357	+ 20
V 768 Oph	2437079.431	0.701895	1059	2436755 — 2437498	356	+ 19
V 777 Oph	2436780.365	0.521392	1425	2436755 — 2437498	354	+ 17
375. 1934 Oph	2436758.478	0.513010	27758	2423573 — 2437813	1	+ 26
HV 10913 Oph	2436757.411	0.538160	1381	2436755 — 2437498	354	+ 22
CF Ser	2437454.398	0.53886	15991	2419556 — 2438173	337	+ 32
CG Ser	2420634.471	0.56096889	27272	2418884 — 2434183	339	+ 33
CO Ser	2436758.387	0.4451630	40996	2419920 — 2438170	338	+ 29
VX Vir	2419865.337	0.5430115	26254	2419481 — 2433737	259	+ 59
BL Vir	2419511.404	0.6686395	22393	2419482 — 2434455	264	+ 58
BM Vir	2419511.476	0.6718147	26266	2419482 — 2437128	263	+ 63
BQ Vir	2419858.356	0.6359125	23547	2419482 — 2434456	266	+ 60
BU Vir	2420247.47	0.5813090	29537	2419498 — 2436668	270	+ 60
BX Vir	2426070.544	0.8375589	21031	2419513 — 2437128	273	+ 64

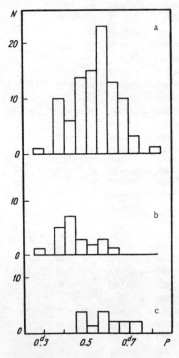

FIGURE 58. Frequency of periods for constant-period stars:

(a) total number of stars, (b) stars of first group, (c) stars of second group.

The first part of Table 108 contains data on bright stars that we call "galactic field stars", whereas the second part is concerned with stars from three selected areas: Aquarius, Ophiucus — Serpens, and Virgo. These three areas were scanned in great detail by Tsesevich using Simeise, Moscow, and Odessa photographs; the name "cloud" is tentatively attached to each of these areas.

SOME CONCLUSIONS

Study of the 90 constant-period stars leads to a number of remarkable, quite general conclusions.

Frequency of Periods

The main characteristic of a variable is its period. We give here the frequency of periods for constant-period variables (Figure 58). In the table below, the first column gives the range of periods, the second column gives the count of stars in the

94

corresponding interval, without the stars in the "clouds", and the third column gives the total number of constant-period stars in that range:

δP	n_1	n	δP	n_1	n
0.25 — 0.30	1	1	0.55 — 0.60	17	23
0.30 — 0.35	0	0	0.60 — 0.65	8	13
0.35 — 0.40	10	10	0.65 — 0.70	7	10
0.40 — 0.45	6	6	0.70 — 0.75	2	3
0.45 — 0.50	11	14	0.75 — 0.80	0	0
0.50 — 0.55	11	15	0.80 — 0.85	0	1

Analysis of the bimodal distribution of Figure 58 shows that one group of constant-period stars has relatively short periods, whereas another group is characterized by longer periods. Stars in "clouds" have longer periods, and this confirms our subdivision.

Distribution According to Spectral Properties

Table 109 lists some spectral data for constant-period stars, Table 110 same data for stars with periods varying in proportion to time. The tables include the spectral types at the minimum as determined by G. Preston from the Ca II lines, the difference ΔS of SpH and SpCa II at the minimum, the radial velocities, and the galactic longitudes and latitudes. Asterisk (*) marks objects not studied by Tsesevich; the corresponding data were assembled from various sources.

TABLE 109. Constant-period stars

Star	Period	Sp Ca II	ΔS	v_r, km/sec	l	b	Group
TV Lib*	0.270	F3	2	— 60	321°	+38°	1
V 559 Sco	0.351	—	—	—	331	+27	—
AA Aql	0.362	F5	0	— 85	11	—26	1
SS Cnc	0.367	F3	2	+ 5	167	+28	1
TZ Aur	0.392	F4	2	— 10	144	+22	1
V 445 Oph	0.397	F5	1	— 15	336	+27	1
TW Her	0.400	F3	2	— 15	23	+24	1
W Crt*	0.412	F2	3	+ 65	245	+41	1
AR Per	0.426	F6	0	— 10	123	— 1	1
CZ Lac	0.432	F4	1	—120	68	— 5	1
KX Lyr*	0.441	F5	0	— 60	36	+19	1
U Tri	0.447	F3	2	— 60	106	—27	1
ST Oph	0.450	A8	6	— 45	351	+15	2
CG Peg*	0.467	F4	2	+ 5	45	—22	1
DX Del	0.473	F4	2	— 45	26	—20	1
XZ Dra*	0.476	F2	3	— 25	63	+22	1
AV Ser	0.488	—	(6)	— 55	339	+35	2
ET Peg	0.490	F2	4	—	58	—28	2
Y Lyr	0.503	F4	1	—110	40	+20	1
AO Peg	0.547	F4	1	+115	38	—24	1
W CVn	0.552	A9	7	+ 20	36	+70	2
UY Cyg	0.561	F2	3	— 5	42	—10	1
SW Dra	0.570	F3	3	— 30	93	+48	1
RX Eri	0.587	A7	9	+ 70	182	—32	2
S Com	0.587	A7	7	— 35	187	+87	2
ST Com	0.599	F0	5	—100	320	+80	2
AE Dra*	0.603	F1:	(4)	—	51	+25	1
AT And	0.617	F2	3	—250	78	—18	2
BH Peg*	0.641	F1	5	—260	54	—39	2
AV Vir	0.657	F0	6	+ 35	296	+70	2
SU Dra	0.660	A6	10	—185	100	+49	2
VY Ser*	0.714	A7	9	+ 5	334	+43	2

We see from Tables 109 and 110 that the stars clearly fall into two groups. The results of this analysis are listed in Table 111.

TABLE 110. Stars with time-proportional periods

Star	Period	SpCaII	ΔS	v_r, km /sec	l	b	Group
AV Peg	0.390^d	F5	0	—100	46°	—25°	1
SW And*	0.442	F6	0	— 22	84	—33	1
VX Her	0.455	A9	5	—375	3	+38	2
AN Ser	0.522	F5	0	— 60	351	+44	1
V341 Aql	0.578	F2	3	—135	14	—23	1?
SV Eri	0.714	A6	9	— 5	162	—52	2

TABLE 111. Averaged data

Range of P	$\overline{S_p}$	$\overline{\Delta S}$	$\overline{v_r}$	Maximum v_r	Number of stars
$0^d.270$—$0^d.447$	F 4.4	1.2	49	—120	14
0.450—0.547	F 2.2	3.0	93	—375	10
0.552—0.714	A 9.5	6.8	88	—250	14

Thus, as the period increases, the spectral type becomes earlier, and the difference in hydrogen and ionized calcium spectral types progressively increases; a highly important point is that the spread of ΔS and v_r also increases.

My interpretation of this effect is the following. The short-period stars constitute a homogeneous group with relatively normal spectra and smaller radial velocities. This group also includes stars of longer periods, which are difficult to separate from the entire ensemble.

Proceeding from the ΔS criterion, I arrived at the following division into groups (which is unfortunately not particularly reliable). The stars of the "short-period" group are marked by 1, and the "long-period" stars are marked by 2. The average galactic latitudes are the following:

for 23 stars of group 1, +25°
for 13 stars of group 2, +47°.

We see that the two groups actually differ in the average galactic latitude The frequency of periods for the stars of the two groups separately is the following:

δP	Number of stars		δP	Number of stars	
	group 1	group 2		group 1	group 2
$0^d.25$—$0^d.30$	1	0	$0^d.50$— $0^d.55$	2	1
.30— .35	0	0	.55— .60	3	4
.35— .40	5	0	.60— .65	1	2
.40— .45	7	0	.65— .70	0	2
.45— .50	3	4	.70— .75	0	2

The division into two groups is obvious. Following the same approach we tried to divide the stars from Table 108 into three groups. This could not be accomplished, however, since their spectral characteristics are not known. We can therefore calculate only their average galactic latitudes:

δP	n	$\overline{|b|}$
0.270—0.447^d	4	19°.5
0.450—0.547	12	35.8
0.552—0.714	20	36.7

These data also confirm our previous findings.

Chapter 3

REGULAR VARIATION OF PERIODS

As recently as a quarter of century ago RR Lyrae stars were commonly regarded as an extraordinary phenomenon, whereas now many astronomers are of the opinion that each of these stars has a variable period. Detailed analysis of observations indeed shows that the periods of numerous stars are variable, and that different stars differ in the actual variation of their periods.

"TWIN " STARS

RV Coronae Borealis and RR Geminorum are located in two entirely different parts of the sky. They were both discovered on Moscow plates by L. P. Tserasskaya [Ceraski]. Their light curves belong to different subtypes (RRc and RRa, respectively), their periods are different, $0^d.332$ and $0^d.397$, and yet these stars are remarkably alike in their overall behavior.

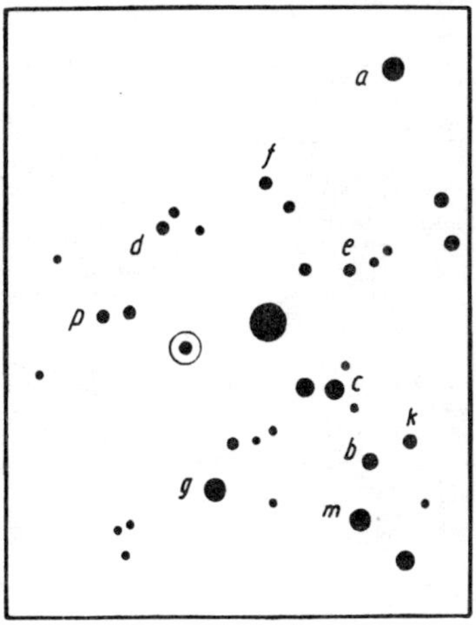

FIGURE 59. Comparison stars for RV Coronae Borealis.

RV Coronae Borealis

One of the most remarkable stars of subtype RRc. Although it was observed by numerous astronomers, and among them such an experienced observer as S. N. Blazhko, its peculiar properties were established only in the result of extensive cooperation between various researchers. Tsesevich discovered extreme instability of the star's period, but the old and new observations could not be linked up. At Tsesevich's request G. S. Filatov estimated the stellar magnitude from the photographs of the Dushanbe collection and P. Ahnert from the Sonneberg plates. These findings shed light on the behavior of this star over a period of fifty years.

97

The starting material for the present study thus included all the data published in various journals, original observations of Prof. S.N. Blazhko, Filatov's magnitude estimates, information on maxima supplied by P. Ahnert, the data obtained by R.I. Chuprina from Odessa photographs, and the maxima determined by G.A. Lange from visual observations.

The various data were processed as follows. A "seasonal" light curve was constructed using the approximate equation of the period, applicable to the relevant short interval of time. If sufficiently numerous observations were available, the average light curve was computed. When only few observations were on hand, the individual observations were plotted on a graph and a smooth curve was drawn from which the maxima were then determined. The complete list of the maxima is given in Table 114. The original observations of P. Ahnert were not available. The table, however, lists the maxima communicated by Ahnert.

Since the unpublished observation series are highly valuable, they are reproduced here in full; some details of reduction are given whenever justified.

Observations of S.N. Blazhko. Comparison stars in visual observations (Figure 59): $b = 0^s.0$; $c = 4^s.9$; $d = 9^s.3$; $e = 14^s.0$; $f = 15^s.8$. Blazhko's observation series (Table 115) was partitioned into "seasons"; a brief characteristic of his observations will be found in Table 113. All the average seasonal curves (Table 112) were plotted and the maxima were determined graphically. In addition to the average maxima, the individual maxima were also determined from Blazhko's observations carried out in different evenings (Table 115) and completely covering the maximum region of the curve. These maxima are also listed in Table 114.

TABLE 112. Average light curves (Blazhko's observations)

No. 1 JD 2420754—JD 2420978

φ_1	s	n	φ_1	s	n	φ_1	s	n
$0^P.010$	6.9	5	$0^P.193$	7.2	5	$0^P.890$	10.5	4
.030	5.0	5	.246	10.4	4	.912	10.0	4
.060	6.9	5	.296	11.2	3	.956	7.1	4
.080	6.6	5	.696	13.4	2	.972	6.6	4
.126	6.4	6	.823	12.1	5	.988	4.4	4
.160	7.1	5	.863	11.7	5			

No. 2 JD 2422918—JD 2422943

φ_2	s	n	φ_2	s	n	φ_2	s	n
$0^P.019$	9.2	5	$0^P.798$	11.7	5	$0^P.913$	8.1	5
.078	8.4	5	.830	11.3	5	.945	7.1	3
.146	8.4	4	.858	10.7	5	.980	6.8	3
.726	12.2	5	.893	8.7	5			

No. 3 JD 2423244—JD 2423249

φ_3	s	n	φ_3	s	n	φ_3	s	n
$0^P.023$	6.4	3	$0^P.198$	8.0	2	$0^P.956$	7.4	2
.061	6.6	2	.258	9.3	2	.978	4.9	2
.091	6.9	2	.886	11.7	3	.999	4.9	2
.136	7.4	3	.928	11.3	2			

TABLE 112 (continued)

No. 4 JD 2423687—JD 2423722

φ_3	s	n	φ_3	s	n	φ_3	s	n
$0^P.067$	6.8	2	$0^P.273$	8.1	3	$0^P.896$	11.3	1
.139	7.4	3	.829	13.2	2	.950	7.0	3
.199	7.6	3	.873	11.4	2	.998	6.4	2

No. 5 JD 2424370—JD 2424376

φ_3	s	n	φ_3	s	n	φ_3	s	n
$0^P.002$	6.8	3	$0^P.152$	7.8	2	$0^P.932$	7.4	2
.026	7.0	3	.266	8.6	4	.961	6.6	4
.066	7.1	2	.872	11.0	2	.987	7.0	3
.105	7.2	3	.897	9.1	2			

No. 6 JD 2425101—JD 2425122

φ_4	s	n	φ_4	s	n	φ_4	s	n
$0^P.015$	6.5	5	$0^P.766$	14.5	2	$0^P.940$	7.9	5
.043	6.6	5	.833	12.6	5	.967	6.7	5
.113	7.1	5	.864	12.0	5	.990	6.3	5
.196	8.1	5	.891	11.1	5			
.664	13.9	2	.918	9.7	5			

No. 7 JD 2426549—JD 2426597

φ_5	s	n	φ_5	s	n	φ_5	s	n
$0^P.053$	7.8	3	$0^P.659$	12.7	4	$0^P.933$	8.1	4
.148	7.3	4	.753	14.2	2	.951	7.0	3
.268	7.6	3	.906	10.8	4	.988	7.1	2

No. 8 JD 2426809—JD 2426974

φ_5	s	n	φ_5	s	n	φ_5	s	n
$0^P.007$	5.8	5	$0^P.307$	8.2	3	$0^P.939$	6.3	5
.042	6.7	5	.529	11.6	4	.959	6.4	5
.072	5.7	5	.656	12.2	2	.974	5.8	4
.104	7.6	5	.768	11.9	2	.995	6.4	3
.152	6.7	5	.842	11.6	4			
.218	7.4	4	.906	8.5	5			

No. 9 JD 2427151—JD 2427365

φ_6	s	n	φ_6	s	n	φ_6	s	n
$0^P.019$	6.8	5	$0^P.308$	9.7	3	$0^P.947$	7.0	5
.058	6.5	5	.747	13.3	5	.966	6.8	4
.079	6.8	5	.824	12.7	5	.976	7.3	4
.115	7.5	5	.859	11.3	5	.993	7.6	3
.157	7.7	5	.889	8.5	5			
.220	8.2	5	.916	7.7	5			

No. 10 JD 2428071—JD 2428121

φ_7	s	n	φ_7	s	n	φ_7	s	n
$0^P.006$	6.3	5	$0^P.232$	7.1	3	$0^P.866$	12.4	5
.037	7.3	5	.338	8.2	3	.919	9.5	5
.072	6.9	5	.582	13.8	2	.950	7.5	5
.160	7.8	5	.752	13.1	3	.978	6.8	5

TABLE 113. A brief characteristic of Blazhko's observations

JD	Result	Equation used
2419506—9509	Graph through individual points	$\text{Max} = 2420743.413 + 0.3316424 \cdot E;$
2419860—20241	Ditto	$P^{-1} = 3.015296$ (φ_1)
2420754—20978	Average curve No. 1	
2422202—22255	Graph through individual points	$\text{Max} = 2422247.336 + 0.331584 \cdot E;$
2918—2943	Average curve No. 2	$P^{-1} = 3.015827$ (φ_2)
3244—3249	Average curve No. 3	
3687—3722	Average curve No. 4	$\text{Max} = 2423244.398 + 0.3315566 \times$
4370—4376	Average curve No. 5	$\times E; P^{-1} = 3.0160763$ (φ_3)
5101—5122	Average curve No. 6	$\text{Max} = 2424370.365 + 0.3315454 \times$
6123	Individual curve	$\times E; P^{-1} = 3.0161782$ (φ_4)
6239—6266	Graph through individual points	$\tilde{\text{Max}} = 2426239.452 + 0.3315838 \times$
6549—6597	Average curve No. 7	$\times E; P^{-1} = 3.0158289$ (φ_5)
6809—6974	Average curve No. 8	
7151—7365	Average curve No. 9	$\text{Max} = 2427255.749 + 0.3315565 \times$
7943—7944	Graph through individual points	$\times E; P^{-1} = 3.0160772$ (φ_6)
8071—8121	Average curve No. 10	$\text{Max} = 2428078.333 + 0.3316190 \times$
8305	Individual curve	$\times E; P^{-1} = 3.0155088$ (φ_7)

TABLE 114. List of maxima

Source	Max hel JD	E	O — B
Blazhko (phot.)	2417823.33	— 41864	+0.578
	8202.28:	— 40721	+ .496:
	8888.34	— 38652	+ .451
	9160.27	— 37832	+ .460
Blazhko (seas. curve)	9509.431	— 36779	+ .433
	9861.239	— 35718	+ .401
Blazhko (ind. max.)	20743.417	— 33058	+ .492
	0754.354	— 33025	+ .486
	0772.256	— 32971	+ .481
	0773.264	— 32968	+ .494
	0774.257	— 32965	+ .492
Blazhko (max. No. 1)	0795.487	— 32901	+ .499
Blazhko (ind. max.)	0960.304	— 32404	+ .505
	0961.300	— 32401	+ .506
	1161.292	— 31798	+ .536
	1471.368	— 30863	+ .555
	1484.297	— 30824	+ .551
Blazhko (seas. curve)	2247.328	— 28523	+ .543
Blazhko (ind. max.)	2247.340:	— 28523	+ .555
	2255.290	— 28499	+ .547
	2929.402:	— 26466	+ .492
	2931.402	— 26460	+ .502
Blazhko (max. No. 2)	2931.385	— 26460	+ .485
Blazhko (ind. max.)	2939.350	— 26436	+ .492
	2943.335	— 26424	+ .497
	3244.396	— 25516	+ .455
Blazhko (max. No. 3)	3244.395	— 25516	+ .454
Blazhko (ind. max.)	3249.375	— 25501	+ .460
	3708.243	— 24117	+ .377
Blazhko (max. No. 4)	3708.574	— 24116	+ .376
Blazhko (ind. max.)	4370.365	— 22120	+ .270
Blazhko (max. No. 5)	4375.330	— 22105	+ .261
Blazhko (ind. max.)	4376.332	— 22102	+ .268
Tsesevich	5064.264:	— 20027	+ .106:
Blazhko (ind. max.)	5107.388	— 19897	+ .120
	5111.372	— 19885	+ .125
	5112.363	— 19882	+ .121
Blazhko (max. No. 6)	5112.360	— 19882	+ .118
Blazhko (ind. max.)	5113.354	— 19879	+ .117
	5122.311:	— 19852	+ .120:
	6123.378:	— 16833	+ .051:
	6239.387:	— 16483	— .004:

TABLE 114 (continued)

Source	Max hel JD	E	O — B
Blazhko (max. No. 5)	242 6266.290	— 16402	+0.039
Blazhko (ind. max.)	6266.295:	— 16402	+ .044:
Lause /301/	6418.517	— 15943	+ .056
	6444.403	— 15865	+ .076
	6485.504	— 15741	+ .057
	6545.537	— 15560	+ .069
Blazhko (ind. max.)	6588.282	— 15431	+ .036
	6589.268	— 15428	+ .027
Blazhko (max. No. 7)	6589.275	— 15428	+ .034
Blazhko (ind. max.)	6597.275:	— 15404	+ .075:
	6810.415	— 14761	— .011
	6825.352:	— 14716	+ .003:
	6831.357:	— 14698	+ .039:
	6835.305	— 14686	+ .012
Blazhko (max. No. 8)	6837.298	— 14680	+ .011
Blazhko (ind. max.)	6837.303	— 14680	+ .016
	6955.310	— 14324	— .031
	6962.288	— 14303	— .016
	6974.241	— 14267	— .001
Florya /125/	7255.732	— 13418	— .049
Blazhko (ind. max.)	7333.312	— 13184	— .068
Blazhko (max. No. 9)	7333.335	— 13184	— .043
Blazhko (ind. max.)	7341.275	— 13160	— .062
	7342.271	— 13157	— .061
	7345.276	— 13148	— .040
	7346.288:	— 13145	— .023:
	7348.263	— 13139	— .038
Ahnert /174/	7462.674	— 12794	— .033
Blazhko (ind. max.)	7943.390	— 11344	— .154
Blazhko (seas. curve)	7943.725	— 11343	— .150
Blazhko (ind. max.)	7944.388	— 11341	— .151
Lange	7985.204	— 11218	— .123
	7988.195	— 11209	— .116
Blazhko (ind. max.)	8078.330	— 10937	— .180
	8083.318	— 10922	— .166
	8098.223	— 10877	— .183
Blazhko (max. No. 10)	8098.239	— 10877	— .167
Blazhko (ind. max.)	8110.169	— 10841	— .175
	8121.177:	— 10808	— .211:
Ahnert /174/	8285.596	— 10312	— .171
	8304.491	— 10255	— .178
	8306.479	— 10249	— .180
Gur'ev /27/	8358.231	— 10093	— .159
Payne — Gaposchkin /329/	8616.891	— 9313	— .156
Ahnert /174/	8626.497	— 9284	— .167
	8630.512	— 9272	— .131
	8656.397	— 9194	— .112
	9015.492	— 8111	— .153
	9023.504	— 8087	— .099
	9103.409	— 7846	— .113
	9109.395	— 7828	— .096
	9421.441	— 6887	— .097
	9495.382	— 6664	— .105
	9496.379	— 6661	— .103
	9619.709	— 6289	— .132
	9790.474	+ 5774	— .137
	9808.437	— 5720	— .082
	30104.530	— 4827	— .128
	0409.648	— 3907	— .093
	0496.536	— 3645	— .087
	0497.508	— 3642	— .110
	0499.491	— 3636	— .116
	0513.424	— 3594	— .111
	0578.443	— 3398	— .088
	0587.425	— 3371	— .060
	0590.411	— 3362	— .058
	0734.668	— 2927	— .052
	0798.645	— 2734	— .076
	1530.556	— 527	— .033
	1610.462	— 286	— .045
	1705.348	0	.000
	1960.421	+ 769	+ .064

TABLE 114 (continued)

Source	Max hel JD	E	O — B
Filatov	2431992.178	+ 865	−0.014 d
Ahnert /174/	2943.700	+ 3734	+ .114
	2946.691	+ 3743	+ .121
	2950.655	+ 3755	+ .105
	2954.656	+ 3767	+ .126
	3030.547	+ 3996	+ .078
	3038.521	+ 4020	+ .094
Filatov	3062.412	+ 4092	+ .109
Ahnert /174/	3124.443	+ 4279	+ .128
	3126.470	+ 4285	+ .165
Batyrev /3/	3126.469	+ 4285	+ .164
Ahnert /174/	3154.346	+ 4369	+ .186
	3364.573	+ 5003	+ .171
Filatov	3418.347	+ 5165	+ .224
Ahnert /174/	3687.687	+ 5977	+ .295
	3797.492	+ 6308	+ .337
Filatov	3860.461	+ 6498	+ .300
Ahnert /174/	3889.343	+ 6585	+ .331
	4119.520	+ 7279	+ .370
	4120.539	+ 7282	+ .395
	4122.528	+ 7288	+ .393
	4132.515	+ 7318	+ .432
Filatov	4177.571	+ 7454	+ .389
Ahnert /174/	4217.401	+ 7574	+ .425
	4451.592	+ 8280	+ .498
	4453.595	+ 8286	+ .512
	4455.590	+ 8292	+ .517
	4534.460	+ 8530	+ .463
	4538.426	+ 8542	+ .450
Alaniya /1/	4539.461	+ 8545	+ .490
Ahnert /174/	4540.467	+ 8548	+ .501
Filatov	4578.605	+ 8663	+ .504
Ahnert /174/	4628.323	+ 8813	+ .480
	4628.355	+ 8813	+ .512
	4631.339	+ 8822	+ .512
	4635.317	+ 8834	+ .511
	4780.590	+ 9272	+ .537
Ahnert /174/	4890.419	+ 9603	+ .603
Filatov	4958.459	+ 9808	+ .662
Ahnert /174/	4962.416	+ 9820	+ .640
	5128.583	+10321	+ .670
	5196.585	+10526	+ .691
	5216.491	+10586	+ .700
	5220.466	+10598	+ .696
	5224.508	+10610	+ .759
	5391.291	+11113	+ .741
	5542.574	+11569	+ .809
	5549.581	+11590	+ .852
Filatov	5658.336	+11918	+ .839
Ahnert /174/	5962.512	+12835	+ .927
	5966.514	+12847	+ .949
	5972.458	+12865	+ .924
	5976.425	+12877	+ .912
Batyrev /3/	5988.353	+12913	+ .902
	5991.346	+12922	+ .910
	5997.315	+12940	+ .910
	6001.294	+12952	+ .910
Filatov	6076.269	+13178	+ .941
Ahnert /174/	6263.672	+13743	+ .984
	6279.653	+13791	+1.047
	6285 627	+13809	+1.052
	6287.601	+13815	+1.036
Filatov	6408.342	+14179	+1.070
Ahnert /174/	6722.451	+15126	+1.143
	6724.463	+15132	+1.165
Chuprina	6740.411	+15180	+1.196
Filatov	6800.412	+15361	+1.175
Ahnert /174/	6805.363	+15376	+1.152
	6808.363	+15385	+1.167
	6810.338	+15391	+1.153

TABLE 114 (continued)

Source	Max hel JD	E	O — B
Ahnert /174/	2436813.348	+15400	+1.178
	6814.334	+15403	+1.169
Lange	7902.282	+18683	+1.431
	7903.262	+18686	+1.416
	8152.353	+19437	+1.466
	8221.347	+19645	+1.485
	8222.326	+19648	+1.469
	8223.330	+19651	+1.479
	8228.313	+19666	+1.487

TABLE 115. Observations of S. N. Blazhko

JD hel	s	JD hel	s	JD hel	s
241...		242...		242...	
9506.474	3.3	0774.231	8.3	2204.325	10.5
9509.407	2.9	.244	7.1	.339	10.5
.413	3.5	.252	5.9	.363	11.6
9860.258	6.4	.260	6.4	.403	12.8
.266	11.2	.271	7.3	2235.339	14.3:
.273	11.5	.278	7.5	2247.293	9.3
.288	10.6	.287	7.1	.298	8.6
.301	11.1	.301	7.5	.316	7.1
.316	11.5	.317	8.0	.334	6.7:
.444	10.6	.321	8.2	2255.227	12.2
9861.302	11.1	.340	8.8	.248	10.7
.308	11.5	0780.279	7.3	.258	9.3
.314	9.3	0795.152	6.4	2918.405	12.0
.328	10.9	.164	6.2	2925.427	12.2
.377	10.4	.169	6.8	.436	12.8
9878.299	10.5	.177	7.1	.450	12.8
.314	10.5	.185	7.1	2929.320	11.6
.400	14.0	.195	7.1	.339	11.6
9886.290	12.1	.207	7.1:	.371	8.8
.310	12.4	.225	9.3	.372	6.0
.417	13.1	.237	9.3	.376	7.1
242...		0959.322	2.9	.387	6.7
0035.275	14.7	.336	4.9	2931.275	12.8
.303	14.4	.358	3.9	.313	14.0
0241.295	4.9:	.390	9.3	.329	12.2
0754.262	13.8	0960.253	12.8	.339	11.4
.295	13.3	.265	12.2	.344	11.2
.318	9.3	.273	10.9	.350	10.5
.359	11.7	.279	11.2	.356	9.3
0755.246	13.1	.292	9.3	.362	8.2
.289	12.1	.301	2.1	.368	7.8
.320	11.2	.313	3.7	.376	7.1
0772.204	11.6	.332	4.9	.386	6.6
.251	5.7	.353	6.0	.395	7.1
.258	5.9	.406	9.3	.404	7.1
.271	7.1	0961.288	6.4	.418	7.1
.290	7.1	.292	7.1	.428	7.5
0773.210	12.1	.297	6.6	.440	8.2
.216	11.6	.301	2.9	.447	8.2
.221	11.2	.305	2.9	2939.263	10.5
.228	10.7	.310	3.8	.271	12.1
.246	7.1	.330	6.0	.277	11.6
.254	6.4	.355	6.7	.290	11.2
.271	8.2	.390	14.0:	.299	10.9
.279	8.2	.410	15.0	.307	10.2
.288	7.7	0978.262	7.1	.313	9.3
.325	6.7	.271	7.1	.322	8.0
.328	9.3	.277	3.7	.332	7.5
0774.204	10.9	2202.318	8.0	2943.263	11.6
.208	11.2	.346	9.3	.268	11.6
.214	11.6	2204.280	10.2	.275	11.0
.224	9.3	.307	9.3	.280	10.9

TABLE 115 (continued)

JD hel	s	JD hel	s	JD hel	s
242...		242...		242...	
2943.285	10.6	4376.300	8.0	5122.294	7.7
.295	10.2	.310	7.2	.304	6.6
.302	9.3	.318	7.1	.319	6.6
.320	7.1	.329	7.1	6123.352	11.2
.330	6.7	.341	7.8	.368	7.1
.338	6.7	.353	7.1	.385	7.1
.349	7.3	.365	7.1	.397	7.5
.358	7.5	.425	9.3	.411	8.2
.370	8.0	4794.194	12.1	.433	7.5
.386	8.8	.198	11.3	6239.366	14.0
3244.347	12.6	.203	10.9	.387	6.6
.365	11.2	.208	10.1	.398	9.3
.369	11.2	.212	10.1	.414	8.2
.380	7.8	.222	10.9	6246.336	13.1
.390	4.9	.230	11.6	.348	12.1
.402	6.4	.237	11.8	.358	10.2
.416	7.1	4800.168	11.6	6261.179	12.4
.428	7.1	.179	10.6	.189	13.1
.447	9.3	.195	10.6	.203	14.0
.465	8.6	.232	11.6	.216	14.0
3249.345	11.4	.235	12.4	6266.275	9.3
.350	11.2	5101.306	12.8	.294	5.5
.360	7.1	.317	15.0	.317	7.1
.365	4.9	.334	14.0	6549.358	12.1
.369	4.9	.373	14.0	.365	12.4
.373	4.9	.428	6.2	.276	13.1
.378	6.4	.435	6.0	.380	13.1
.384	6.4	5107.328	12.4	.392	14.0
.394	6.2	.338	12.1	.410	14.5
.402	6.7	.352	11.6	6588.244	11.6
.407	6.2	.363	9.8	.251	10.5
.422	7.1	.368	8.2	.255	9.3
.436	7.5	.374	7.5	.257	8.2
.450	9.3	.384	6.7	.259	7.1
.464	9.3	.390	5.4	.265	7.1
3687.295	12.4	.395	5.9	.273	6.0
.308	11.6	.406	6.4	.287	7.1
.323	11.3	.415	6.9	.319	6.7
.356	7.8:	.431	7.1	.346	7.1
3708.195	14.0	.439	6.9	.387	7.8
.221	8.0	.447	7.5	6589.239	11.2
.229	7.1	.463	7.8	.244	9.8
.237	6.0	5111.316	11.6	.249	7.5
.246	4.9	.324	12.1	.252	7.5
.264	6.6	.330	11.2	.257	6.7
.272	7.1	.337	10.2	.293	7.5
.283	7.1	.340	9.3	.320	8.0
.309	7.7	.353	7.1	.337	7.5
.331	7.1	.365	6.9	.363	7.8
3722.218	8.0	.376	7.1	6597.228	8.2
.225	7.1	.418	7.5	.254	8.7
.234	7.5	5112.296	15.0	.275	7.1
.243	7.5	.305	13.1	6809.273	11.5
.256	8.0	.320	12.1	.283	11.1
.273	9.3	.329	11.6	.324	12.4
4370.320	11.5	.337	11.2	.337	12.0
.328	10.2	.347	8.2	.364	12.4
.341	7.5	.354	6.0	6810.284	11.6
.354	4.9	.362	6.4	.303	12.4
.365	6.8	.369	7.1	.392	11.6
.401	7.2	.383	7.1	.408	7.8
.416	7.8	.400	7.1	.415	4.9
4371.348	6.9	.444	9.3	6825.344	4.9
.354	6.8	5113.310	11.9	.350	2.9:
.360	7.1	.322	10.9	.352	1.9:
.369	6.4	.333	8.2	.360	6.3:
4375.321	7.5	.339	7.8	6831.357	1.9
.333	7.1	.352	5.9	.382	3.4
.338	6.4	.354	6.0	.409	6.5
.345	6.7	.361	6.4	6835.281	8.7
.360	7.1	.373	6.9	.287	7.1
.372	7.4	.396	7.5	.290	5.5
.387	7.8	.425	8.2	.296	6.7
.404	7.8	5122.262	12.1	.298	6.7
.424	8.2	.270	11.6	.308	3.4
.445	9.3	.282	10.2	.323	6.0
4376.291	10.5	.286	8.0	.334	6.5

TABLE 115 (continued)

JD hel	s	JD hel	s	JD hel	s
242...		242...		242...	
6835.348	7.4	7342.233	12.8	7944.353	11.6
.355	7.3	.250	8.7	.362	10.2
.375	7.2	.272	7.1	.369	7.8
.400	7.9	.319	9.3	.376	6.7
.424	8.4	7345.190	14.0	.381	3.9
6837.282	9.3	.208	13.4	.394	4.9
.285	8.4	.224	11.1	.399	3.9
.288	7.4	.238	8.2	.410	4.9
.297	5.7	.259	7.1	.422	6.7
.300	4.9	.272	6.6	.430	7.1
.313	5.7	.290	7.1	.436	7.1
.319	5.7	.316	7.1	.454	8.4
.331	7.4	.345	8.8	8071.279	12.1
.358	7.9	7346.196	12.8	.329	12.1
.396	8.4	.207	12.8	8078.204	14.0
6839.292	7.4	.216	11.6	.243	14.0
.299	7.4	.223	9.3	.285	12.8
.308	6.5	.232	8.2	.302	10.3
.317	7.4	.238	8.2	.313	8.2
.329	7.4	.251	7.1	.317	7.1
.342	7.4	.260	7.1	.325	6.7
.369	7.9	.284	6.0	.333	6.4
6955.268	11.4:	.304	6.0	.344	6.7
.294	10.6	.328	7.1	8079.180	13.5
.310	7.1	7348.200	12.8	8083.283	11.6
.342	7.3	.229	8.0	.299	7.5
.365	7.7	.238	8.0	.303	6.4
6962.241	12.4	.244	7.1	.310	4.9
.254	11.6	.263	6.0	.313	5.8
.267	8.2	7356.291	8.2	.326	5.9
.285	6.0	.331	11.6	8098.163	13.1
.301	7.1	7360.157	8.2	.176	13.5
.314	7.1	.188	8.8	.190	12.1
.336	8.2	.208	7.1	.201	9.3
.360	7.8	.210	7.1	.206	7.1
6974.214	10.5	.224	7.1	.215	6.7
.224	8.2	.242	9.3	.219	6.7
.236	7.1	.260	8.2	.230	7.5
.246	6.0	7364.146	6.0	.238	7.5
.260	7.1	.160	6.8	.243	8.0
.279	7.8	.171	7.1	.250	6.7
7151.339	4.9	.186	6.0	.255	7.5
7211.304	6.8	.208	7.3	.265	8.4
.319	6.9	7365.153	7.7	.282	9.3
.340	6.4	.158	8.2	8110.129	11.6
.348	7.1	.167	7.1	.143	9.3
.366	7.1	.187	7.1	.151	8.2
.391	7.1	.214	7.7	.154	7.1
7333.290	11.6	.238	8.2	.164	6.7
.295	9.3	.268	9.3	.170	6.7
.312	7.1	7533.380	8.2	.180	7.1
.324	7.1	7594.346	8.2	.186	7.1
.345	7.1	.358	10.2	.194	7.1
.361	7.7	.370	9.3	.199	7.1
7341.195	13.5	7943.374	8.8	.221	7.8
.201	13.4	.378	8.2	.241	8.4
.204	12.8	.382	6.0	8121.171	6.7
.238	11.6	.386	4.9	.179	6.7
.246	9.3	.396	4.9	.186	5.9
.252	6.8	.404	5.9	.196	7.1
.265	6.0	.416	7.1	.209	7.1
.278	5.5	.425	7.3	.222	8.2
.333	7.8	.435	8.8	.241	9.3
.363	8.8	.449	10.2		
7342.218	12.8	7944.348	12.1		

Observations of G. S. Filatov. Filatov's observations are not as numerous (generally one photograph per night). The maxima can therefore be found only from the seasonal curves. All the seasonal curves between JD 2431880 and JD 2436833 can be recovered using the equation

$$\text{Max hel JD} = 2431705.348 + 0.331710 \cdot E; \quad P^{-1} = 3.0146815. \tag{65}$$

105

The entire observation series was partitioned into "seasons":

No.	JD	n	No.	JD	n
1	2431880—2432078	22	6	2434536—2434625	17
2	3027— 3116	11	7	5602— 5695	18
3	3381— 3451	11	8	5957— 6105	22
4	3823— 3915	31	9	6334— 6463	14
5	4091— 4277	33	10	6670— 6833	18

Filatov also measured the magnitude of the variable from photographs on panchromatic plates shot without light filter. These observations gave a maximum on the seasonal curve assembled between 2434870 and 2435015. Magnitude scale and comparison stars (see Figure 59):

	Photographic s	For panchromatic plates s
a	0.0	0.0
m	10.8	12.0
c	21.0	22.5
k	31.6	32.9
g	43.2	—

No average curves were computed for lack of sufficient data; all the points were plotted on a graph, a smooth curve was drawn through these points, and the maximum was determined. Filatov's observations are listed in Tables 116—117.

TABLE 116. Photographic observations (Filatov)

JD hel	s	JD hel	s	JD hel	s
243...		243...		243...	
1880.477	4.9	3418.408	10.8	4125.344	26.8
1887.341	6.3	3437.267	7.2	4149.267	26.9
1888.361	10.8	3445.266	21.0	4150.287	26.8
1889.364	5.9:	3450.234	21.0	.310	24.9
1907.316	7.2	3451.278	27.4	4152.363	14.5
1908.316	6.3:	.301	33.6	4153.327	14.5
1909.323	17.9	3823.318	5.9:	.349	6.9
1932.237	23.9:	3825.277	6.3	4163.343	10.8
1938.242	26.3	.300	6.3	.414	24.9
1971.403	24.9	3829.291	4.9	4177.218	6.0:
1981.406	26.3	3830.254	6.9:	.262	13.6
1991.289	27.4	3832.257	5.9	4180.329	17.9
1992.211	6.0:	.279	5.9	4181.281	16.4
1993.315	27.4	3850.205	17.3	.329	17.3
1995.290	31.6:	3851.240	6.3	4188.329	27.8
1996.326	21.0	3856.294	35.6	4189.282	34.6:
1999.319	27.0	3859.211	25.2	.304	26.8
2000.307	29.0	3860.189	4.9	4192.302	28.7
2003.237	16.5	3861.237	23.1	.325	35.4
2024.196	27.4	3862.194	21.0	4205.226	23.9
2029.248	26.3	3864.235	26.0	4207.220	26.3
2078.142	6.9:	3865.191	13.9	4211.261	26.0
2687.326	5.8	.214	26.8	4212.257	28.5
3027.313	4.3	3868.255	17.9	4213.277	29.0
.344	10.8:	3881.160	26.8	4214.249	35.6
3033.307	16.9	.182	33.5	.274	37.8
3040.363	26.8	3883.162	26.9	4216.273	37.6
3056.292	34.5	3884.186	19.0	4218.240	38.9
3062.191	25.0	.209	26.8	4222.258	25.2
.345	17.9	3885.218	27.8	4269.158	6.5
3064.358	15.9	3887.167	26.8	4277.132	13.9
3067.310	26.3	3891.184	31.6	4536.372	27.4
3115.199	7.8	3892.166	35.7	4540.246	14.8:
3116.187	10.8	.189	26.8	4544.345	14.9
3381.332	26.3	3911.128	21.0	4549.357	17.9
3382.313	26.3	3914.130	21.0	4550.362	10.8
.370	24.2	3915.131	25.8	4567.298	5.0:
3416.312	16.9	4091.355	16.5	4572.281	4.0:
3418.358	6.9	4103.373	16.4	4574.256	4.9

TABLE 116 (continued)

JD hel	s	JD hel	s	JD hel	s
243...		243...		243...	
4575.287	10.8	5693.207	7.9	6368.281	7.9
4576.284	16.4	5695.214	21.0	6369.327	15.9
4578.278	6.5	5957.338	27.4	6376.342	21.0
4579.306	6.5	5987.367	3.2	6377.344	36.6
4590.229	6.5	6007.272	4.3	6408.294	10.8
4593.227	14.9	6008.286	4.9	6432.194	5.9
4595.220	15.9	6037.228	10.8	6450.163	4.9
4605.216	6.5	6042.275	33.6	6453.172	10.8
4625.140	14.9	6047.190	21.0	6460.151	16.8
5330.197	4.0	6065.192	26.3	6463.149	7.6
5338.139	4.0	6069.199	28.5	6670.395	14.9
5361.136	5.9	6070.164	34.6	6671.427	5.9
5602.387	15.3	6071.160	35.6	6675.443	24.9
5604.322	14.8	6072.216	31.6	6688.346	5.9
5609.307	7.2	6074.169	35.6	6699.337	16.9
5629.304	26.8	6076.161	35.6	6700.401	27.4
5633.306	26.8	.249	6.9	6734.375	6.9
5637.313	24.9	6077.192	36.6	6748.227	27.4
5638.384	23.9	6080.202	31.6	6751.316	5.0
5656.222	25.8	6081.188	25.2	6756.312	5.4
5658.324	3.6	6096.152	6.5	6757.310	6.5
5659.318	3.6	6097.140	6.5	.353	14.9
5661.239	35.6	6098.163	6.3	6676.255	0.0
5664.259	8.1	6105.161	13.9	6786.334	7.6
5665.227	25.8	6334.318	25.0	6811.144	16.9
5689.208	3.2	6339.272	27.2	6812.151	27.0
5691.267	6.9	6362.203	4.0::	6814.168	3.0
5692.189	5.0	6364.334	25.4	6833.181	25.2

TABLE 117. Observations on panchromatic plates (Filatov)

JD hel	s	JD hel	s	JD hel	s
243...		243...		243...	
3736.420	16.2	4931.311	18.7	4983.171	36.9
3889.165	26.5	4946.242	19.7	4985.159	28.7
4124.336	27.7	4947.265	36.9	4987.168	27.7
4125.320	24.2	4948.259	27.7	4989.182	37.9
4870.355	27.2	4951.269	8.0	5003.143	22.5
4920.294	16.8	4954.223	28.2	5004.142	32.9
4922.270	18.3	4958.214	22.5	5008.125	30.1
4923.292	16.2	4961.225	36.9	5011.123	27.7
4924.249	22.5	4975.201	29.1	5015.126	22.5
4925.266	18.7	4976.191	35.9	6108.156	14.1
4930.278	28.2	4980.171	32.9	6109.136	8.0

New observations of G.A. Lange. Lange observed this star visually in 1962 and 1963 using a 5-in. refractor. Magnitude scale and comparison stars:

	1962	spring 1963	summer 1963
	s_1	s_2	s_3
f	0.0	0.0	0.0
d	9.4	9.5	10.0
p	15.8	20.3	18.0

Seven individual maxima were derived from these observations. Lange's new observations are listed in Table 118.

TABLE 118. New observations (Lange)

JD hel	s_1	JD hel	s_1	JD hel	s_1
243...		243...		243...	
7902.274	9.4	7902.354	14.0	7903.276	6.6
.278	5.9	7903.246	12.6	.283	9.4
.285	4.7	.253	9.4	.297	12.6
.292	6.6	.256	6.6	.324	13.1
.298	7.5	.261	5.6	.362	16.0
.306	9.4	.264	5.2		
.318	12.6	.269	6.3		

JD hel	s_2	JD hel	s_2	JD hel	s_2
243...		243...		243...	
8150.236	18.1	8150.321	14.9	8150.332	7.6
.307	17.1	.328	9.5	.338	5.7

JD hel	s_2	JD hel	s_2	JD hel	s_2
243...		243...		243...	
8150.345	4.8	8152.330	12.7	8152.368	6.3
8152.288	18.1	.337	9.5	.373	9.5
.302	17.6	.342	7.1	.379	11.7
.315	16.7	.346	5.9	.386	13.1
.322	14.9	.352	4.8	.394	13.8
.326	13.8	.362	5.7	.408	15.4

JD hel	s_3	JD hel	s_3	JD hel	s_3
243...		243...		243...	
8221.308	15.0	8222.321	6.0	8223.342	6.0
.313	14.0	.324	5.0	.345	6.0
.322	13.0	.332	6.0	.350	7.0
.331	10.0	.338	7.0	.358	7.0
.337	6.0	.344	8.0	.370	7.3
.342	5.0	.351	10.0	.376	9.1
.348	4.0	.367	13.0	.383	10.0
.352	5.0	.376	14.0	.403	13.0
.360	6.0	.392	15.0	8228.302	10.0
.365	7.0	8223.305	15.3	.309	7.0
.371	8.0	.312	13.0	.311	6.0
.379	10.0	.316	10.0	.316	6.0
.388	12.0	.319	8.0	.322	7.0
.392	13.0	.322	7.0	.328	8.0
8222.308	14.0	.325	5.8	.341	10.0
.313	10.0	.331	5.0	.357	13.0
.318	7.0	.337	5.0		

Variation of the period. Table 114 lists all the maxima. The column O — B lists the residues relative to the ephemeris

$$\text{Max hel JD} = 2431705.348 + 0.3316118 \cdot E. \qquad (66)$$

The tabulated data are plotted in Figure 60. Figure 61 gives the averag
light curve No. 8. The variation of the O — B residues is astonishing. Give
such extreme variations of the period, one can easily lose all track of
epochs. The wealth of observations and the requirement of smoothness
imposed on the O — B curve nevertheless enabled us to establish the correc
count of epochs. It is clearly seen that the deviation from the linear ephemeris reaches five and more periods, and the variation of the period is

quite irregular. At the beginning of observations, the period was large. Then it became smaller, again increased, and remained almost constant for over 20 years. According to Lange's latest observations, the period decreased again.

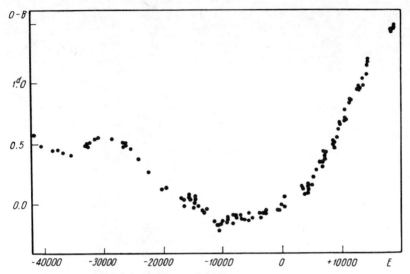

FIGURE 60. O—B residues for RV Coronae Borealis.

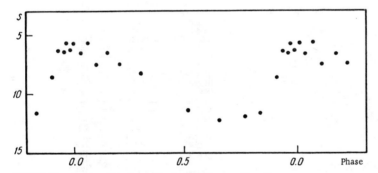

FIGURE 61. Average light curve No. 8 of RV Coronae Borealis.

RR Geminorum

Discovered by Tserasskaya [Ceraski] /216/ some 60 years ago. An object of extensive repeated observations. Observed by Tsesevich for several seasons. The maxima are listed in Table 119.

TABLE 119. List of maxima

Source	Max hel JD	E	O — A
Blazhko /199/	2415103.29	— 11122	—0^d.148
	5484.31	— 10163	— .104
	6198.265	— 8366	— .031
	6200.240	— 8361	— .043
	6215.335	— 8323	— .044
	6221.295	— 8308	— .043
	6228.431	— 8290	— .057
	6240.361	— 8260	— .045
	6422.323	— 7802	— .030
	6463.247	— 7699	— .024
	6550.254	— 7480	— .018
Graff /244/	6872.460	— 6669	+ .007
	6877.243	— 6657	+ .023
Blazhko /199/	6879.215	— 6652	+ .009
Graff /244/	6880.422	— 6649	+ .024
	6884.382	— 6639	+ .011
	6886.356	— 6634	— .001
	6903.429	— 6591	— .010
Blazhko /199/	6906.233	— 6584	+ .013
	6907.422	— 6581	+ .010
	6910.211	— 6574	+ .018
	6914.185	— 6564	+ .020
	6919.347	— 6551	+ .017
	6921.327	— 6546	+ .011
Graff /244/	7115.590	— 6057	+ .012
Luizet /312/	7228.440	— 5773	+ .039
	7249.481	— 5720	+ .026
Blazhko /199/	7262.201	— 5688	+ .033
	7277.291	— 5650	+ .027
	7296.366	— 5602	+ .033
	7300.335	— 5592	+ .029
Luizet /312/	7300.345	— 5592	+ .039
	7304.335	— 5582	+ .047
	7315.426	— 5554	+ .024
	7613.410	— 4804	+ .061
	7636.451	— 4746	+ .060
	7646.386	— 4721	+ .064
	7704.390	— 4575	+ .067
Blazhko /199/	8024.219	— 3770	+ .099
	8029.357	— 3757	+ .072
	8035.322	— 3742	+ .079
	8062.338	— 3674	+ .081
	9006.319	— 1298	+ .163
Luizet /312/	9133.434	— 978	+ .154
	9139.393	— 963	+ .154
	9141.366	— 958	+ .141
	9160.430	— 910	+ .136
	9448.462	— 185	+ .152
	9460.391	— 155	+ .163
	9487.422	— 87	+ .180
	9491.373	— 77	+ .158
	9497.350	— 62	+ .176
	9512.426	— 24	+ .156
	9742.457	+ 555	+ .172
	9744.452	+ 560	+ .180
	9767.473	+ 718	+ .160
	9806.408	+ 616	+ .163
	9833.429	+ 784	+ .170
	9872.370	+ 882	+ .179
	9893.421	+ 935	+ .175
	9895.406	+ 940	+ .174
Waterfield /364/	20178.659	+ 1653	+ .178
Luizet /312/	0243.430	+ 1816	+ .195
Waterfield /364/	0461.901	+ 2366	+ .171
	0465.981	+ 2376	+ .189
	0506.791	+ 2479	+ .170
Luizet /312/	0560.426	+ 2614	+ .175
Waterfield /364/	1134.851	+ 4060	+ .157
	1227.796	+ 4294	+ .142
Graff /244/	2763.526	+ 8160	+ .051
	2765.514	+ 8165	+ .052
Waterfield /364/	2941.870	+ 8609	+ .023
Dubyago /163/	4256.338	+ 11918	— .054
Martynov /163/	4256.335	+ 11918	— .057
Waterfield /364/	4523.684	+ 12591	— .066
Ivanov /40/	4586.464	+ 12749	— .054

TABLE 119 (continued)

Source	Max hel JD	E	$O - A$
Ivanov /241/	2414598.373	+ 12779	$- 0^d.063$
	4602.342	+ 12789	— .067
Martynov /163/	4857.387	+ 13431	— .065
	4938.392	+ 13635	— .102
Ivanov /40/	4967.387	+ 13708	— .107
Martynov /163/	4967.408	+ 13708	— .086
	5139.421	+ 14141	— .088
Dubyago /163/	5150.533	+ 14169	— .100
Martynov /163/	5150.534	+ 14169	— .099
Dubyago /163/	5160.457	+ 14194	— .107
Martynov /163/	5160.455	+ 14194	— .109
Ivanov /40/	5173.578	+ 14227	— .096
Dubyago /163/	5175.546	+ 14232	— .114
Martynov /163/	5175.548	+ 14232	— .112
	5185.488	+ 14257	— .104
	5205.333	+ 14307	— .122
Dubyago /163/	5221.232	+ 14347	— .113
Martynov /163/	5221.232	+ 14347	— .113
Dubyago /163/	5221.639	+ 14348	— .104
Martynov /163/	5221.640	+ 14348	— .103
Dubyago /163/	5234.342	+ 14380	— .113
Martynov /163/	5234.341	+ 14380	— .114
Dubyago /163/	5238.330	+ 14390	— .098
Martynov /163/	5238.328	+ 14390	— .100
Dubyago /163/	5266.140	+ 14460	— .096
Martynov /163/	5266.131	+ 14460	— .105
	5300.302	+ 14546	— .099
Dubyago /163/	5301.489	+ 14549	— .104
	5321.334	+ 14599	— .122
Martynov /163/	5321.341	+ 14599	— .115
Dubyago /163/	5323.328	+ 14604	— .114
Martynov /163/	5323.322	+ 14604	— .120
	5326.512	+ 14612	— .108
Dubyago /163/	5327.289	+ 14614	— .126
Ivanov /163/	5327.297	+ 14614	— .118
Martynov /163/	5327.305	+ 14614	— .110
Dubyago /163/	5358.300	+ 14692	— .101
Martynov /163/	25358.294	+ 14692	— .107
Dubyago /163/	5514.407	+ 15085	— .119
Dubyago /163/	5518.383	+ 15095	— .116
	5531.490	+ 15128	— .118
Martynov /163/	5531.493	+ 15128	— .115
Tsesevich	6250.496	+ 16938	— .160
	6739.501	+ 18169	— .186
Blazhko /23/	7142.326	+ 19183	— .186
	7150.267	+ 19203	— .191
	7175.293	+ 19266	— .192
	7179.266	+ 19276	— .192
Lange /63/	7340.552	+ 19682	— .195
Tsesevich	7386.644	+ 19798	— .186
Blazhko /23/	7531.242	+ 20162	— .192
Kleissen /297/	7532.423	+ 20165	— .202
Blazhko /23/	7533.231	+ 20167	— .189
Lange /23/	7804.156	+ 20849	— .198
Gur'ev /27/	8180.389	+ 21796	— .173
Blazhko	8248.289	+ 21967	— .205
Gur'ev /27/	9183.110	+ 24320	— .146
Tsesevich	31103.310	+ 29153	+ .079
Batyrev /4/	3003.277	+ 33935	+ .331
	3026.319	+ 33993	+ .332
	3032.279	+ 34008	+ .333
	3034.267	+ 34013	+ .335
	3038.243	+ 34023	+ .338
Alaniya /2/	4806.289	+ 38473	+ .561
Mandel /72/	6308.177	+ 42253	+ .792
Batyrev /4/	6486.544	+ 42702	+ .788
	6488.529	+ 42707	+ .785
	6646.272	+ 43104	+ .816
	6704.271	+ 43250	+ .815
Lange	7764.311	+ 45918	+ .955
	7943.499	+ 46369	+ .977
	7945.483	+ 46374	+ .975
	7962.569	+ 46417	+ .979
	7974.485	+ 46447	+ .977

111

The O−A residues (Figure 62) were calculated from

$$\text{Max hel JD} = 2419521.804 + 0.39726364 \cdot E. \qquad (67)$$

This plot is highly reminiscent of the corresponding plot for RV Coronae Borealis. By 1962 the value of the O−A residues exceeded double the period! Yet the fluctuations are very smooth and gradual.

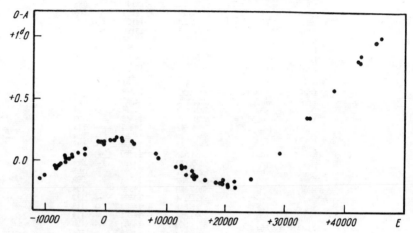

FIGURE 62. O−A residues for RR Geminorum.

The period of RR Geminorum apparently remained constant (or nearly constant) over the last 17,000 epochs. The corresponding equation

$$\text{Max hel JD} = 2431103.3125 + 0.3973158 \cdot E \qquad (68)$$

is effective after JD 2431000. All the O−B residues, except one, fall within the margin of observation error. Thus, no variation of the period could be detected during the last 20 years.

SZ Hydrae is closely related to RR Geminorum and RV Coronae Boreali Its main property is that the period changed gradually, and then remained constant for 20,000 cycles.

SUDDEN CHANGES OF PERIOD ("JUMPS"
OF RESIDUES)

Analysis of the available material shows that not all the periods vary smoothly and gradually. The periods of certain stars change abruptly, so the O−C diagrams are an assembly of straight lines. These objects are fairly numerous.

AX Aquarii

Observed by Tsesevich on the photographs of the Simeise collection. Light-variation amplitude very small, although quite significant magnitude enhancement is observed:

Max JD	E	O—A	O—B
2421466.38	—10180	—0d.01	—
3673.31	— 4494	—.01	—
3678.32	— 4481	—.05	—
4736.43	— 1755	+.01	—
6568.47	+ 2965	+.05	—
6948.43	+ 3944	+.03	—
7367.24	+ 5023	+.04	+0d.02
30236.44	+12415	+.14	—.02
2771.48	+18946	+.28	—.02
3178.32	+19994	+.35	+.03
4625.34	+23722	+.40	+.01

The O—A residues were calculated from the equation (GCVS, 1958)

$$\text{Max JD} = 2425417.60 + 0.388135 \cdot E; \ P^{-1} = 2.576423. \tag{69}$$

We see that starting with $E = 5023$, the equation no longer fits the observations. Around that epoch, the period increased discontinuously, and the new equation became effective

$$\text{Max hel JD} = 2425417.52 + 0.388155 \cdot E; \ P^{-1} = 2.576290, \tag{70}$$

as is evident from the O—B column. Reduction of the observations to one period according to (69) and (70) in the respective intervals confirmed this conclusion. Note, however, that the first four observations (2420394 — 0396) do not fit equation (69). At that time the maximum apparently lagged and the period varied. There are unfortunately too few observations for that epoch. Comparison stars in the observations of Table 120 (Figure 63): $u = -3^s.1; \ a = 0^s; \ b = 8^s.5.$

TABLE 120. Simeise observations (Tsesevich)

JD hel	s	JD hel	s	JD hel	s
242...		242...		242...	
0394.328	+1.2	5530.333	+2.0	8780.373	0.0
	+1.0		+2.4	243...	
0396.295	+1.9	6210.391	+2.0	0236.443	—3.1
	+1.0		+2.1		—5.6
1463.435	0.0	6235.323	+1.9	2771.476	—3.0
	+1.2		+2.1	2775.476	+1.9
1466.380	—4.1	6568.466	—6.1		+1.0
	—3.1		—8.1	3157.422	—6.0
3649.409	—1.6	6948.432	—6.1	3178.322	—6.1
	+2.1		—6.1		—9.1
3673.307	—1.6	7302.472	0.0	3860.485	0.0
	—7.1	7362.284	0.0		+1.9
3674.300	—1.0		—2.1	3867.506	+4.7
	+2.0:	7367.243	—6.1		+3.8
3678.315	—4.1:		—4.1	3897.395	+3.9
	—4.0:	7685.430	—2.0		+2.8
4033.380	0.0		—5.1	3913.301	+2.1
	+2.1	8404.426	+1.9		+1.9
4736.426	—4.1		+2.8	4223.477	+1.1
4740.471	+2.0	8433.323	—2.0		+2.1
	—3.0:	8460.273	—3.1	4242.414	0.0
4744.409	+2.0:		+ 1.1		+2.0
4773.410	—2.0:	8461.300	0.0	4596.464	—1.6
5485.469	0.0		—2.3		—1.0
5503.420	—4.1	8776.368	+2.0	4625.335	—7.1

113

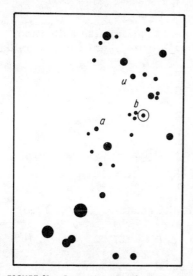

FIGURE 63. Comparison stars for AX Aquarii.

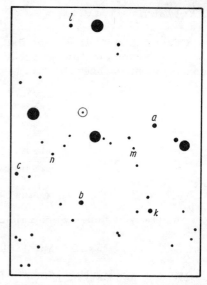

FIGURE 64. Comparison stars for BN Virginis.

BN Virginis

Discovered and studied by Boyce /206/. Tsesevich studied the star from Simeise photographs, which gave tentatively

$$\text{Max hel JD} = 2426853.25 + 0.391286 \cdot E; \ P^{-1} = 2.555741 ; \qquad (71)$$

the light curve of Table 121 is based on this equation. The seasonal light curves gave the following maxima:

Max hel JD	E	O—C	O—D
2419865.424	0	$0^d.000$	$+0^d.002$
26033.488	$+ 15764$	$-.011$	$-.006$
8257.512	$+ 21448$	$.000$	$+.006$
32650.373	$+ 32675$	$+.006$	—

The O—C residues were calculated from

$$\text{Max hel JD} = 2419865.424 + 0.391276 \cdot E; \ P^{-1} = 2.555740705. \qquad (72)$$

TABLE 121. Average light curve

φ	s	n	φ	s	n	φ	s	n
$0^p.005$	5.1	6	$0^p.431$	17.8	6	$0^p.763$	16.4	4
.053	8.6	4	.464	15.1	6	.870	11.6	4
.205	12.8	5	.548	18.0	6	.898	12.4	5
.284	14.4	6	.617	19.9	5	.964	4.4	4
.391	19.1	6	.687	19.3	5			

114

Comparison stars in the observations of Table 122 (Figure 64): $k = 0^s.0$; $a = 10^s.0$; $b = 15^s.3$; $c = 21^s.1$.

TABLE 122. Simeise observations (Tsesevich)

JD hel	s	JD hel	s	JD hel	s
241...		242...		242...	
9511.414	12.1	6084.396	6.3	8637.398	10.0
9513.375	16.5	6397.562	18.2	8992.488	18.2
	14.2		17.6		13.2
9863.356	19.2	6423.459	18.2	9339.460	12.6
9865.424	5.5	6424.458	10.0	9362.331	21.1
	7.7		10.0		17.6
9922.333	7.8	7510.552	3.0	243...	
	13.2:		1.8	0112.354	17.3:
242...		7513.478	15.3		15.3:
0608.392	23.1		19.9	2650.373	2.9
0929.568	17.6	7541.354	18.6		5.3
3848.528	15.3	7898.478	15.3	2996.403	21.1
	13.5		22.1		18.8
4231.516	21.1	8246.514	9.0	3002.339	22.1
4620.363	19.7		12.1		24.6
	17.6	8249.534	22.1	3358.478	18.2
6030.541	19.3		17.6		18.8
	21.1	8257.512	5.0	3734.402	18.6
6033.509	8.6		4.0		17.2
	11.8	8276.418	12.3	3771.314	13.0
6060.483	6.9		14.0		14.1
	6.0	8278.340	20.0	4118.398	7.5
6070.507	12.1		11.3		12.1
	13.2	8280.329	17.6	4555.482	21.1
6084.396	7.5	8637.398	9.1		

When the reduction of the bulk of the material had been completed, Tsesevich estimated the star's magnitude from new Moscow photographs (Table 123) taken with the 16-in. astrograph. Comparison stars (see Figure 64): $l = 0^s.0$; $a = 8^s.4$; $m = 12^s.1$; $n = 21^s.8$. The new observations reliably give the maximum Max hel JD = 2437099.278, which yields $O-C = +0^d.103$ for $E = 44045$. This signifies that the period is subject to considerable changes. If the period indeed changed abruptly, we have the two equations

$$\text{Max hel JD} = 2419865.422 + 0.39127585 \cdot E \text{ for } E < 21500; \qquad (73)$$

$$\text{Max hel JD} = 2432650.373 + 0.3912845 \, (E - 32675) \text{ for } E > 30000. \qquad (74)$$

The period increased by $0^d.00000865$. The new observations are listed in Table 123.

TABLE 123. Moscow observations (Tsesevich)

JD hel	s	JD hel	s	JD hel	s
243...		243...		243...	
3052.341	11.2	7052.331	9.6	7102.316	14.0
3061.458	8.4	7078.321	13.9	7103.317	10.2
3352.429	13.3	7079.295	5.6	7106.315	3.1
3358.483	14.1	7080.297	13.1	7113.315	13.9
3388.379	9.6	7087.301	12.1	7128.301	6.3
4077.423	11.4	7099.323	5.9		
7051.338	13.2	7100.318	13.2		

115

V 696 Ophiuchi

Discovered and investigated by Hughes—Boyce /275/. The original period $0^d.405$. Tsesevich estimated the magnitude of this variable from 73 photographs taken with the 16-in. astrograph of the Southern Section of the Shternberg Astronomical Institute, derived a tentative equation of the elements, and plotted two seasonal light curves from which two certain maxima were determined:

$$\text{Max hel JD} = 2436733.426 \text{ and Max hel JD} = 2437079.455.$$

The corresponding equation is

$$\text{Max hel JD} = 2436733.426 + 0.393215 \cdot E; \ P^{-1} = 2.5443138. \tag{75}$$

This equation was used to plot the average light curve (Figure 65, Table 124).

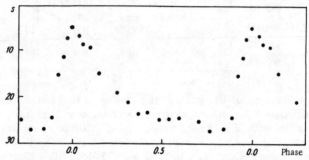

FIGURE 65.   Light curve of V 696 Ophiuchi.

TABLE 124.   Average light curve

φ	s	n	φ	s	n	φ	s	n
$0^P.033$	6.9	3	$0^P.416$	23.1	4	$0^P.840$	26.9	4
.051	8.3	3	.481	24.9	5	.881	24.1	5
.090	9.1	4	.537	24.6	5	.914	15.3	1
.146	14.7	5	.598	24.5	4	.943	11.6	2
.243	19.1	4	.708	25.0	5	.967	7.3	3
.309	21.1	5	.766	27.1	4	.992	4.8	2
.364	23.7	5						

Comparison stars (Figure 66) for the observations of Table 125: $u = 0^s.0$; $k = 9^s.6$; $a = 18^s.5$; $b = 22^s.2$; $c = 25^s.9$.

Simeise and Odessa photographs (Tables 127, 128) gave the seasonal curves and the maxima (Table 126).

TABLE 125. Observations at the Southern Section of the Shternberg Institute

JD hel	s	JD hel	s	JD hel	s
243...		243...		243...	
6702.495	21.0	7050.485	20.6	7077.467	11.9
6703.496	20.7	.506	24.0	.512	8.6
6720.444	3.8	.527	24.3	7078.370	16.2
.489	8.7	.549	25.9	.414	24.3
6721.432	24.3	7051.384	19.6	.458	24.7
.479	25.4	.450	25.9	.503	27.9
6722.382	15.3	.472	25.9	7079.345	25.9
.429	8.4	.493	24.5	.392	25.9
.475	19.4	.514	11.2	.440	8.1
6724.419	10.7	.536	6.1	.488	9.6
6728.382	14.6	7052.443	22.2	7080.347	20.1
.428	21.0	.464	24.0	.402	22.2
.475	21.5	.486	24.3	.450	26.9
6729.442	24.3	.507	24.3	7087.434	20.6
.483	5.8	.527	25.0	7116.365	24.3
6733.439	5.6	7053.491	8.1	7135.305	6.7
6734.432	24.0	.522	8.7	7136.301	25.9
.478	26.9	7072.396	7.5	7137.304	16.2
6747.336	25.0	.443	15.9	7377.545	7.5
.380	25.9	7074.392	9.6	7378.582	25.9
.418	25.0	.438	17.9	7486.317	20.6
6749.442	26.9	.484	21.3	7488.313	28.9:
6750.422	22.2	7077.378	25.9	7490.325	26.9
6751.424	26.9	.423	28.9	7493.318	21.3
7047.547	26.9				

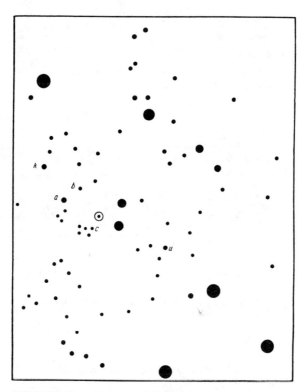

FIGURE 66. Comparison stars of V 696 Ophiuchi.

117

TABLE 126. List of maxima

Source	Max hel JD	E	O — A	O — B
Tsesevich (Simeise)	2424312.376	— 24134	—0^d.001	—
	5385.482	— 21405	+ .004	—
	6853.371	— 17672	— .003	—
	33802.382	0	.000	0.000^d
Tsesevich (Moscow)	6733.426	+ 7454	— .028	+ .001
	7079.455	+ 8334	— .034	—.001

TABLE 127. Simeise observations (Tsesevich)

JD hel	s	JD hel	s	JD hel	s
241...		242...		243...	
9920.449	[18.1	5385.465	12.1	2686.452	11.0
	[25.9	5388.489	[25.9		8.4
242...			28.9	3031.495	[25.9
0253.498	[18.1	5739.496	[24.1		[25.9
0630.489	25.9:	5744.497	15.1	3034.505	24.9
	[18.1		18.1		18.1
0634.464	23.1:	5745.470	25.9	3802.362	9.9
0636.497	25.9		[18.1		9.9
	24.8	5764.404	[18.1	3803.364	[18.1
1008.428	[25.9		[18.1		25.9
	[25.9	5770.348	25.9	3806.452	[25.9
1369.461	27.9		[18.1		25.9
	[25.9	6486.488	9.4	3823.344	26.4
3193.372	[25.9	6853.359	10.4		[25.9
	24.9	7931.500	30.9	3824.334	[26.1
4296.366	[25.9	7961.432	16.6		[25.9
	[25.9	7963.448	10.9	3829.331	24.6
4312.376	9.9	243...		3834.326	24.3
	8.5	0137.478	19.1		20.7
4652.452	[18.1		23.3	4131.497	12.1
5382.442	20.7	0161.380	[18.1		
5385.465	20.1		[18.1		

TABLE 128. Odessa observations

JD hel	s	JD hel	s	JD hel	s
243...		243...		243...	
6744.358	12.1	6758.389	[25.9	7111.370	18.1
6756.344	10.9	7075.464	24.9	.396	24.2
.401	24.6	7079.467	10.6	.424	25.0
6758.363	[25.9	.361	19.1:		

The O — A residues were calculated from the equation

$$\text{Max hel JD} = 2433802.382 + 0.39322138 \cdot E, \qquad (76)$$

which was valid between JD 2424312 and JD 2433802. The O — B residues were computed from the equation

$$\text{Max hel JD} = 2433802.382 + 0.39321745 \cdot E, \qquad (77)$$

which is valid since JD 2433802 to this day. The period apparently decreased suddenly by 0^d.00000393.

DM Cygni

Discovered by Tserasskaya [Ceraski] /200/ and carefully studied by Martynov. His summary list of maxima, adopted in the present work, was based on 825 observations /73/. Tsesevich observed the star in 1931, 1932, and 1957; he also established the power scale of the comparison stars (Figure 67). The photovisual magnitudes were borrowed from Martynov's paper (the power scale is converted to stellar magnitudes using the relation $m' = 9.51 + 0.07404\ s$):

Star	s	m	m'	m'−m
k	0.0	—	9.51	—
r	5.6	—	9.92	—
a	13.7	10.49	10.52	+.03
b	21.9	11.22	11.13	−.09
f	24.8	11.36	11.35	−.01
c	26.7	11.42	11.49	+.07
d	35.9	—	12.17	—

The stellar magnitudes m' show a good fit with the magnitudes m, and in what follows the luminosity of the variable is therefore expressed in powers, which can be readily converted to stellar magnitudes if desired.

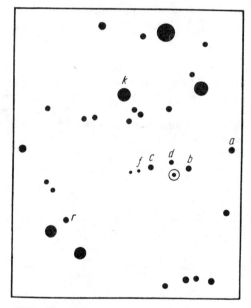

FIGURE 67. Comparison stars for DM Cygni.

Using Martynov's equation

$$\text{Max hel JD} = 2425887.5597 + 0.41985715 \cdot E;$$
$$P^{-1} = 2.38176246,$$

(78)

Tsesevich determined the seasonal light curves (Table 129).

TABLE 129. Seasonal light curves

φ	s	n	φ	s	n	φ	s	n
1931,	**2426538—6578**							
0.009[p]	11.7	5	0.360[p]	20.6	5	0.798[p]	27.0	5
.025	11.9	5	.390	21.4	5	.829	27.2	5
.046	12.0	5	.436	22.0	5	.856	25.9	5
.067	12.7	5	.470	22.6	5	.888	27.1	5
.090	12.8	5	.503	23.2	5	.906	25.1	5
.130	13.7	5	.540	23.1	5	.915	24.1	5
.159	15.1	5	.595	22.2	5	.934	18.0	5
.200	16.6	5	.649	23.8	5	.955	15.5	5
.242	17.9	5	.674	24.5	5	.972	14.8	5
.299	19.1	5	.732	25.2	5	.985	12.1	6
.324	20.6	5	.767	25.3	5	.995	12.1	6
1932,	**2426943—6984**							
.014	11.2	5	.348	25.2	5	.938	20.6	5
.047	11.8	5	.451	25.3	1	.949	17.9	3
.088	14.3	5	.688	26.1	3	.958	17.9	3
.128	17.1	3	.848	28.4	3	.971	13.7	2
.197	16.6	2	.919	23.4	3	.981	12.4	4
.242	21.4	4						
1957,	**2436027—6057**							
.011	15.0	3	.191	15.6	3	.692	25.5	2
.029	14.5	3	.218	16.4	3	.744	25.8	4
.046	10.8	5	.245	17.9	3	.795	25.6	4
.070	11.2	4	.276	18.3	3	.852	26.7	5
.094	10.3	4	.308	18.5	1	.895	27.5	5
.126	11.9	6	.406	19.5	1	.935	26.2	3
155	12.9	4	.503	24.0	2	.988	21.5	5

The maxima listed in Table 130 were determined using this equation. The old observations of Tsesevich have been published in full by Ustinov and Odynskaya /72/, and Table 131 therefore includes only the new observations. The summary table also lists the maxima from other sources and the new unpublished findings of Lange. Torondzhadze, mainly concerned with the motion of this star, published magnitude estimates from old Moscow photographs. Tsesevich's seasonal light curves revealed three additional maxima. They are also listed in the table.

TABLE 130. List of maxima

Source	Max hel JD	E	O — A	O — B
Torondzhadze /116/	2415290.409	—25240	+0.038[d]	—
	8570.320	—17428	+.031	—
Esch	25923.246	+ 85	—.002	—
Martynov /73/	242.5968.176	+ 192	+0.004[d]	—
	5973.214	+ 204	+.003	—
	6200.353	+ 745	.000	—
Tsesevich	6558.496	+ 1598	+.005	—
Martynov /73/	6635.325	+ 1781	.000	—
	6681.092	+ 1890	+.002	—
	6920.411	+ 2460	+.003	—
	6928.384	+ 2479	—.002	—
	6934.263	+ 2493	—.001	—
	6938.458	+ 2503	—.004	—
	6943.497	+ 2515	—.003	—
	6952.315	+ 2536	—.002	—
	6954.417	+ 2541	.000	—
	6957.356	+ 2548	.000	—
	6959.455	+ 2553	.000	—
	6962.394	+ 2560	.000	—
	6963.232	+ 2562	—.002	—

TABLE 130 (continued)

Source	Max hel JD	E	O — A	O — B
Tsesevich	242 6963.240	+ 2562	$+0^d.006$	—
Martynov /73/	6975.411	+ 2591	+.001	—
	6981.289	+ 2605	+.001	—
	6983.390	+ 2610	+.003	—
	7092.131	+ 2869	+.001	—
Dombrovskii /35/	7315.073	+ 3400	—.001	—
Martynov /73/	7334.386	+ 3446	—.001	—
	7336.486	+ 3451	—.001	—
	7337.327	+ 3453	+.001	—
	7340.262	+ 3460	—.003	—
	7342.365	+ 3465	.000	—
	7344.463	+ 3470	—.001	—
	7345.302	+ 3472	—.002	—
	7683.288	+ 4277	—.001	—
	7687.489	+ 4287	+.002	—
Selivanov /95/	7688.327	+ 4289	.000	—
Martynov /73/	8041.431	+ 5130	+.004	—
	8046.466	+ 5142	+.001	—
	8047.301	+ 5144	—.004	—
Gur'ev /27/	8076.695	+ 5214	.000	—
	8434.415	+ 6066	+.002	—
Martynov /73/	8508.305	+ 6242	—.003	—
	8509.148	+ 6244	.000	—
	8746.366	+ 6809	—.001	—
	8749.309	+ 6816	+.003	—
	8751.404	+ 6821	—.001	—
	8754.347	+ 6828	+.003	—
Gur'ev /27/	9224.585	+ 7948	+.001	—
Torondzhadze /116/	30585.354:	+11189	+.013:	—
Satyvaldyev	3557.527	+18268	+.017	+.001
Born /204/	3872.430	+19018	+.027	+.010
	3896.358	+19075	+.023	+.006
Sofronievitsch /204/	3896.360	+19075	+.025	+.008
Born /203/	3898.447	+19080	+.013	—.004
Sofronievitsch /204/	3898.444	+19080	+.010	—.007
Born /204/	3917.340	+19125	$+0^d.012$	$—0^d.005$
Sofronievitsch /204/	3917.340	+19125	+.012	—.005
	4001.314	+19325	+.015	—.003
Born /204/ \	4001.312	+19325	+.013	—.005
Alaniya /1/	4579.445	+20702	+.002	(—.017)
Tsarevskii /142/	4680.225	+20942	+.017	—.003
Tsesevich	6041.414	+24184	+.029	+.005
Satyvaldyev	6259.732	+24704	+.021	—.003
Lange	7851.415	+28495	+.026	—.003
	7854.357	+28502	+.029	.000

TABLE 131. New visual observations (Tsesevich)

JD hel	s	JD hel	s	JD hel	s
243...		243...		243...	
6027.399	24.3	6041.453	13.7	6052.329	10.7
.411	24.0	.466	18.1	.343	10.7
.421	24.3	.472	16.8	.352	11.9
.427	24.3	.480	17.8	.365	13.7
.435	24.3	.489	17.5	.381	16.4
.447	24.6	.493	17.8	.403	18.3
.457	24.6	.507	18.7	.431	18.5
.463	24.8	6043.357	28.7	.472	19.5
.468	25.5	.385	30.4	.511	23.6
.475	25.5	.441	29.1	6053.354	24.5
.487	24.8	.458	28.4	6057.297	25.7
.493	25.3	6051.369	29.0	.333	18.8
6028.426	11.7	.395	29.1	.339	23.9
6041.326	28.7	.417	28.7	.356	11.0
.340	29.0	.444	24.9	.361	10.4
.381	23.7	.452	22.9	.372	10.0
.392	17.2	.458	18.3	.379	9.7
.399	15.4	.463	17.0	.395	11.1
.406	11.7	.470	15.8	.409	11.7
.410	10.7	.476	12.2	.420	12.2
.421	10.2	.483	11.2	.433	14.7
.427	10.5	.495	13.2	.443	18.0
.434	11.2	.514	13.7	.458	18.3
.439	12.2	6052.307	10.7		
.445	12.6	.320	9.8		

It is clearly seen from Table 130 that equation (78) fits the observations between $E = 0$ and $E = 7948$, whereas after that

$$\text{Max hel JD} = 2433872.420 + 0.41985842 \, (E - 19018), \qquad (79)$$

and it is for this equation that the O−B residues were computed. It thus seems that the period increased by $0^d.00000127$. Before that the period was apparently shorter. Between $E = -25000$ and $E = 0$ the period probably was $0^d.4198557$. The jump in the period is $0^d.0000015$.

VZ Herculis

Observed visually for over 30 years. The good distribution of observations over the years is suitable for studying the slow variation of the period. Table 134 lists the maxima, mainly those from seasonal light curves. The parts of the curve near the maximum were determined from the new observations of Tsesevich (Tables 132, 133). The age was calculated from the equation

$$\text{Max hel JD} = 2425004.457 + 0.44032518 \cdot E; \; P^{-1} = 2.27104886. \qquad (80)$$

The residues O−A are plotted in Figure 68.

TABLE 132. Seasonal light curve 1951 (Tsesevich)

φ	m	n	φ	m	n	φ	m	n
$0^P.907$	11.56	3	$0^P.044$	10.32	2	$0^P.196$	10.90	6
.950	11.53	2	.064	10.36	3	.212	10.90	3
.006	10.43	1	.086	10.43	4	.246	11.07	3
.032	10.30	2	.149	10.51	5	.276	11.17	3

TABLE 133. Seasonal light curve 1956 (Tsesevich)

φ	m	n	φ	m	n	φ	m	n
$0^P.927$	11.59	3	$0^P.021$	10.46	3	$0^P.077$	10.49	4
.958	11.57	2	.030	10.26	2	.119	10.49	3
.982	11.27	3	.045	10.26	3	.176	10.88	3
.009	10.79	4	.058	10.32	3	.223	11.06	2

TABLE 134. List of maxima

Source	Max hel JD	E	O − A	O − B
Leiner /305/	2422388.496	− 5941	$+0^d.002$	—
Tsesevich /157/	4786.500	− 495	+.001	—
	4970.550	− 77	−.004	—
	5086.357	+ 186	−.002	—
	5325.455	+ 729	.000	—
	6066.520	+ 2412	.000	—
	6185.847	+ 2683	−.001	—
	6546.476	+ 3502	+.003	—
	6967.425	+ 4458	+.002	—

TABLE 134 (continued)

Source	Max hel JD	E	O — A	O — B
Florya /125/	242 7274.328	+ 5155	—0d.001	0d.000
Solov'ev /108/	7973.567	+ 6743	+.004	+.001
Parenago /90/	9089.353	+ 9277	+.009	.000
Tsesevich /157/	30612.440	+ 12736	+.015	—.001
	0941.363	+ 13483	+.016	—.002
Alaniya /1/	3856.324	+ 20103	+.033	—.001
Tsesevich /157/	3897.278	+ 20196	+.037	+.002
	5693.367	+ 24275	+.044	.000
Lange	6781.407	+ 26746	+.044	—.006
	6792.413	+ 26771	+.042	—.008
	6793.303	+ 26773	+.051	+.001
	7818.383	+ 29101	+.057	+.001
	7848.335:	+ 29169	+.067	+.011:
	7851.408	+ 29176	+.058	+.002
	7855.373	+ 29185	+.060	+.004
Tsesevich	7869.468	+ 29217	+.064	+.009
Lange	7873.427	+ 29226	+.061	+.005
	7878.270:	+ 29237	+.060	+.004
	7885.320	+ 29253	+.065	+.009
	7900.285	+ 29287	+.059	+.003
Tsesevich	7903.365	+ 29294	+.057	.000

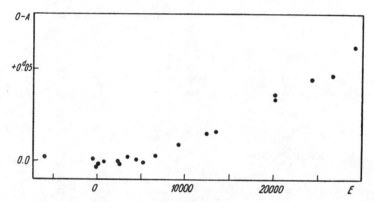

FIGURE 68. O—A residues for VZ Herculis.

Analysis of Table 134 shows that the period changed abruptly near $E = 5100$. Two equations of the period are thus valid: for $-6000 < E < 5100$ (O—A)

$$\text{Max hel JD} = 2425004.4590 + 0.44032394 \cdot E, \qquad (81)$$

or $+5100 < E < 24275$ (O—B)

$$\text{Max hel JD} = 2425004.4456 + 0.44032631 \cdot E. \qquad (82)$$

Table 135 lists Tsesevich's new observations. Comparison stars and stellar magnitudes borrowed from /157/: $a = 9^m.87$; $c = 10^m.59$; $e = 11^m.01$; $f = 11^m.69$.

123

TABLE 135. New observations (Tsesevich)

JD hel	m	JD hel	m	JD hel	m
243...		243...		243...	
5601.326	10.75	3885.378	10.43	7857.428	11.58
.328	10.75	.390	10.26	.466	11.56
.341	10.25	.396	10.23	7869.400	11.57
.344	10.28	.409	10.48	.425	11.49
.348	10.35	3886.284	10.31	.430	11.56
.353	10.76	.285	10.35	.439	11.44
.358	10.59	.289	10.44	.442	11.28
5691.387	11.42	.298	10.33	.445	11.09
.407	11.53	.311	10.44	.446	10.87
.447	11.55	.341	10.87	.450	10.82
5692.406	11.57	.382	11.16	.454	10.81
.446	11.59	3887.293	11.35	.456	10.39
.478	10.65	3889.295	11.55	.460	10.28
.481	10.35	.311	11.57	.463	10.31
.483	10.32	3890.279	10.33	.467	10.23
.490	10.25	.286	10.44	.474	10.25
.496	10.29	.292	10.83	.478	10.31
.505	10.32	.300	10.93	.483	10.39
.513	10.42	.308	10.96	.489	10.73
5693.297	11.59	.342	11.09	7900.281	10.29
.311	11.59	3892.304	11.54	.287	10.33
.318	11.59	.316	11.49	.300	10.30
.331	11.57	.317	11.55	.320	10.69
.333	11.57	.343	11.55	.326	10.87
.340	11.38	.351	11.61	.345	10.95
.342	11.28	.360	11.55	7903.271	11.56
.346	11.15	3897.278	10.33	.279	11.56
.353	10.91	.283	10.42	.283	11.56
.356	10.74	.292	10.41	.288	11.57
.359	10.38	.308	10.46	.294	11.59
.364	10.21	.339	10.49	.298	11.56
.376	10.32	.346	10.78	.301	11.57
.387	10.28	.355	10.85	.306	11.56
.402	10.35	.372	11.24	.313	11.62
.414	10.70	.380	11.26	.319	11.57
.420	10.82	.404	11.39	.323	11.58
.430	10.83	3920.248	10.90	.329	11.58
.435	10.94	.250	10.94	.332	11.58
.445	11.01	.254	10.92	.336	11.57
.453	11.12	.258	10.92	.340	11.43
3883.346	11.54	.267	10.96	.342	11.40
.354	11.46	.273	11.01	.344	11.32
.374	11.54	3922.238	11.57	.346	11.31
3884.338	11.51	.283	11.58	.350	11.09
.349	11.49	3941.229	11.55	.353	10.78
.361	11.57	3948.234	11.54	.356	10.43
.379	11.57	.251	11.54	.362	10.23
3885.281	11.58	3951.208	11.54	.368	10.23
.291	11.54	7854.388	11.52	.374	10.23
.301	11.54	.398	11.52	.380	10.29
.308	11.51	.434	11.50	.388	10.35
.312	11.54	7857.353	11.49	.405	10.42
.317	11.58	.378	11.50	.413	10.69
.333	11.54	.391	11.58	.420	10.91
.342	11.59	.402	11.54	.431	11.01
.359	11.49				

RV Capricorni

Some 20 years ago Tsesevich discovered a pronounced Blazhko effect for this star. Numerous observations by various astronomers were examined by Tsesevich and Ustinov /117/. Klepikova /42/ continued the project and confirmed the main results. Nearly ten years have passed since, and new observations require revision of the previous results. Table 136 lists the maxima supplementing Table 3 of /117/.

TABLE 136. New maxima

Source	Max hel JD	E	O − A	ω	Q	O − N
Batyrev /13/	2433883.279	+ 21325	−0^d.017	45°.6	−0^d.034	+0^d.017
Tsesevich	3886.401	+ 21332	− .029	50.6	−.048	+ .005
Batyrev /13/	3891.326	+ 21343	− .029	58.5	−.050	+ .005
	3922.282	+ 21412	+ .032	108.8	+.009	+ .066
	3927.204	+ 21423	+ .029	116.8	+.007	+ .063
	3948.203	+ 21470	− .017	150°.8	−0^d.029	+ .018
Klepikova /42/	4601.443	+ 22929	− .050	130.1	−.069	.000
	4606.369	+ 22940	− .059	138.2	−.076	.000
	4610.392	+ 22949	− .056	144.7	−.070	− .006
	4623.377	+ 22978	− .055	165.8	−.061	− .006
	4624.278	+ 22980	− .050	167.2	−.056	.000
	4628.297:	+ 22989	− .061	173.7	−.064	− .011
	4631.431	+ 22996	− .061	178.7	−.062	− .011
	4632.327:	+ 22998	− .061	180.2	−.061	− .011
	4649.344	+ 23036	− .058	207.8	−.047	− .008
	4650.235	+ 23038	− .063	209.2	−.051	− .012
	4654.244	+ 23047	− .083	215.7	−.069	− .033
	4658.278	+ 23056	− .079	222.3	−.063	− .029
	4675.285	+ 23094	− .087	249.9	−.064	− .026
	4684.243	+ 23114	− .084	264.3	−.060	− .033
Lange /68/	7137.435	+ 28593	− .135	282.6	−.121	− .030
	7142.366	+ 28604	− .129	290.7	−.106	− .025
	7159.368	+ 28642	− .142	318.2	−.126	− .037
Tsesevich	7159.379	+ 28642	− .131	318.2	−.115	− .026
Lange /68/	7164.306	+ 28653	− .129	326.2	−.115	− .024
	7167.436	+ 28660	− .133	331.2	−.121	− .028
	7172.371	+ 28671	− .123	339.3	−.114	− .018
Tsesevich	7172.401	+ 28671	− .093	339.3	−.084	+ .012
Lange /68/	7176.413	+ 28680	− .111	345.8	−.105	− .006
Tsesevich	7176.411	+ 28680	− .113	345.8	−.108	− .008
Lange /68/	7186.289	+ 28702	− .086	1.9	−.087	+ .019
	7198.361	+ 28729	− .103	21.5	−.112	+ .002
Tsesevich	7198.401:	+ 28729	− .063	21.5	−.072:	+ .042
Lange /68/	7225.245	+ 28789	− .084	64.9	−.106	+ .022
	7497.464:	+ 29397	− .100	146.5	−.113	+ .012
	7498.372:	+ 29399	− .087	148.0	−.100	+ .025
	7501.488	+ 29406	− .105	153.0	−.116	+ .007
	7511.341	+ 29428	− .103	168.9	−.108	+ .009
	7519.401	+ 29446	− .103	182.0	−.102	+ .010
	7520.304	+ 29448	− .095	183.5	−.093	+ .017
	7524.320	+ 29457	− .109	190.0	−.105	+ .004
	7528.350	+ 29466	− .109	196.5	−.102	+ .004
Lange	7871.339:	+ 30232	− .099	32.7	−.112	+ .021
	7878.497	+ 30248	− .105	44.4	−.122	+ .015
	7883.422	+ 30259	− .105	52.3	−.124	+ .015
	7874.322	+ 30261	− .101	53.8	−.121	+ .020
	7887.450	+ 30268	− .107	58.8	−.128	+ .013
	7900.438	+ 30297	− .104	79.9	−.128	+ .017
	7909.392	+ 30317	− .105	94.5	−.129	+ .016
	7910.292	+ 30319	− .101	95.9	−.125	+ .020
	7931.332	+ 30366	− .105	130.0	−.124	+ .016
	7932.229	+ 30368	− .104	131.4	−.122	+ .018
	7962.210	+ 30435	− .122	180.1	−.122	.000

Table 136 gives the O − A residues calculated from the equation

$$\text{Max hel JD} = 2424334.947 + 0.44775374 \cdot E. \qquad (83)$$

We see that the O − A residues are negative and progressively increase in magnitude, reaching 2.5 hrs. This confirms our earlier conclusion that the period of this star abruptly diminished (see Figure 3 in /117/). This development, however, is not easy to investigate since the Blazhko effect is close to 220 days and none of the seasonal observation series covers the entire cycle of this effect. Therefore the formula ω = 1°.621645 (JD − 2426973.3) was used to calculate the inequality phases, which are also listed in Table 136. Assuming a constant inequality period, Tsesevich estimated the average ephemeris correction O − A for the first season of observations, having plotted the seasonal variation in O − A vs. the phase ω.

From each O—A residue we subtracted $0^d.0243 \sin \omega$ to obtain the Q in Table 136. The averaged values of Q are listed in Table 137.

TABLE 137. Average values of $Q = (O-A) - 0^d.0243 \sin \omega$

$\bar{E}$	$\bar{Q}$	n	$\bar{E} - 21384$	$\bar{Q}_c$	$\bar{Q} - \bar{Q}_c$
21384	$-0^d.024$	6	0	$-0^d.034$	$+0^d.010$
23010	$-$ .062	14	$+$ 1626	$-$.050	$-$.012
28675	$-$ .107	14	$+$ 7291	$-$.105	$-$.002
29431	$-$ .105	8	$+$ 8047	$-$.112	$+$.007
30306	$-$ .123	11	$+$ 8922	$-$.121	$-$.002

The variation of $\bar{Q}$ against $\bar{E}$ is almost linear,

$$\bar{Q} = -0.034 - 0.00000973 \, (\bar{E} - 21384).$$

From this equation we calculated the values of $\bar{Q}_c$. The new maxima can be linked with the period

$$0.44775374 - 0.00000973 = 0.44774401.$$

The O—N residues were calculated from

$$\text{Max hel JD} = 2433883.262 + 0.44774401 \, (E - 21325). \tag{84}$$

To check the validity of the above inequality, we plotted the residues O—N as a function of the phase angle ω (Figure 69). Since the error in the maxima is often very considerable, the observations apparently show a good fit with the previous results.

Thus, the period of the Blazhko effect did not change, although the star's period diminished abruptly by $0^d.00000973$. Tsesevich's new visual observations are listed in Table 138. Comparison stars /117/: $a = 10^m$ $b = 10^m.66$; $c = 11^m.20$; $d = 11^m.98$.

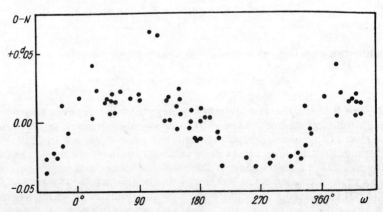

FIGURE 69. The O—N plot for RV Capricorni.

TABLE 138. New visual observations (Tsesevich)

JD hel	m	JD hel	m	JD hel	m
243...		243...		243...	
3853.404	11.63	7158.347	11.79	7172.405	10.78
3883.353	11.02	.364	11.84	.410	10.88
.359	10.98	.398	11.81	.415	10.81
.380	11.02	.426	11.81	.423	10.98
3884.353	10.96	.437	11.79	.433	10.88
.424	11.78	7159.312	11.79	.439	10.93
.445	11.81	.342	11.65	.445	10.95
3885.305	11.79	.355	11.59	.456	11.02
.316	11.82	.363	11.13	7175.388	11.79
.342	11.84	.374	10.75	7176.329	11.77
.400	11.84	.376	10.88	.339	11.79
.430	11.84	.384	10.95	.353	11.77
3886.296	11.88	.388	10.91	.376	11.77
.358	11.79	.401	10.95	.383.	11.70
.381	10.97	.411	10.98	.396	11.09
.385	10.59	.418	10.97	.402	10.59
.389	10.54	.424	10.97	.419	10.56
.395	10.51	.435	10.97	.429	10.77
.409	10.81	.440	11.99	7193.344	11.70
.420	10.52	.446	11.04	.351	11.72
3887.310	10.51	7161.354	11.65	.361	11.74
.314	10.51	7162.417	11.81	.368	11.65
.328	10.56	.449	11.81	.383	11.67
3889.304	11.86	.467	11.70	7196.250	11.11
3890.292	11.84	7165.323	11.80	.261	11.20
.300	11.88	.352	11.72	.277	11.61
.308	11.87	.360	11.72	.331	11.70
.343	11.88	7170.339	11.74	7197.302	11.77
.351	11.88	.348	11.79	.329	11.77
3892.354	11.68	.355	11.79	.382	11.77
.359	11.66	.367	11.80	7198.282	11.81
3897.290	11.11	.374	11.79	.355	11.10
.299	11.49	.383	11.74	.364	11.05
.310	11.61	.398	11.80	.370	10.96
.347	11.67	.422	11.84	.378	10.88
.372	11.75	.448	11.84	.382	10.82
.380	11.81	7172.316	11.72	.399	10.73
.387	11.82	.330	11.70	.410	10.73
.408	11.81	.342	11.62	.426	10.77
3920.251	11.76	.364	11.09	7204.287	11.00
5011.266	11.41	.380	10.97	.294	11.07
.275	11.57	.387	10.96	.352	11.57
.293	11.67	.392	10.81		
7158.326	11.72	.399	10.78		

CP Aquarii

Discovered by Hoffmeister /265/ and investigated by Selivanov /99/.
Solov'ev /104/ examined the Simeise photographs and derived the improved
equation

$$\text{Max hel JD} = 2427634.324 + 0.463407 \cdot E. \qquad (85)$$

Batyrev's visual observations /16/ confirmed Solov'ev's findings.

The star was also observed by Alaniya /2/ and by the Sonneberg
astronomers Busch, Barthel, and Häussler /210/. Recently it was
observed by Tsarevskii and Lange. Table 139 lists the maxima. It does
not include Solov'ev's maxima, since Tsesevich has revised all the
Simeise photographs, which are now much more numerous than at that time
(Table 140). The resulting seasonal curves have given more reliable
maxima. Comparison stars (Figure 70): $a = 0^s.0$; $b = 11^s.2$; $c = 14^s.3$.

TABLE 139. List of Maxima

Source	Max hel JD	E	O — A	O — B
Tsesevich (Simeise)	2419625.237	— 17283	—0^d.033	+0^d.002
	24727.372	— 6273	—.009	+.006
	6209.333	— 3075	—.023	—.014
Selivanov /99/	7685.300	+ 110	—.008	—.004
Tsesevich (Simeise)	8753.472	+ 2415	+.011	+.011
Busch et al. /210/	33831.501	+ 13373	+.026	—
Tsesevich (Simeise)	3838.421	+ 13388	—.005	—
Busch et al. /210/	3947.354	+ 13623	+.027	—
	4211.454	+ 14193	—.015	—
Batyrev /16/	4251.310	+ 14279	—.012	—
	4277.259	+ 14335	—.013	—
Busch et al. /210/	4334.259	+ 14458	—.012	—
Batyrev /16/	4606.282	+ 15045	—.009	—
	4618.337	+ 15071	—.003	—
	4624.364	+ 15084	.000	—
	4625.292	+ 15086	+.001	—
	4631.314	+ 15099	—.001	—
Busch et al. /210/	4636.395	+ 15110	—.018	—
	4662.386	+ 15166	+.022	—
	4714.248	+ 15278	—.017	—
	4958.468	+ 15805	—.013	—
Alaniya /2/	4980.306	+ 15852	+.045	—
Busch et al. /210/	4990.444	+ 15874	—.012	—
	5306.479	+ 16556	—.020	—
	5371.387	+ 16696	+.009	—
	5391.331	+ 16739	+.028	—
	5455.235	+ 16877	—.018	—
	5996.511	+ 18045	—.001	—
	6015.508	+ 18086	—.004	—
	6138.297	+ 18351	—.018	—
	6164.274	+ 18407	+.008	—
	6434.450	+ 18990	+.018	—
	6460.399	+ 19046	+.016	—
	6480.310	+ 19089	+.001	—
Tsarevskii	7142.507	+ 20518	—.011	—
	7144.364	+ 20522	—.007	—
Tsarevskii, Chuprina	7153.633	+ 20542	—.007	—
Tsarevskii	7161.522:	+ 20559	+.004:	—
	7164.308	+ 20565	+.010	—
	7189.313	+ 20619	—.009	—
	7193.491	+ 20628	—.002	—
	7196.281:	+ 20634	+.008:	—
Tsarevskii, Chuprina	7233.337	+ 20714	—.009	—
Lange /67/	7543.349	+ 21383	—.015	—
Tsarevskii, Chuprina	7548.453	+ 21394	—.009	—
Lange	7872.381	+ 22093	—.003	—
	7878.406	+ 22106	—.002	—
	7884.429	+ 22119	—.003	—
Tsarevskii, Chuprina	7884.423	+ 22119	—.009	—
Lange	7885.353	+ 22121	—.006	—
	7898.330	+ 22149	—.005	—

TABLE 140. Simeise observations (Tsesevich)

JD hel	s	JD hel	s	JD hel	s
241...		242...		242...	
8966.223	10.2	0746.381	11.2	1094.296	3.9
	10.6	0786.244	8.6		4.5
9625.352	7.5	0806.213	7.8	1101.379	12.4
	7.2		13.2		12.8

TABLE 140 (continued)

JD hel	s	JD hel	s	JD hel	s
242...		242...		242...	
1137.291	6.0	5883.370	14.2	9112.387	16.3
1431.324	11.7	5889.287	8.7		16.3
	9.7		6.8	9132.350	13.7
2552.353	12.7	6180.480	10.2		15.7
	12.7		13.2	9512.296	15.3
3283.389	− 1.5	6187.503	12.7		9.2
	1.1	6192.488	13.2	9518.296	12.2
3620.412	10.6		14.2	9522.284	− 1.0
	8.8	6208.314	14.2		− 1.0
4357.474	15.8	.411	7.5	9548.281	1.1
	17.3	.314	14.2		0.0
4415.337	12.7	.411	6.9	243...	
	15.2	6209.324	− 1.5	0199.490	9.2
4702.502	13.2	6220.379	15.7	0208.495	12.8
4727.472	8.4	6921.468	11.2		10.1
4770.268	13.2		17.8	0224.467	8.8
4787.347	15.8	7686.329	7.8	3154.330	14.2
	13.2		6.5		14.2
5448.462	− 1.0	8011.500	6.8	3510.295	4.4
	2.0		6.7		3.1
5494.280	2.2	8020.491	7.8	3838.456	0.0
	6.0		9.0	3859.390	7.1
5508.332	7.5	8024.435	14.2		6.1
	4.4		13.7	3894.336	12.2
5525.285	15.2	8369.420	9.3		10.1
5527.326	9.7		8.4	3898.356	9.2
5829.477	7.8	8398.340	13.2		6.6
	5.1		15.8	4212.492	8.8
5837.451	11.2	8753.453	− 1.0		8.6
5860.353	7.8	8847.277	10.6	4607.368	11.2
	7.7		12.2		5.6
5883.370	12.2	9111.378	9.2		

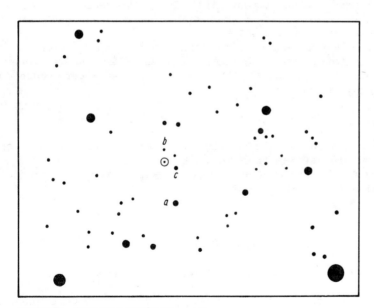

FIGURE 70. Comparison stars for CP Aquarii.

The O —A residues were calculated from the equation

$$\text{Max hel JD} = 2427634.333 + 0.463407 \cdot E. \tag{86}$$

Chuprina reduced Tsarevskii's observations and obtained average seasonal light curves from which 4 maxima were determined. The variation of the O —A residues shows that the period increased abruptly by $0^d.0000018$. The first five maxima give

$$\text{Max hel JD} = 2427634.329 + 0.4634088 \cdot E \tag{87}$$
$$(-17283 < E < +2415).$$

After that,

$$\text{Max hel JD} = 2437884.429 + 0.4634070 \cdot E. \tag{88}$$

BN Aquarii

Discovered by Hoffmeister /263/ and tentatively designated as 94.1931 Elements determined by Tsesevich:

$$\text{Max hel JD} = 2426578.360 + 0.469632 \cdot E. \tag{89}$$

Parenago /87/ studied the Simeise photographs and obtained the average light curve. Between 1933 and 1935 Tsesevich published several maxima /149 — 151/.

From all the observations of Tsesevich, Odynskaya derived a number of average seasonal maxima /81/. Tsarevskii determined the normal maximum from his visual observations /141/. It markedly deviates from the ephemeris maximum.

Tsesevich revised all the material available on this star and estimated its magnitude from the Simeise plates. In 1961 — 1962 new visual observations were carried out. The observation results were reduced using Odynskaya's equation

$$\text{Max hel JD} = 2426577.897 + 0.469641 \cdot E; \quad P^{-1} = 2.12928599. \tag{90}$$

The average light curves were also obtained from this equation. The maxima are listed in Table 141.

The O —A plot shows that the period jumped at a certain point. The equation for JD 2421101 — JD 2428964 (O —B)

$$\text{Max hel JD} = 2426577.901 + 0.46963401 \cdot E; \tag{91}$$

the equation after JD 2428964 (O —C)

$$\text{Max hel JD} = 2426577.833 + 0.46964739 \cdot E. \tag{92}$$

The jump occurred at $E = 5080$. The observations are listed in Tables 142, 143. Comparison stars (Figure 71): visual scale, $a = 0^s.0$; $b = 3^s.5$; $c = 15^s.7$; $d = 20^s.4$; $f = 31^s.7$; photographic scale, $a = 0^s.0$; $b = 8^s.3$; $c = 14^s.6$; $e = 25^s.2$.

TABLE 141. List of Maxima

Source	Max hel JD	E	O — A	O — B	O — C
Tsesevich (Sim.)	2421101.503	− 11661	$+0^d.090$	$+0^d.004$	$+0^d.228$
	4412.415	− 4611	+.033	−.004	+.126
Tsesevich,	6577.888	0	−.009	−.013	+.055
Odynskaya	6957.370	+ 808	+.003	+.005	+.062
Tsesevich (Sim.)	7362.169	+ 1670	−.028	−.021	+.025
Tsesevich,	7393.190	+ 1736	−.004	+.004	+.049
Odynskaya	7659.963	+ 2304	+.013	+.025	+.062
Tsesevich (Sim.)	9518.296	+ 6261	−.023	+.016	+.001
Tsesevich, Odynskaya	30592.385	+ 8548	−.003	+.052	+.006
Tsesevich (Sim.)	3152.415	+ 13999	+.014	+.107	−.012
Tsarevskii	6116.376	+ 20310	+.070	+.208	+.005
	6811.450	+ 21790	+.076	+.224	.000
	7168.405	+ 22550	+.103	+.257	+.023
Tsesevich (visual)	7547.391	+ 23357	+.089	+.248	+.004
Tsarevskii	7553.494	+ 23370	+.087	+.246	+.001
Tsesevich (visual)	7872.390	+ 24049	+.097	+.261	+.007

TABLE 142. Simeise observations (Tsesevich)

JD hel	s	JD hel	s	JD hel	s
241...		242...		242...	
9652.344	14.6	3996.502	22.3	6984.351	22.4
	13.6		23.3		21.4
9987.508	14.6	4385.490	23.3	7304.472	8.4
9991.470	15.7		23.1		10.6
	17.5	4412.445	12.5	7329.408	14.6
242...		4414.368	12.7	7343.411	10.6
0006.355	7.3		12.7		7.3
	8.8	4732.493	22.3	7356.261	18.1
0398.338	24.0	4734.474	25.2		20.8
	20.6	.550	14.6	7362.284	12.9
0717.475	15.7	.512	22.8	7367.243	24.0
	18.5	5124.414	8.9		22.4
0721.441	23.1		9.2	8040.305	9.3
0726.507	21.7	5129.443	23.3		9.3
0743.350	19.9		22.3	8394.496	10.6
0748.405	9.3	5479.451	10.6		10.6
	9.3		11.7	8397.496	21.4
0754.369	21.7	5482.466	22.8	8403.459	13.3
	21.7		22.8		12.7
0777.282	23.1	5497.395	17.2	8776.368	13.0
	24.1	5854.425	22.3		13.7
1075.500	22.9		22.8	8779.373	22.4
	23.7	5855.441	23.1		22.0
1101.478	10.8		23.7	8786.40:	24.0
	10.4	5856.437	21.4		24.1
1102.419	9.2		22.4	8811.349	18.6
1105.450	20.9	5865.474	9.7		17.6
1108.449	22.3		10.1	8817.338	17.6
	22.3	6209.432	19.9		18.6
1112.429	18.8		22.3	9132.458	21.7
	17.5	6216.388	12.8		22.4
1134.330	13.6		14.6	9491.491	14.6
	14.1	6222.483	12.5		14.6
1464.502	9.3		12.8	9498.533	15.8
	7.4	6240.371	17.3		15.7
1466.469	13.2	6248.330	15.7	9514.394	22.3
	11.7		14.6		22.3
1483.359	9.3	6264.233	12.0	9518.403	12.3
1485.334	20.8		12.0	9524.339	22.3
	20.4	6266.280	19.9		20.8
1493.314	16.1		18.8	9880.439	8.5
	14.6	6268.242	20.4		11.7
3647.406	7.9		20.8	243...	
	9.0	6566.431	23.3	0236.443	11.8
3664.287	11.7		20.8		11.7
3731.218	22.0	6976.324	20.9	2771.476	18.8
	24.1		21.4		16.7

131

TABLE 142 (continued)

JD hel	s	JD hel	s	JD hel	s
243...		243...		243...	
2772.487	8.3	3215.245	21.1	4242.414	20.5
	7.3		23.8		19.4
2775.476	21.4	3869.467	21.4	4246.373	17.8
	18.8		21.4		15 6
3152.373	14.6	3896.400	11.9	4273.294	22.0
	14.6		10.8		22.0
3157.422	22.0	3915.339	20.7	4601.471	20.4
3187.337	19.4		20.8		20.4
	19.4				

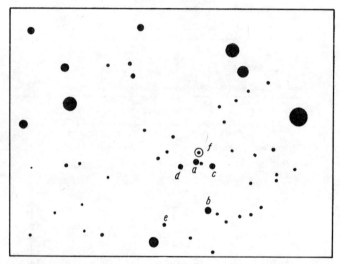

FIGURE 71. Comparison stars for BN Aquarii.

TABLE 143. New visual observations (Tsesevich)

JD hel	s	JD hel	s	JD hel	s
243...		243...		243...	
7543.353	27.9	7547.295	27.9	7547.429	13.2
.376	27.9	.299	28.3	.434	14.0
7545.299	28.9	.308	28.6	.438	13.7
.328	27.6	.322	28.6	.442	13.7
.352	28.3	.336	28.9	7549.286	12.9
.376	28.3	.351	29.0	.293	12.9
.382	27.6	.357	27.6	.334	14.2
.387	27.9	.361	26.5	.355	14.2
393	29.0	.364	25.6	.362	17.6
.398	27.6	.367	24.8	.371	17.8
.402	28.3	.370	19.9	.390	19.2
.421	28.3	.374	17.5	.398	19.9
.430	27.6	.376	13.0	.413	22.3
.439	28.3	.381	12.2	.423	24.5
.445	28.3	.384	11.6	.438	25.8
.450	27.6	.397	11.1	7857.420	17.2:
7547.260	27.6	.404	11.2	7869.383	28.6
.273	27.6	.416	12.2	.427	28.6
.284	27.9	.420	12.2	.448	29.5
.290	27.8	.424	13.0	.458	28.7

TABLE 143 (continued)

JD hel	s	JD hel	s	JD hel	s
243...		243...		243...	
7871.362	28.6	7872.372	23.5	7872.475	14.1
.388	29.2	.373	18.6	.481	18.4
.404	29.4	.375	16.9	.498	18.0
.428	26.0	.377	13.0	.508	19.2
.432	18.5	.380	12.3	.514	24.5
.435	13.5	.385	10.1	7873.336	11.6
.439	10.9	.390	10.6	.351	11.3
.446	10.1	.400	9.8	.369	12.3
.453	9.1	.403	10.2	.381	13.1
.465	9.4	.416	11.3	.392	13.5
.481	10.2	.423	11.3	.407	14.5
.486	11.3	.441	11.6	.436	19.2
.494	10.8	.453	13.9	.469	22.9
7872.371	26.3	.466	13.7		

BW Virginis

Discovered and investigated by Boyce /206/, who obtained

$$\text{Max hel JD} = 2426840.50 + 0.47066 \cdot E. \tag{93}$$

Studied by Tsesevich from Simeise and Odessa photographs, which gave the improved equation

$$\text{Max hel JD} = 2426840.471 + 0.4706635 \cdot E \tag{94}$$

and "seasonal" light curves. These curves were used to determine the maxima (Table 144).

TABLE 144. List of maxima

Source	Max hel JD	E	O—A	O—B	O—C
Tsesevich (Sim.)	2420247.403	—14008	—0^d.014	0^d.000	—
	4621.302	— 4715	+.009	.000	—
Boyce	6840. 50	0	+.029	+.008	+0^d.023
Tsesevich (Sim.)	8252.463	+ 3000	+.001	—	.000
Tsesevich (Odessa)	36667.428	+20879	—.026	—	+.001

The period most probably jumped, so that the observations are represented by two equations:

$$\text{Max hel JD} = 2420247.403 + 0.4706660 \cdot E \text{ before } E = 14000; \tag{95}$$

$$\text{Max hel JD} = 2428252.463 + 0.4706619 \cdot E_1 \text{ after } E = 14000. \tag{96}$$

The period apparently decreased by 0^d.0000041.

Comparison stars for observations of Tables 145, 146 (Figure 72): $k = -8^s.8$; $a = 0^s.0$; $b = 11^s.0$; $c = 21^s.3$.

133

TABLE 145. Odessa observations (Tsesevich)

JD hel	s	JD hel	s	JD hel	s
243...		243...		243...	
6286.456	[15	6347.347	6.6	6660.488	10.8
6288.452	16.8	6607.572	[11.0	6661.466	11.0
6304.398	17.0	6608.596	3.4	6663.484	17.0
6344.357	17.2	6612.636	15.5	6667.417	—2.0
6345.349	13.0	6613.589	16.0	6668.442	7.2

TABLE 146. Simeise observations (Tsesevich)

JD hel	s	JD hel	s	JD hel	s
241...		242...		242...	
9498.285	15.6	6060.483	[17.0	8637.398	17.0
	19.0	6070.507	7.0		18.0
9513.375	[11.0		7.7	8981.450	14.0
.463	15.0	6084.396	16.0		15.0
9922.333	14.0		[19.0	8992.488	[16.0
	13.0	6397.562	9.0		[11.0
242...			8.0	9339.460	16.0
0215.41:	6.4	6420.483	19.4	9348.504	15.0
	5.0		17.5		17.2
0247.420	4.0	6423.459	5.9	9721.446	from 11.0 to 2.6
	5.5	7510.552	18.0		from 11.0 to 3.3
0929.568	14.0		17.7	243...	
	14.0	7513.478	from 18.0 to 11.0	2650.373	7.3
0950.563	2.0			3031.322	16.2
	2.4		8.0		18.2
0958.347	18.0	7896.347	12.0	3034.324	18.2
	19.0		[16.0		17.5
0977.322	18.0	7898.472	15.0	3358.478	17.9
	18.0		14.0		18.7
0978.376	5.0	8246.514	18.0	3734.402	14.0
	2.0	8252.498	2.2		12.0
4621.287	6.0		4.0	3737.426	15.0
	7.7	8257.512	16.0	3771.314	19.0
6033.509	13.0		13.0		18.2
	[16.0	8278.340	5.7	4455.482	14.4
6060.483	[15.0		6.4		11.0
				4456.515	15.0

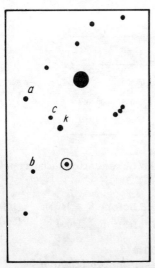

FIGURE 72. Comparison stars for BW Virginis.

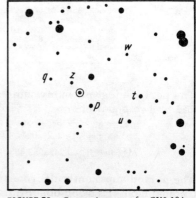

FIGURE 73. Comparison stars for CVS 194.

CVS 194 = 192.1935 Arietis

Discovered by Morgenroth /322/. Studied by Tsesevich from Odessa and Simeise photographs. A tentative equation of the period was derived after a few trials and the seasonal light curves were constructed. The maxima were then determined (Table 147).

TABLE 147. List of maxima

Source	Max hel JD	E	O — A	E_1	O — B
Tsesevich (Sim.)	2420463.362	0	$+0^d.009$	—	—
	4466.477	+ 8488	−.003	—	—
	7811.205	+15580	−.017	—	—
	32824.575	+26210	+.012	—	—
Tsesevich (Odessa)	6823.596	+34689	+.151	0	$-0^d.010$
	7207.544	+35503	+.199	+ 814	+ .010
	7591.470	+36317	+.224	+1628	+ .009
	7914.535	+37002	+.228	+2313	− .010

The O—A residues were calculated from

$$\text{Max hel JD} = 2420463.353 + 0.4716219 \cdot E; \; P^{-1} = 2.120342588; \quad (97)$$

the O—B residues from

$$\text{Max hel JD} = 2436823.606 + 0.4716555 \cdot E_1; P^{-1} = 2.120191538. \quad (98)$$

Table 147 shows that the period changes suddenly by $0^d.0000336$. Reduction to one period shows that the Simeise observations (Table 148) fit equation (97), and Odessa observations (Table 148) fit (98). The average light curve was·calculated from the Odessa observations (the age calculated from (98)):

φ	s	n	φ	s	n	φ	s	n
$0^P.030$	−1.4	3	$0^P.398$	+ 8.2	4	$0^P.875$	+2.5	4
.091	−0.2	3	.537	+ 9.9	4	.918	+1.0	2
.146	+1.2	5	.681	+10.8	5	.988	−1.6	3
.220	+4.8	5	.753	+ 8.2	4			
.300	+7.0	5	.824	+10.3	4			

Comparison stars (Figure 73): $t = -13^s.6$; $w = -7^s.3$; $u = 0^s.0$; $p = +8^s.6$; $q = +18^s.4$; $z = +19^s.4$.

TABLE 148. Simeise observations

JD hel	s	JD hel	s	JD hel	s
242...		242...		242...	
0446.221	10.6	0475.209	1.9	4439.357	12.6
0447.441	6.4		4.9	4447.357	8.6
	8.6	0476.458	11.6	4466.494	4.0
0461.336	12.6	1513.567	15.1		5.7
0462.390	1.2		10.8	4467.369	9.6
	1.0	4435.368	9.6		12.9
0463.385	−2.7		7.4	5150.583	11.6

135

TABLE 148 (continued)

JD hel	s	JD hel	s	JD hel	s
242...		242...		243...	
5150.583	9.6	7718.555	14.0	2824.556	−1.8
5151.570	12.1		10.8		+1.0
	9.6	7779.440	11.6	2831.520	17.2
5180.221	3.9:		13.6		14.5
	3.7:	7781.413	15.5	2832.487	10.6
5889.535	7.1		11.6		12.6
6280.550	8.6	7810.218	6.7	2836.487	3.4
	9.6		6.3		2.6
6306.254	11.6	7811.223	0.0	3210.410	0.0
	13.6		−1.7	3566.290	14.4
6978.478	12.6	8083.500	6.0		14.5
	12.6		7.2	3921.544	13.0
7005.550	8.6	8084.557	10.8		13.5
7011.432	8.6		11.5	4300.411	12.5
	7.5	8841.415	7.1:		14.5
7011.538	10.6		0.0	4717.345	13.6
7012.466	11.6:	8845.455	11.1		
7042.262	15.8		14.6		
	12.6				

TABLE 149. Odessa observations

JD hel	s	JD hel	s	JD hel	s
243...		243...		243...	
6546.328	2.1	7230.510	13.6	7909.564	6.9
6815.521	1.9	.537	13.6	7911.542	6.7
6817.542	1.0	7337.230	−1.0	7912.537	8.1
6823.572	−1.8	7549.516	−2.4	7913.517	12.6
6825.592	2.0	7555.493	[8.6	7914.562	−1.1
6843.541	7.6	.545	10.6	7939.457	10.6
6844.520	10.6	7560.489	5.1	.497	[6.0
6869.456	1.2	7561.518	7.5	7942.484	6.6
6901.368	0.0	7578.492	12.6	7943.480	8.6
6904.352	1.7	7582.444	6.0	7946.527	7.5
6959.235	13.6	7583.494	2.5	7959.408	1.4
6960.248	10.6	7591.466	−1.0	7960.390	7.4
7167.557	4.8	7605.408	6.0	7963.439	11.6
7196.569	1.2	7606.407	11.6	7968.443	7.5
7197.509	10.2	7636.314	−0.5	7969.442	8.6
7198.529	3.8	7643.337	−2.0	7973.450	2.1
7207.554	−2.1	7645.303	0.0	7974.443	−1.8
7227.477	4.3	7690.233	10.6		

UU Virginis

Discovered by Cannon /215/ and studied by Belyavskii. Observed by
Florya /129/, Solov'ev /108/, Alaniya /1/, Lange. Magnitude estimated
by Tsesevich from Odessa and Simeise photographs. Magnitude also
estimated from 737 Harvard photographs, as reported by S. Gaposchkin /238/
and C. Payne-Gaposchkin /329/. Reduction of all the observations gave
the summary list of maxima (Table 150).

The O−A residues were calculated from the equation

$$\text{Max hel JD} = 2419505.314 + 0.47560558 \cdot E.$$ (99)

The variation of the O−A residues shows that the period changed probably
in a jump. The least squares method gave the combined equation (O−B)

$$\text{Max hel JD} = 2419505.309 + 0.47560624 \cdot E$$ (100)

and the linear relations (O−C, O−D)

$$\text{Max hel JD} = 2419505.309 + 0.47560590 \cdot E \text{ до } E = 17\,000; \qquad (101)$$
$$\text{Max hel JD} = 2419505.310 + 0.47560624 \cdot E \text{ после } E = 17\,000. \qquad (102)$$

Comparison stars: photographic observations (Tables 152, 153), Simeise scale $a = 0^s.0$; $b = 8^s.1$; $c = 16^s.8$; Odessa scale $a = 0^s.0 = \text{BD} + 0°2897$; $b = 9^s.3 = \text{BD} + 0°2901$; $c = 17^s.6 = \text{BD} + 0°2898$; visual observations (Table 151) $b = \text{BD} + 0°2901 = -7^s.6$; $d = \text{BD} + 0°2896 = 0^s.0$; $c = \text{BD} + 0°2898 = 18^s.0$.

TABLE 150. List of maxima

Source	Max hel JD	E	O—A	O—B	O—C	O—D
Gaposchkin /238/	2416253.585	− 6837	−0^d.014	−0^d.004	−0^d.006	—
Belyavskii /186/	9505.314	0	.000	+.005	+.005	—
Tsesevich (Sim.)	9865.347	+ 757	.000	+.004	+.004	—
Florya /129/	26878.623	+15503	−.004	−.010	−.004	—
Tsesevich (visual)	7485.501	+16779	+.001	−.005	+.001	—
Solov'ev /108/	7607.743:	+17036	+.012:	+.006:	(+.012:)	+0^d.005
Tsesevich (Sim.)	7896.438	+17643	+.015	+.008	(+.014)	+.007
Solov'ev /108/	7918.779	+17690	+.002	−.004	(+.002)	−.005
	8312.104	+18517	+.001	−.006	(+.001)	−.007
Alaniya /1/	34509.257	+31547	+.014	−.002	(+.009)	−.003
Tsesevich (Odessa)	7020.462	+36827	+.021	+.002	(+.015)	+.001
Lange	7781.432	+38427	+.022	+.002	(+.015)	+.001

TABLE 151. Visual observations (Tsesevich)

JD hel	s	JD hel	s	JD hel	s
242...		242...		242...	
7459.402	1.5	7485.474	1.2	7508.310	−2.8
.434	1.0	.482	−2.9	.326	−5.0
.450	1.9	.489	−4.3	.333	−3.4
.466	2.0	.494	−4.6	.339	−2.5
.475	2.5	7490.331	1.6	7509.315	−2.5
.493	2.7	.466	8.0	.349	1.0
.515	3.4	7492.470	11.4	.377	4.0
.559	4.9	.492	11.4	.390	3.6
7460.297	−6.3	.504	11.4	.414	4.5
.311	−3.3	.510	10.6	7510.185	2.1
.322	−4.0	7508.161	6.5	.195	3.6
.339	−3.3	.203	10.4	.199	1.0
.383	−1.1	.213	10.2	.202	−2.2
.489	5.4	.224	11.0	.211	−4.8
.527	6.5	.240	11.9	.223	−4.2
7485.442	10.5	.263	11.6	.275	−1.3
.453	10.8	.281	9.6	.301	4.0
.460	4.6	.289	3.6	.350	4.5
.466	3.4	.291	2.3	.388	10.1
.470	2.6	.302	−2.9	.440	11.1

TABLE 152. Simeise observations (Tsesevich)

JD hel	s	JD hel	s	JD hel	s
241...		241...		241...	
9481.395	12.0	9513.272	11.6	9857.385	12.0
	11.0		12.4		12.9
9482.436	6.8	9527.304	7.7	9859.356	12.0
	9.3		9.0		12.0
9505.298	4.3	9835.474	6.5	9863.356	13.9
	2.0		4.8		13.9
9511.414	12.7	9841.456	14.9	9865.424	4.6
	12.9		15.3		4.3

TABLE 152 (continued)

JD hel	s	JD hel	s	JD hel	s
242...		242...		242...	
0191.486	13.2	6030.541	13.3	8612.512	14.4
	11.6	6033.509	3.8		14.2
0597.401	4.5		3.0	8624.339	12.9
	4.5	6060.483	13.7		7.3
0606.277	17.8:		14.9	8631.399	7.1
	14.6	6397.562	13.9		9.3
0608.392	7.1		12.4	8635.336	9.3
	5.7	6420.483	13.1		13.3
0929.452	7.1		14.2	8981.450	13.3
	12.0	6423.459	4.4		11.0
0929.568	12.6	6424.458	4.6	8992.488	12.2
	12.7		2.5		11.7
0983.367	13.6	6769.356	4.5	9339.460	4.0
	13.5		11.4		4.5
0991.288	7.7	7513.478	14.2	9362.331	9.2
	6.3		14.9		7.1
1665.507	4.9	7886.374	13.3	243...	
	6.1		9.3	2650.373	13.5
3521.336	4.3	7896.347	10.0		7.1
	4.0		9.3	2996.403	4.8
3847.491	11.0	7896.461	3.8		2.7
	13.9		4.0	3002.339	11.7
3848.528	3.6	7899.472	8.1		10.3
	5.2		6.7	3026.340	4.4
4231.516	6.0	7904.518	4.5		4.9
	6.1		3.6	3035.296	18.8
4234.283	4.5	7923.415	11.0	3734.402	12.7
	4.0	7928.361	5.7		13.1
4621.287	13.6		4.8	3737.426	5.4
	14.7	8246.514	4.8		8.1
4964.399	4.9		5.9	4476.305	7.3
	5.5	8249.534	7.1		10.6
6030.541	14.9		9.3		

TABLE 153. Odessa observations (Tsesevich)

JD hel	s	JD hel	s	JD hel	s
243...		243...		243...	
6288.424	12.1	6668.417	11.8	7373.454	5.3
6304.371	10.5	6698.348	15.7	7377.485	13.4
6306.399	14.5	6959.569	2.8	7378.436	10.3
6335.340	10.5	6971.549	5.5	7396.410	10.3
6338.354	13.9	6995.550	8.8	7397.418	14.0
6345.322	14.0	6997.570	13.0	7398.408	11.8
6607.530	10.8	7015.453	12.1	7400.409	13.4
6608.546	11.7	7016.426	11.2	7405.380	10.3
6612.593	10.2	7017.428	12.4	7406.405	12.9
6613.548	9.3	7020.471	1.9	7424.341	5.2
6660.426	1.0	7044.331	6.5	7425.322	5.9
6661.402	1.5	7046.409	11.4	7426.340	11.1
6663.458	10.3	7052.373	5.2		
6667.389	10.3	7326.638	13.2		

VV Librae

Studied by Ashbrook /181/, who derived the following equation:

$$\text{Max hel JD} = 2424268.50 + 0.47812 \cdot E. \tag{103}$$

Examination of Simeise photographs (Table 154) shows that the period changed suddenly near JD 2425000. This is evident from the following approximate maxima:

Max hel JD	E	O —A	O—B
2420245.37	−12231	$-0^{d}\!.15$	$-0^{d}\!.018$
0988.44	−10677	−. 10	+.019
4268.47	− 3817	−. 04	−.004
6093.523	0	−.010	−.018
32699.378	+13816	.000	−.170

The O —A residues were calculated from

$$\text{Max hel JD} = 2426093.533 + 0.4781301 \cdot E, \qquad (104)$$

and the O — B residues from

$$\text{Max hel JD} = 2426093.541 + 0.4781418 \cdot E. \qquad (105)$$

The period jumped by $+0^{d}.0000117$. Comparison stars (Figure 74): $n = 0^{s}.0$; $m = 6^{s}.8$; $p = 17^{s}.9$; $q = 23^{s}.4$.

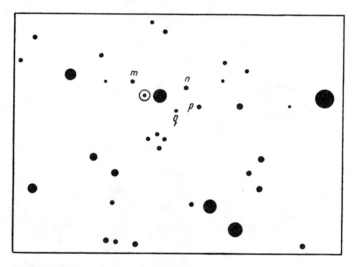

FIGURE 74. Comparison stars for VV Librae.

TABLE 154. Simeise observations (Tsesevich)

JD hel	s	JD hel	s	JD hel	s
241...		242...		242...	
9887.473	6.4	4268.474	1.0	5028.336	5.6
	5.6		0.0	5035.365	21.6
242...		5002.412	5.1		17.0
0245.438	7.9		5.4	5732.475	17.0
	6.8	5005.417	11.4		18.9
0249.458	17.9		12.9	5743.390	16.9
0988.435	1.4	5008.494	20.1		19.7
	2.7		16.9	5743.485	19.9
3550.424	6.4	5027.436	6.8	5768.340	18.9
	5.6		9.8		20.9

139

TABLE 154 (continued)

JD hel	s	JD hel	s	JD hel	s
242...		242...		243...	
5772.366	16.1	6483.363	16.9	2674:458	16.7
	16.9		17.9		15.4
5798.362	18.9	6833.514	19.3	2699.369	5.1
	16.9		20.1		3.0
5800.358	17.0	6857.397	16.9	3035.468	8.6
	17.0		26.4:	3419.412*	from 17.9
6093.511	1.9	7932.472	3.7		to 6.8
	0.0	8696.393	17.9		from 17.9
6095.480	5.9		22.6		to 5.8
	6.0				

* A rise in brightness seen in a photograph obtained by Metcalf's method.

BR Aquarii

Discovered by Hoffmeister and observed repeatedly. Studied by Tsesevich /149—151/, Parenago /87/, Lange /56—58/. Solov'ev /108/. Alaniya /1/, Tsesevich and Odynskaya /83/, and Tsarevskii. To determine the variation of its period, Tsesevich selected all the published maxima and also determined the magnitude from all the Simeise and old Harvard plates. At Tsesevich's request, Yu. E. Migach estimated the magnitude from the 265 photographs in the Odessa collection.

A complete list of maxima is given in Table 155.

TABLE 155. List of maxima

Source	Max hel JD	E	O — A
Tsesevich (Harv.)	2415488.458	—23014	—0^{d}.041
	6551.514	—20808	—.018
	7448.797	—18946	.000
	8756.627	—16232	+.001
	20002.311	—13647	+.019
Tsesevich (Sim.)	0773.337:	—12047	+.033:
Tsesevich (Harv.)	0975.709	—11627	+.015
	2248.355:	— 8986	+.009:
	3575.460	— 6232	+.010
Tsesevich (Sim.)	4432.254	— 4454	+.016
Tsesevich (Harv.)	4556.562	— 4196	.000
	5884.654	— 1440	+.024
Tsesevich /149-151/	6578.538	0	—.003
	6595.407	+ 35	.000
	6946.216	+ 763	—.001
Tsesevich, Okunev /149-151/	6952.490	+ 776	+.008
	6983.324	+ 840	+.002
	6984.290	+ 842	+.004
Tsesevich (Sim.)	7336.552	+ 1573	+.010
Lange /56-58/	7339.435	+ 1579	+.002
	7800.120	+ 2535	+.007
Solov'ev /108/	8397.649	+ 3775	+.002
Gur'ev /27/	8480.052	+ 3946	+.003
Tsesevich, Odynskaya /83/	30954.491	+ 9081	—.024
Tsesevich (Sim.)	3894.465	+15182	—.015
Alaniya /1/	3952.306	+15302	+.001
Migach	4624.498	+16697	—.033
	5743.415	+19019	—.047
Tsarevskii	6085.542	+19729	—.057
Migach	6129.393	+19820	—.057
	6488.395	+20565	—.058
	6847.378:	+21310	—.077:

TABLE 155 (continued)

Source	Max hel JD	E	O — A
Lange /65/	2437195.310:	+22032	— 0^d.064:
	7196.258	+22034	—.080
Migach	7222.293	+22088	—.066
Lange /68/	7496.473	+22657	—.078
	7497.445	+22659	—.069
	7522.489	+22711	—.083
	7524.422	+22715	—.078
	7525.389	+22717	—.074
Migach	7582.264	+22835	—.062
Lange	7636.220	+22947	—.076
	7908.471	+23512	—.089
	7909.434	+23514	—.090
	7910.402	+23516	—.086
	7925.336	+23547	—.090
Migach	7939.324	+23576	—.076

The O — A residues calculated from

$$\text{Max hel JD} = 2426578.541 + 0.4818824 \cdot E \qquad (106)$$

are shown in Figure 75. We see from the figure that the O — A plot is fairly complex. It could have been represented by two linear sections, assuming a jump of period, had it not been for a highly reliable point at $E = 9081$. This point is obtained from the average curve. To check the data, Tsesevich determined the individual maxima from the 1942 — 1943 observations:

Max hel JD	E	O—A
2430959.319	+9091	—0^d.014
0960.285	+9093	— .013
0969.428	+9112	— .025
0970.403	+9114	— .014

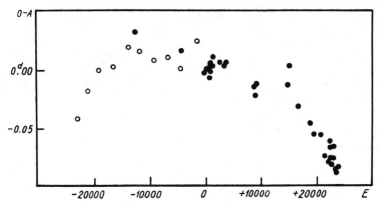

FIGURE 75. O—A residues for BR Aquarii (o Harvard observations).

They bear out the significance of the large negative deviation for this epoch. Anyhow, after JD 2434624 we have

$$\text{Max hel JD} = 2434624.500 + 0.4818749 \cdot E, \qquad (107)$$

141

so that the period dropped abruptly by $0^d.0000075$. Comparison stars for the Simeise observations of Table 156 (Figure 76): $u = 0^s.0$; $k = 8^s.6$; $a = 14^s.4$; $b = 23^s.7$.

TABLE 156. Simeise observations (Tsesevich)

JD hel	s	JD hel	s	JD hel	s
242...		242...		242...	
0045.324	15.4	4432.300	9.4	8080.301	19.6
	13.6	4434.340	12.5	8097.229	18.5
0394.426	19.0	4443.223	20.6		18.5
	20.6		20.6	8404.522	15.9
0398.425	5.7	4469.298	15.6		15.9
	6.4		18.8	8434.393	11.8
0418.380	19.3	4761.474	12.7		11.5
	18.1		11.9	8845.452	20.0
0420.374	20.0	4762.454	11.8	9516.477	19.5
	16.7		7.6		19.7
0749.395	20.0	4799.409	10.5	9522.400	7.0
0754.369	17.5		11.5		7.4
	13.1	5481.495	17.5	9524.442	14.4
0772.262	21.0	5499.407	19.0		16.8
	20.2		19.0	9904.306	19.0
0773.366	6.7	5510.407	16.5	9904.416	20.6
0777.371	19.3	6189.492	20.0	243 . . .	
	18.6	6216.479	19.8	2807.476	16.3
0786.342	5.5	6219.482	18.5		14.4
1134.431	17.2		18.8	3151.454	7.5
	14.4	6237.428	6.7		10.0
1137.383	19.6	6242.377	18.5	3155.484	19.5
	19.0	6249.355	24.7		19.6
1464.410	8.6		24.7	3233.246	18.2
	8.6	6951.488	13.2		18.8
1483.455	19.6		11.9	3235.242	9.6
1510.327	20.2	7308.522	20.3		6.8
	19.0		20.0	3894.426	9.6
3322.322	20.3	7336.467	20.0		9.8
	19.3		20.8	4241.432	7.2
3673.409	14.4	7337.408	20.0		6.9
4060.411	16.7		20.3	4246.462	18.1
	18.2	8051.486	20.2		20.6
4418.296	5.4	8067.453	20.3	4253.427	8.6
	7.5		19.6		12.6
4419.398	16.9	8070.326	21.0	4256.505	18.4
	17.3		19.0		18.5
4421.506	18.4	8077.420	18.8	4653.306	20.2
	20.0		19.6		20.8
4432.300	6.0	8080.301	19.6		

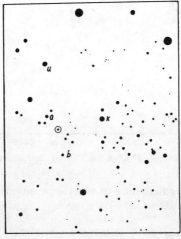

FIGURE 76. Comparison stars for BR Aquarii.

VV Pegasi

Discovered by Tserasskaya /219/. Initially observed by Zinner /376/ and Blazhko /195/, later by Lange /58/, Solov'ev /108/, Batyrev, and others.

Batyrev /7/ summarized all the published data, but his summary is not complete. It does not include the maxima that can be derived from Blazhko's observations (Tsesevich reconstructed the most complete set of data from Blazhko's archives) and Zinner's maxima (although the latter are not quite accurate). The complete summary is given in Table 157.

TABLE 157. List of maxima

Source	Max hel JD	E	O — A	O — B	O — C
Zinner /376/	2419704.426:	— 821	$+0^d.095$:	$+0^d.088$:	—
	9979.38:	— 262	+ .045::	+.039:	—
	20037.402:	— 139	— .003:	—.010:	—
	0042.291:	— 129	+ .002:	—.004:	—
	0043.282:	— 127	+ .016:	+.010:	—
	0058.408	— 96	+ .002	—.004	—
	0059.389	— 94	+ .007	.000	—
	0062.347:	— 88	+ .034:	+.028:	—
	0064.256::	— 84	— .010::	—.016:	—
	0066.225:	— 80	+ .005:	—.001:	—
	0068.161::	— 76	— .012::	—.018:	—
	0105.297	0	+ .007	+.001	—
Blazhko /195/	1195.356	+ 2232	+ .006	+.001	—
	1466.403	+ 2787	+ .003	—.002	—
	1470.310	+ 2795	+ .003	—.002	—
	1471.283	+ 2797	— .001	—.005	—
	3232.373	+ 6403	— .003	—.005	—
	3249.472	+ 6438	+ .003	+.001	—
	3275.358	+ 6491	+ .005	+.003	—
	4389.350	+ 8772	+ .006	+.005	—
Tsesevich	5120.454	+10269	+ .008	+.008	—
	5121.422	+10271	— .001	—.001	—
Blazhko	5126.302	+10281	— .004	—.005	—
	5150.232	+10330	— .005	—.005	—
	6595.348	+13289	.000	+.001	—
	6596.324	+13291	— .001	+.001	—
Tsesevich	6598.275	+13295	— .003	—.002	—
	6600.230	+13299	— .002	.000	—
	6601.213	+13301	+ .004	+.006	—
	6602.183	+13303	— .002	—.001	—
	6603.163	+13305	+ .001	+.002	—
Lange /58/	7332.796	+14799	— .003	—.001	—
Blazhko	7356.235	+14847	— .006	—.004	—
	7712.269	+15576	.000	+.003	—
Lange	7780.153	+15715	— .001	+.002	—
Solov'ev /108/	8427.255	+17040	.000	+.004	—
Batyrev /7/	33217.271	+26848	+ .003	—	$+0^d.003$
	3218.243	+26850	— .002	—	— .002
	3264.158	+26944	+ .006	—	+ .005
	3504.432	+27436	— .003	—	— .005
	3507.371	+27442	+ .006	—	+ .003
	3508.342	+27444	.000	—	— .002
	3509.323	+27446	+ .005	—	+ .002
	3510.301	+27448	+ .006	—	+ .003
	3532.275	+27493	+ .003	—	.000
	3599.181	+27630	+ .001	—	— .003
Domke, Pohl /203, 204/	3929.325	+28306	$+0^d.001$	—	—.006
Alaniya /1/	4630.269	+29741	+ .123	—	—
Lange /65/	7176.309	+34936	+ .038	—	—.001
	7189.290	+34981	+ .042	—	+.002
Lange /68/	7494.5245	+35606	+ .0400	—	—.0024
	7496.4887	+35610	+ .0506	—	+.0082
	7497.4549	+35612	+ .0401	—	—.0023
	7498.4314	+35614	+ .0398	—	—.0026
	7518.4592	+35655	+ .0441	—	+.0015

TABLE 157 (continued)

Source	Max hel JD	E	O — A	O — B	O — C
Lange /68/	2437519.4371	+35657	+0^d.0453	—	+0^d.0026
	7520.4122	+35659	+.0436	—	+.0010
	7521.3883	+35661	+.0430	—	+.0003
	7522.3645	+35663	+.0424	—	—.0003
	7523.3394	+35665	+.0405	—	—.0021
Lange	7878.4013	+36392	+.0515	—	+.0053
	7898.4218	+36433	+.0485	—	+.0021
	7900.3768	+36437	+.0500	—	+.0035
	7902.3209	+36441	+.0405	—	—.0059
	7903.2993	+36443	+.0422	—	—.0043

The O —A residues were calculated from Batyrev's equation:

$$\text{Max hel JD} = 2420105.290 + 0.48837821 \cdot E. \tag{108}$$

The variation of these residues (Figure 77) shows that the period jumped relatively recently. The observations can be represented by two equations: prior to $E = 18000$ (O—B)

$$\text{Max hel JD} = 2420105.296 + 0.48837765 \cdot E; \tag{109}$$

and after $E = 26000$ (O—C)

$$\text{Max hel JD} = 2433217.268 + 0.48838307 (E - 26848). \tag{110}$$

The period thus increased by 0^d.00000542. The residues are fairly large and show a systematic variation. The star possibly reveals Blazhko effect, although more extensive observation series are required to establish this fact conclusively.

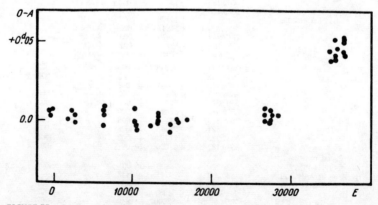

FIGURE 77. O—A residues for VV Pegasi.

BV Virginis

Observed by Tsesevich on the photographs of the Simeise, Odessa, and the new Moscow collections. The seasonal curves yielded the following maxima:

Source	Max hel JD	E	O — A	O — B
Tsesevich (Sim.)	2419922.33:	−12381	$+0^d.100$	$+0^d.003$
	26070.474	0	.000	− .014
Boyce /206/	6839.20	+ 1548	+ .009	+ .006
Tsesevich (Sim.)	8278.290	+ 4446	− .010	+ .006
Tsesevich (Odessa)	36613.558	+21231	+ .045	—
Tsesevich (Moscow)	7110.281	+22211	+ .113	—

The O — A residues were calculated from

$$\text{Max hel JD} = 2426070.474 + 0.496587 \cdot E. \tag{111}$$

The period apparently jumped. Two equations apply: before 2428500 (O — B)

$$\text{Max hel JD} = 2426070.488 + 0.4965803 \cdot E; \tag{112}$$

after 2436600

$$\text{Max hel JD} = 2436613.558 + 0.496656 \cdot E_1. \tag{113}$$

Comparison stars used in observations (Tables 158 — 160, Figure 78): Odessa scale $u = 0^s.0$; $m = 11^s.0$; $t = 18^s.8$; $n = 23^s.8$; Moscow scale $m = 0^s.0$; $w = 9^s.1$.

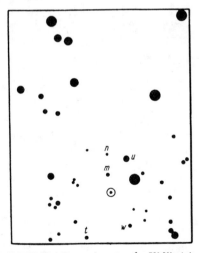

FIGURE 78. Comparison stars for BV Virginis.

TABLE 158.   Simeise observations (Tsesevich)

JD hel	s	JD hel	s	JD hel	s
241...		243...		242...	
9513.375	18.8	7513.478	19.8	0108.447	17.0
9892.326	23.8		17.8	0112.354	17.9
	23.8	7541.354	9.0		17.9
9922.333	11.0	7898.478	7.0	2650.373	18.7
	10.0		11.0		17.4
242...		8246.514	8.0	3358.478	11.0
4620.363	19.8		8.6		14.4
	18.8	8252.498	11.0	3734.402	14.2
6033.509	18.8		11.0		14.4
	[18.8	8257.512	14.0	3771.314	17.5
6060.483	17.9		17.0		21.8
	18.8	8276.418	14.4	4118.398	[23.8
6070.507	12.6		18.7	4120.393	19.8
	9.9	8278.340	8.8	4455.482	13.0
6084.396	10.0		9.2		[18.8
	9.0	8280.329	9.0	4456.516	23.8
6397.562	14.4	8637.398	16.8		20.8
	21.3		16.9	4503.348	20.8
6423.459	[18.8	8992.488	14.0		
7510.552	16.1		15.0		
	19.8	9339.460	11.0		

TABLE 159.   Odessa observations (Tsesevich)

JD hel	s	JD hel	s	JD hel	s
243...		243...		243...	
6304.398	[23.8	6608.596	10.1	6661.466	24.8
6344.357	23.8	6612.636	12.0	6663.484	[18.0
6347.347	16.0	6613.589	9.9	6667.417	21.2
6607.572	12.0	6660.487	18.0	6668.442	21.5

TABLE 160.   Moscow observations (Tsesevich)

JD hel	s	JD hel	s	JD hel	s
243...		243...		243...	
3052.341	8.6	7051.338	6.8	7100.318	—1.0
3061.458	6.4	7052.331	7.1	7102.316	—1.0
3352.429	—1.0	7079.295	7.1	7103.317	—1.0
3358.483	3.4	7080.297	6.4	7106.315	1.1
3388.379	6.8	7087.301	3.4	7113.315	3.8
4077.423	2.0	7099.323	0.0	7128.301	5.7

RZ Ceti

Discovered by Hoffmeister /261/. Studied by Lange /58/, Gur'ev /29/, Solov'ev /108/, Alaniya /2/, and Tsesevich /146/. The period was found to vary, and Tsesevich estimated the magnitude from Harvard photographs of the RH and RB series to obtain a new list of maxima (Table 161). The O—A residues (Figure 79) were calculated from

$$\text{Max hel· JD} = 2426585.448 + 0.510633 \cdot E. \qquad (114)$$

The period became considerably shorter. The following equation is valid for the recent epochs:

$$\text{Max hel JD} = 2433954.379 + 0.510613 \cdot E'. \qquad (115)$$

146

TABLE 161. List of maxima

Source	Max hel JD	E	O — A
Tsesevich (Harv.)	2425960.419	— 1224	—0^{d}.014
Tsesevich /146/	6591.573	+ 12	— .003
Tsesevich (Harv.)	6642.634	+ 112	— .005
Lange /58/	7343.225:	+ 1484	— .002:
Tsesevich (Harv.)	7371.316	+ 1539	+ .004
Tsesevich /146/	7422.383	+ 1639	+ .008
Gur'ev /29/	8096.408	+ 2959	— .003
Tsesevich (Harv.)	8256.273	+ 3272	+ .034
Solov'ev /108/	8427.299	+ 3607	— .002
Tsesevich (Harv.)	9016.617	+ 4761	+ .045
	9544.644	+ 5795	+ .078
	9917.906	+ 6526	+ .067
	30416.801	+ 7503	+ .074
	1182.750	+ 9003	+ .073
	1753.624	+10121	+ .059
	2292.834	+11177	+ .041
	3010.770	+12583	+ .027
	3906.884	+14338	— .020
Alaniya /1/	3954.379	+14431	— .014
Lange	7675.218	+21718	— .157
	7906.481:	+22171	— .211:
	7974.434	+22304	— .172
	7991.271:	+22337	— .186:

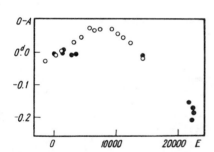

FIGURE 79. O—A residues for RZ Ceti (O Harvard observations).

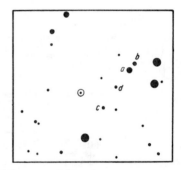

FIGURE 80. Comparison stars for BO Virginis.

BO Virginis

Discovered and studied by Boyce /206/. GCVS gives

$$\text{Max hel JD} = 2426860.20 + 0.52005 \cdot E.$$

Investigation of the Simeise photographs (Tsesevich) gave the improved formula

$$\text{Max hel JD} = 2419498.28 + 0.520058 \cdot E \qquad (116)$$

147

from which the seasonal light curves were constructed and the following maxima determined:

Max hel JD	E	O — C
2419513.388	0	—^{d}.002
26423.414	+13287	+ .003
8992.506	+18227	+ .005
33358.387	+26622	— .008

The period apparently remained constant. The final equation

$$\text{Max hel JD} = 2419513.390 + 0.52005878 \cdot E; \quad P^{-1} = 1.92285957. \quad (117)$$

The ages for the construction of the general average light curve (Table 162) were calculated from this equation.

TABLE 162. General average curve

φ	s	n	φ	s	n	φ	s	n
0^{P}.011	8.2	4	0^{P}.327	13.7	7	0^{P}.884	10.2	4
.062	7.5	5	.396	14.5	8	.939	8.9	3
.096	8.8	4	.538	16.4	6	.967	7.3	6
.131	10.7	6	.619	15.8	5	.987	8.6	4
.166	11.6	6	.734	15.6	4			
.246	15.5	5	.796	18.1	4			

Comparison star for the observations of Table 163 (Figure 80): $a = 0^{s}.0$; $b = 11^{s}.5$; $c = 17^{s}.6$.

TABLE 163. Simeise observations (Tsesevich)

JD hel	s	JD hel	s	JD hel	s
241...		242...		242...	
9482.436	14.5:	6397.562	15.0	8992.488	7.3
9498.285	6.7	6420.483	11.5		8.4
	5.8		11.5	9339.460	14.8
9511.414	11.5	6423.459	6 2		10.4
	15.1	6424.458	8.0	9349.400	18.6
9513.375	8.3		7.7		18.6
	7.1	7510.552	12.4	9362.331	10.4
9863.356	10.4		12.4		10.4
9865.424	8.3	7513.478	6.6	9721.441	15.4
	8.8		8.6	243...	
242...		7898.478	11.5	2996.403	8.0
0215.41:	6.0		11.5		8.4
0608.392	13.5	8246.514	13.9	3002.339	15.1
	17.6		17.6		16.1
0929.568	11.5	8249.534	18.6	3031.322	7.3
4231.516	13.9		18.6		9.4
4620.363	9.9	8257.512	14.6	3034.324	12.4
	9.4		15.1		10.4
6030.541	17.6	8276.418	8.0	3358.478	10.4
	17.6		8.1		11.5
6033.509	15.0	8278.340	17.6	3734.402	8.6
	13.9		19.6		8.3
6060.483	9.9	8624.339	6.0:	3737.426	16.6
	10.4	8631.399	16.4		18.6
6070.507	16.1		15.6	3771.314	8.1
	14.6	8637.398	10.4		7.1
6084.396	11.5		11.5	4455.482	17.6
	11.5	8981.450	15.1		
6397.562	13.0		17.6		

Tsesevich also measured the new Moscow photographs (Table 164) taken with the 16-in. spectrograph. Comparison stars (Figure 80): $a = 0^s.0$; $d = 12^s.2$; $c = 17^s.6$. These observations clearly bracket a maximum, and the seasonal light curve gives with full certainty Max hel JD = 2437078.313. For $E = 33775$ this gives an ephemeris correction $O-C = -0^d.062$. The period thus changed quite abruptly. The observations can be represented by two equations:

$$\text{Max hel JD} = 2419513.388 + 0.52005916 \cdot E; \quad E < 20000; \tag{118}$$

$$\text{Max hel JD} = 2433358.387 + 0.52005117(E-26622); \quad E > 24000. \tag{119}$$

TABLE 164. Moscow observations (Tsesevich)

JD hel	s	JD hel	s	JD hel	s
243...		243...		243...	
3052.341	15.3	7052.331	3.3	7100.318	12.2
3061.458	16.3	7078.321	1.6	7102.316	8.5
3358.483	10.2	7079.295	11.6	7103.317	3.5
3388.379	19.1	7080.297	15.8	7106.315	18.6
4077.423	17.6	7087.301	15.4	7113.315	12.2
7051.338	6.7	7099.323	15.4	7128.301	7.1

EZ Lyrae

Discovered by Hoffmeister /262/ in 1930. Studied by Jacchia /286/, who obtained for the period

$$\text{Max hel JD} = 2426577.406 + 0.52534 \cdot E. \tag{120}$$

Parenago /88/ obtained magnitude estimates from old Moscow photographs, derived an improved formula, and detected a suspicion of variable period. His equation was

$$\text{Max hel JD} = 2426577.406 + 0.5252599 \cdot E. \tag{121}$$

Also observed by Zverev /38/, Dombrovskii /36/, Gur'ev /27/, Florya /36/, Solov'ev /108/, Born /203/, Sofronievitsch /204/, Alaniya /2/. Solov'ev reduced Gur'ev's observations and estimated the magnitude from Dushanbe sky survey photographs. A maximum was determined from the resulting light curve. Tsesevich reduced a small series of his own observations and also added Batyrev's findings. Chuprina reduced Migach's photographic observations and obtained three average seasonal light curves. The average maxima were determined from these curves. Tsesevich revised Parenago's observations and the seasonal curves yielded two certain old maxima. The various maxima are listed in Table 165.

The $O-A$ residues were calculated from (121).

Kurochkin derived the equation

$$\text{Max hel JD} = 2433914.341 + 0.525268 \cdot E. \tag{122}$$

149

TABLE 165. List of maxima

Source	Max hel JD	E	O — A	O — B	O — C	O — D
Tsesevich (Mos.)	2415940.341	−34219	−0^d.027	+0^d.146	+0^d.015	—
	8186.327	−29943	− .052	+ .086	− .020	—
Jacchia /286/	26577.406	−13968	000	+ .008	− .003	—
Dombrovskii /36/	7302.293	−12588	—	+ .026	+ .022	—
Florya /36/	7309.618	−12574	—	− .003	− .006	—
Zverev /38/	7763.450	−11710	—	− .003	− .001	—
Gur'ev /27/	8115.377	−11040	—	− .005	+ .001	—
Solov'ev /108/	8429.478	−10442	—	− .015	− .005	—
	33023.479	− 1696	—	− .007	—	−0^d.006
Tsesevich	3885.449	− 55	—	− .002	—	− .002
Born /203/	3914.339	0	—	− .002	—	− .002
Sofronievitsch /204/	3914.341	0	—	.000	—	.000
	3924.317	+ 19	—	− .004	—	− .004
Born /203/	3924.323	+ 19	—	+ .002	—	+ .002
	3925.373	+ 21	—	+ .001	—	+ .001
	4498.439	+ 1112	—	.000	—	− .001
Sofronievitsch /204/	4498.442	+ 1112	—	+ .003	—	+ .002
Alaniya /2/	4954.392	+ 1980	—	+ .020	—	+ .019
Chuprina (Odessa)	4967.506	+ 2005	—	+ .003	—	+ .001
	6265.441	+ 4476	—	.000	—	− .002
Batyrev	6489.211	+ 4902	—	+ .006	—	+ .003
	6490.257	+ 4904	—	+ .002	—	− .001
	6499.180	+ 4921	—	− .005	—	− .008
	6793.348	+ 5481	—	+ .013	—	+ .010
	6812.244	+ 5517	—	− .001	—	− .004
Chuprina (Odessa)	7501.396	+ 6829	—	.000	—	− .004

It was used to calculate the O − B residues and the epochs E. The variation of the residues O − C and O − D can apparently be represented by two linear relations

$$\text{Max hel JD} = 2433914.270 + 0.5252621 \cdot E; \tag{123}$$

$$\text{Max hel JD} = 2433914.341 + 0.5252686 \cdot E, \tag{124}$$

assuming that near $E \sim 10000$ the period increased abruptly by $0^d.0000065$.

Tsesevich observed this star visually in 1951 (Table 166) using the following comparison stars (Figure 81): $a = 0^s.0$; $b = 10^s.6$; $c = 16^s.0$.

TABLE 166. Visual observations (Tsesevich)

JD hel	s	JD hel	s	JD hel	s
243...		243...		243...	
3853.328	12.8	3884.390	10.2	3889.309	8.2
.333	14.0	.419	9.3	.314	12.4
.341	14.0	.429	9.3	3890.293	8.1
.357	13.4	.442	12.8	.315	9.1
.394	6.9	.454	10.7	.349	10.6
.399	4.9	3885.287	12.8	3892.312	7.4
.402	6.2	.306	12.8	.345	8.5
3854.364	14.7	.393	12.8	3895.416	8.3
.370	13.7	.425	11.7	3897.293	12.9
.381	14.8	.443	7.4	.326	11.6
.408	13.3	3886.309	12.8	.339	12.2
.419	13.1	.354	12.8	.349	12.8
.429	13.7	.385	13.8	.386	13.1
3855.465	12.8	.430	14.2	.413	12.8
3859.351	12.8	.443	14.2	.446	13.5
3884.352	14.0	3887.300	13.5		

5443

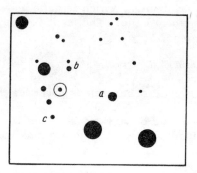

FIGURE 81. Comparison stars for EZ Lyrae.

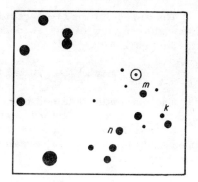

FIGURE 82. Comparison stars for BH Aquarii.

BH Aquarii

Discovered by Hughes /274/, who determined its period. Hughe's elements were improved by Tsesevich /155/, who obtained

$$\text{Max hel JD} = 2426948.43 + 0.525805 \cdot E. \tag{125}$$

The Simeise observations revealed times of enhanced brightness (Table 167). These observations are listed in Table 168.

TABLE 167. Times of enhanced brightness

Max hel JD	E	O — A	O — B	O — C
2420394.328	—12465	$+0.^{d}057$	$-0.^{d}036$	$+0.^{d}011$
4033.380	— 5544	+.013	—.020	—.014
4736.426	— 4207	+.058	+.037	+.035
5115.494	— 3486	+.020	+.006	—.001
5530.333	— 2697	—.001	—.009	—.019
6264.330	— 1301	—.028	—.023	—.042
6948.432	0	+.002	+.018	—.009
8404.426	+ 2769	+.042	+.082	+.039
33860.480	+13146	—.183	—.051	—.157

TABLE 168. Simeise observations

JD hel	s	JD hel	s	JD hel	s
242...		242...		242...	
0394.328	7.8	5530.333	5.4	8404.426	9.8
	6.9		3.9		7.0
1463.434	17.9	6210.391	19.9	8460.273	19.4
	16.9		16.7		17.9
2554.423	17.0	6235.323	20.9	8461.300	19.9
3649.409	19.9		20.9		17.9
	20.9	6236.359	21.9	8780.373	9.8
4033.380	4.9		16.7	243...	
	4.9	6264.330	7.8	3860.485	5.9
4736.426	6.3		7.8		3.8
4745.513	[17.9	6568.466	18.9	3867.506	19.9
	[17.9		19.4		19.9
4773.410	19.9	6948.432	1.1	3897.395	17.9
5115.494	4.1		1.3		17.9
5485.469	21.9	7685 430	19.9	3913.301	20.4
	22.9		22.9		19.9

The spread of the O−A residues is fairly high, which is quite under-standable in view of the few observations available. The least squares method gave for O−B

$$\text{Max hel } JD = 2426948.414 + 0.5257962 \cdot E; \quad P^{-1} = 1.90187757. \tag{126}$$

The maximum obtained from the last eight observations precedes the calculated maximum by 0.1 period.

In 1961 Tsesevich carried out a few visual observations (Table 169). Reduction to one period revealed considerable deviation of the maximum from ephemeris:

$$\text{Max hel } JD = 2437571.401,$$

which gives O−B = $-0^d.199$ for $E = 20204$.

The light variation of the star is thus described by two equations

$$\text{Max hel JD} = 2426948.441 + 0.5258022 \cdot E \ (\text{before } E = 8000); \tag{127}$$

$$\text{Max hel JD} = 2433860.480 + 0.5257751 \cdot E_1. \tag{128}$$

The period contracted abruptly by $0^d.0000271$. Comparison stars (Figure 82): photographic scale, $m = 0^s.0$; $n = 8^s.9$; $k = 17^s.9$; visual scale $m = 0^s.0$; $n = 11^s.8$; $k = 18^s.8$.

TABLE 169. Visual observations (Tsesevich)

JD hel	s	JD hel	s	JD hel	s
243...		243...		243...	
7544.372	21.8	7549.412	16.8	7582.344	24.8
7545.301	21.8	7571.393	8.6	.353	22.8
.364	22.8	.399	8.2	.360	23.8
7547.291	21.8	.406	8.8	.383	22.8
.310	21.8	.414	9.4	.393	[18.8
.338	22.8	.427	9.7	7583.282	23.8
.347	21.8	.442	10.7	.304	23.8
7549.399	15.8	7582.321	23.8		

CS Serpentis

Discovered by Hoffmeister /267/ and studied by Tsesevich /157/. The variation of the period was investigated from Simeise, Odessa, Moscow, and Harvard photographs. The list of maxima is given in Table 170. The residues were calculated from the equation

$$\text{Max hel JD} = 2431176.430 + 0.5267959 \cdot E. \tag{129}$$

The plot of the O−A residues shows that the period markedly changed (Figure 83).

TABLE 170. List of maxima

Source	Max hel JD	E	O — A
Tsesevich (Harv.)	2416228.786	—28375	+0^d.190
	7242.861	—26450	+.183
	8137.875	—24751	+.170
	9140.897	—22847	+.173
	21004.650	—19309	+.122
Tsesevich (Sim.)	1008.36	—19302	+.144
Tsesevich (Harv.)	1916.549	—17578	+.137
Tsesevich (Sim.)	3193.46	—15154	+.095
Tsesevich (Harv.)	4213.886	—13217	+.117
	5222.682	—11302	+.099
	6102.951	— 9631	+.092
	7227.635	— 7496	+.067
	8323.878	— 5415	+.048
	9416.986	— 3340	+.054
	30902.521	— 520	+.025
Tsesevich (Sim.)	1176.430	0	.000
Tsesevich (Harv.)	2334.876	+ 2199	+.022
	3390.572	+ 4203	+.019
Tsesevich (Sim.)	3445.36	+ 4307	+0.020
	3771.41	+ 4926	—.017
Tsesevich (Mos.)	6749.420	+10579	+.016
Tsesevich (Odessa)	7044.410	+11139	.000
Tsesevich (visual)	7112.38	+11268	+.014

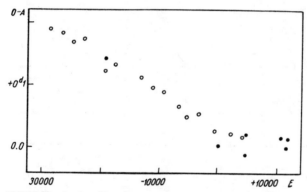

FIGURE 83. O—A residues for CS Serpentis (o Harvard observations).

Comparison stars (Figure 84): photographic observations (Tables 171—173) $s = -19^s.1$; $u = 0^s.0$; $k = 5^s.6$; $a = 12^s.2$; $b = 18^s.0$; new visual observations (Table 174) $u = 0^s.0$; $b = 11^s.4$; $c = 16^s.1$.

TABLE 171. Simeise observations (Tsesevich)

JD hel	s	JD hel	s	JD hel	s
242...		242...		242...	
1008.290	12.2	4645.374	2.8	3393.481	10.4
.363	1.6	5066.341	6.5		10.9
2486.396	11.1		5.6	3445.356	3.7
	9.4	5390.316	7.3	3447.352	13.4
2491.352	5.6	7595.384	2.8		13.2
	4.5		9.7	3747.514	10.4
3193.463	—1.5	7597.360	10.9		10.3
	0.0	7607.356	12.2	3771.408	3.7
4297.363	11.1		9.4		3.4
	10.3	9376.505	5.6	4129.463	14.1
4645.374	3.4		7.5		15.1

153

TABLE 172. Odessa observations (Tsesevich)

JD hel	s	JD hel	s	JD hel	s
243...		243...		243...	
6702.460	9.7	7019.551	8.6	7378.572	6.3
6703.469	11.5	7020.563	9.2	7400.556	−2.0
6722.427	8.6	7028.553	4.6	7402.509	11.2
6726.385	4.0	7029.568	8.3	7404.526	12.2
6728.409	0.0	7044.461	0.0	7405.522	9.8
6729.407	−2.0	7046.503	−2.0	7406.529	9.4
6730.384	8.9	7052.465	5.6	7453.381	10.6
6732.387	11.3	7071.456	7.0	7458.367	10.4
7015.579	7.8	7080.372	4.0		
7016.565	3.4	7373.606	3.1		

TABLE 173. Moscow observations (Tsesevich)

JD hel	s	JD hel	s	JD hel	s
242...		243...		243...	
5738.458	9.8	6702.448	16.2	6751.374	12.0
6097.441	18.0	6703.448	12.6	6752.320	13.7
8653.445	13.1	6715.352	10.8	6762.386	13.1
8654.434	18.0	6716.367	11.4	6961.589	10.8
8655.448	14.0	.416	10.8	.640	12.0
8656.393	12.6	6721.379	12.0	6971.589	9.5
8665.417	15.2	6723.314	14.4	6972.611	11.4
243...		.360	14.0	7050.440	9.8
4131.438	9.0	.405	12.6	7051.428	6.3
4133.423	4.3	.451	12.0	7052.420	2.8
6376.365	10.0	.491	11.4	7072.420	−1.0
.417	−3.4	6724.474	11.4	7078.392	12.6
.456	−4.1	6729.393	−3.5	7080.378	4.0
6377.443	11.3	6749.402	−2.7	7100.345	2.8
6395.410	−1.7	6750.327	9.5	7102.408	−2.5
6396.342	12.6	.372	11.4	7103.413	−1.3
6398.328	13.7	6751.330	14.8	7106.351	14.4

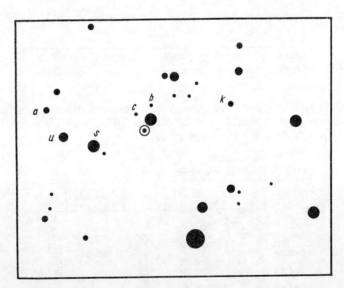

FIGURE 84. Comparison stars for CS Serpentis.

TABLE 174.   Visual observations (Tsesevich)

JD hel	s	JD hel	s	JD hel	s
243...		243..		243...	
6018.362	7.4:	6022.378	8.6	7112.457	8.9
.372	7.0	7112.428	7.3	7113.334	14.0
.381	8.3	.432	8.5	.352	13.5
6019.355	7.9	.435	8.3	.363	12.6
6020.340	6.1	.447	8.9		

RY Piscium

Discovered and investigated by Belyavskii [Beljawsky] /187, 188/ and Dubyago /224/. Also observed by Lange /56 — 58/ and Gur'ev /27/. S. Gaposchkin found variations of the period, although their exact nature is not clear.

Tsesevich estimated the magnitude from Simeise photographs; Lange made visual observations, and Migach measured 165 photographs taken with the seven-camera astrograph. Average seasonal curves were constructed and the maxima determined. The results were supplemented by the maxima of Dubyago, Lange, and Gur'ev (Table 175).

TABLE 175.   List of maxima

Source	Max hel JD	E	O — A	O — B	O — C
Tsesevich (Sim.)	2419677.268	—11008	$+0^d.103$	$+0^d.002$	—
	20775.369	— 8935	+.080	—.002	—
	4053.267	— 2747	+.027	+.002	—
	5148.192	— 680	+.006	.000	—
Dubyago /224/	5508.400	0	.000	.000	—
Lange /56-58/	7328.542	+ 3436	.000	—	$+0^d.003$
Tsesevich (Sim.)	7336.483	+ 3451	—.005	—	—.002
Tsesevich (visual)	7665.449	+ 4072	+.001	—	+.002
Lange /56-58/	7767.154	+ 4264	—.002	—	—.001
Gur'ev /27/	8486.525	+ 5622	.000	—	—.002
Tsesevich (Sim.)	9934.268:	+ 8355	—.001:	—	—.009:
	33188.402	+14498	+.020	—	.000
Migach	4623.453:	+17207	+.041·	—	+.014:
	6128.406	+20048	+.039	—	+.007
	6485.422	+20722	+.019	—	—.014
	6824.468	+21362	+.040	—	+.005
	7198.459	+22068	+.044	—	+.007
	7606.349	+22838	+.044	—	+.006
Lange	7902.465	+23397	+.042	—	+.003
	7910.402	+23412	+.033	—	—.006
Migach	7946.436	+23480	+.046	—	+.007
Lange	7961.258	+23508	+.036	—	—.004
	7962.302	+23510	+.020	—	—.019

The O — A residues were calculated from

$$\text{Max hel JD} = 2425508.400 + 0.529727 \cdot E. \tag{130}$$

The O — A residues (Figure 85) can be fitted with two equations:

$$\text{Max hel JD} = 2425508.400 + 0.5297178 \cdot E; \quad P^{-1} = 1.88779762$$
$$\text{for } E < 0; \tag{131}$$

155

$$\text{Max hel J D} = 2425508.390 + 0.5297291 \cdot E; \ P^{-1} = 1.88775735$$
$$\text{for } E > 0. \qquad (132)$$

The period thus increased abruptly by $0^d.0000113$. The spread of the individual points at the maximum suggests a Blazhko effect. Comparison stars used with Simeise photographs (Figure 86): $m = 0^s.0$; $a = 5^s.5$; $k = 10^s.7$; $b = 16^s.0$. The Simeise photographic observations (Table 177) were used to derive the average light curve (Table 176).

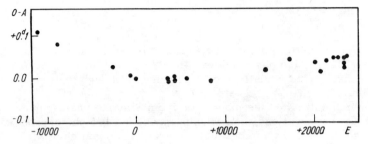

FIGURE 85. O—A residues for RY Piscium.

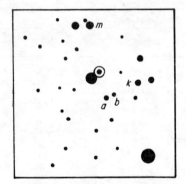

FIGURE 86. Comparison stars for RY Piscium.

TABLE 176. Average light curve

φ	s	n	φ	s	n	φ	s	n
$0^p.022$	1.8	4	$0^p.378$	13.2	5	$0^p.712$	14.3	6
.078	4.0	5	.392	13.6	6	.739	13.0	6
.114	5.9	6	.406	13.4	5	.787	14.7	5
.130	7.8	6	.428	13.3	6	.799	14.2	6
.143	7.2	6	.445	12.1	6	.832	12.7	7
.159	6.8	6	.468	13.0	6	.853	14.2	6
.199	9.2	5	.502	13.9	6	.898	11.0	6
.242	9.7	6	.578	13.4	6	.928	6.4	5
.287	10.4	5	.600	13.9	6	.943	7.1	4
.311	10.8	6	.638	13.7	6	.966	5.4	5
.341	11.0	5	.661	14.4	6	.980	7.7	5
.369	13.1	6	.698	14.4	5			

TABLE 177. Simeise observations

JD hel	s	JD hel	s	JD hel	s
241...		242...		242...	
9656.382	14.8	4058.343	14.5	8051.486	9.4
	13.6		13.4	8067.453	12.8
9657.271	10.7	4060.412	14.8		13.9
9658.262	8.6		14.7	8815.304	14.4
	9.3	4403.320	13.9		12.2
9677.248	1.2		12.0	8845.342	12.0
	1.3	4416.371	13.4		10.7
9715.318	13.9		10.7	8847.349	5.5
	14.2	4469.298	12.8		9.3
242...			14.8	9142.486	12.0
0014.484	13.9	4746.532	15.0		9.7
	14.6		14.9	9173.438	14.0
0016.404	9.3	4762.544	4.1		12.2
0017.515	11.9	4799.314	13.6	9191.304	14.2
	10.7		13.0		12.6
0043.447	9.7	5148.405	13.4	9520.446	14.9
0065.390	12.8		14.0		13.9
	17.0	5151.467	10.7	9522.503	14.9
0100.214	9.0		6.8		14.9
	9.0	5499.407	3.1	9524.442	12.0
0105.322	3.5		2.4	9903.445	15.1
	4.6	5509.466	−1.0	9904.416	13.6
0398.426	13.9	5513.488	14.8	9934.314	4.4
	13.9	5528.443	13.9		3.3
0418.299	6.9		14.8	9936.241	14.8
	4.4	5534.433	8.3		12.5
0422.279	13.8		4.6	243...	
	12.8	5543.422	6.8	2796.476	8.3
0749.395	5.5		5.5		7.6
0750.520	4.4	5552.244	14.0	2807.436	14.2
0754.559	13.7	5559.322	6.8	.516	6.8
	13.9		5.5	.436	13.9
0772.263	8.6	5860.499	13.6	.516	8.1
	7.6		13.8	2854.319	9.5
0775.479	9.0	5891.316	13.6		10.7
	10.0		14.8	2861.291	13.6
0805.452	16.0	6216.478	14.7		13.6
	14.8	6219.482	13.4	3158.484	13.8
0838.183	13.7		14.8		13.0
	14.6	6240.481	2.8	3180.448	8.6
1137.384	13.4	6247.492	8.6		10.7
	13.9	6266.398	6.8	3182.471	14.2
1137.478	13.0		10.7		14.0
	10.7	6269.279	13.4	3184.415	13.4
1510.327	13.6		10.0		14.0
	14.0	6279.365	12.0	3188.353	7.9
1512.275	9.0	6296.226	12.0		7.8
	9.3		8.6	3190.413	14.2
3318.515	5.5	6596.513	6.5		13.6
3322.408	11.8		5.5	3205.292	13.4
	8.8	6597.407	13.9	3210.320	14.2
3668.482	14.0	6602.475	10.7	3233.247	13.6
	13.1		12.5		13.9
3673.409	12.8	6603.498	9.3	3235.243	13.6
.515	2.6		9.3		13.1
	3.1	6948.531	13.6	3266.184	9.7
3675.427	14.9		13.6		12.0
	16.0	6976.495	13.9		13.1
3680.468	8.7		12.5	3898.443	13.6
	9.7	6980.472	8.7		14.8
3699.379	4.1		8.1	3921.366	14.5
3730.305	10.7	7336.467	7.6		14.8
	12.8		7.6	3923.334	13.0
4035.499	13.6	7342.393	5.5		14.0
	12.8		3.7	3952.254	2.0
4053.350	9.4	8051.486	12.7		7.2
	9.0				

CVS 1410 = 120.1936 Hydrae

Investigated by Richter /342/, who obtained

$$\text{Max hel JD} = 2425981.629 + 0.532955 \cdot E. \tag{133}$$

Simeise and Odessa photographs, as well as the Moscow photographs obtained with the 16-in. astrograph, were studied by Tsesevich. The few Simeise photographs between JD 2419451 and JD 2428194 fall on one smooth curve if (133) is used in age calculation. After JD 2432921 the observations cannot be fitted with a single light curve using (133). The seasonal light curves gave the maxima listed in Table 178.

TABLE 178. List of maxima

Source	Max hel JD	E	O — B
Simeise	2434396.423	0	—0^{d}.001
Moscow	4420.403	+ 45	—.003
	6626.246	+4184	+.012
Odessa	6628.378	+4188	+.012
	7016.328	+4916	—.017
Moscow	7293.469	+5436	—.003

The O — B residues were calculated from the equation

$$\text{Max hel JD} = 2434396.424 + 0.5329375 \cdot E;\ P^{-1} = 1.876392635. \tag{134}$$

The period thus contracted by 0^{d}.0000175.

Comparison stars for the observations in Tables 179 — 181 (Figure 87): $a = 0^{s}.0;\ b = 7^{s}.5;\ c = 11^{s}.0;\ d = 14^{s}.5.$

TABLE 179. Simeise observations

JD hel	s	JD hel	s	JD hel	s
241...		242 ..		242...	
9451.353	3.7	0918.287	9.9	4562.300	[11.0
	4.6	0920.361	11.0	8194.520	7.0
9471.369	14.0	1284.274	9.0	243...	
	13.0		10.0	2921.526	14.0
242...		4199.309	14.0	3660.495	11.0
0158.498	11.0		11.0		10.0
	9.6	4225.262	14.0	3686.430	12.5
0180.314	11.0		11.0		14.0
0564.443	13.0	4559.289	7.7	4396.423	1.1
0893.377	13.0		9.0		0.0
0918.287	10.0				

TABLE 180. Odessa observations

JD hel	s	JD hel	s	JD hel	s
243...		243...		243...	
6628.383	3.4	7015.352	7.0	7317.449	[6.0
6954.380	13.0	7016.332	0.0	7377.362	12.0
6959.473	9.6	7017.337	[11.0	7396.308	7.0
6993.409	6.6	7020.355	12.5	7398.303	[8.0
6996.435	[11.0	7039.297	8.0	7400.307	8.0:

158

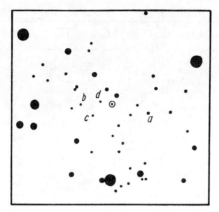

FIGURE 87. Comparison stars for CVS 1410 =
= 120.1936 Hydrae.

TABLE 181. Moscow observations

JD hel	s	JD hel	s	JD hel	s
243...		243...		243...	
4059.367	11.7	6635.426	8.7	7209.416	12.2
4062.253	7.5	6639.322	14.5	7293.469	1.0
4390.448	16.5	.373	15.5	.491	4.3
4420.403	1.1	.422	14.5	.512	3.8
4421.311	12.9	6640.291	15.5	7312.387	12.2
4454.322	14.5	.338	16.0	.409	11.8
4455.283	11.2	.384	11.3	.431	12.9
5187.389	11.5	.431	11.3	.452	13.1
5219.327	5.0	.479	15.5	.474	13.3
6626.246	2.4	6656.303	10.0	.496	14.5
6632.369	13.5	.349	9.8	7638.456	8.4
.425	14.5	6660.301	10.1	.478	11.5
.484	15.5	.345	4.5	.495	9.5
6635.322	4.2	6661.260	10.0	7699.328	10.5
.373	5.3	.305	13.9		

VY Librae

Discovered by Hoffmeister /264/ and studied by Jensch /287/, who
obtained

$$\text{Max hel JD} = 2425653.667 + 0.533948 \cdot E. \qquad (135)$$

Observed by Lange /61/ and Gur'ev /27/, who determined the maxima
from the average light curves. In 1936, Solov'ev determined a maximum
from the Simeise plates. Tsesevich studied in detail a more complete
collection of Simeise photographs.

Observed visually by Tsesevich in 1943. Lange recently repeated this
observation series. In addition to these data, the material is based on
magnitude estimates from Simeise and Odessa photographs (Tsesevich),
which made it possible to construct seasonal curves for the determination
of the maxima. The complete list of maxima is given in Table 182.

TABLE 182. List of maxima

Source	Max hel JD	E	O − B	O − C	E	O − D
Tsesevich (Sim.) 2420247.510		−10125	−0^d.025	−0^d.002	—	—
	5035.396	− 1158	+ .009	− .017	—	—
Jensch /287/	5653.667	0	− .024	− .018	—	—
Lange /61/	7871.672	+ 4154	− 0.011	− 0.012	—	—
Gur'ev /27/	9343.787	+ 6911	+ .028	+ .022	—	—
Tsesevich (visual)	30900.199	+ 9826	+ .001	− .010	—	—
Tsesevich (Sim.)	3773.346	+15207	+ .010	—	0	−0^d.000
Lange /67/	7432.418	+22060	− .018	—	6853	−.003
	7462.329	+22116	− .008	—	6909	+.007
	7463.381	+22118	− .024	—	6911	−.009
Lange	7470.335	+22131	− .011	—	6924	+.004
	7763.469	+22680	− .011:	—	7473	+.006:
	7764.522	+22682	− .026	—	7475	−.009
	7817.392	+22780	− .016	—	7574	+.002

The O − B residues were calculated from

$$\text{Max hel JD} = 2425653.691 + 0.5339413 \cdot E; \quad P^{-1} = 1.87286505. \quad (136)$$

The period in all probability changes. The maxima are better fitted with two equations: for $E \leqslant 9826$ (O − C)

$$\text{Max hel JD} = 2425653.685 + 0.53394304 \cdot E; \quad (137)$$

for $E \geqslant 15207$ (O − D)

$$\text{Max hel JD} = 2433773.346 + 0.53393774 \cdot E'. \quad (138)$$

The period thus possibly jumped by 0^d.00000530.

Comparison stars (Figure 88): visual observations (Table 184): $p = 0^s.0$; $m = 8^s.9$; $q = 15^s.7$; $r = 20^s.7$; photographic observations (Table 185) $A = 0^s.0$; $p = 8^s.2$; $m = 16^s.0$; $n = 24^s.2$.

The average light curve (Table 183) was derived from the visual observations.

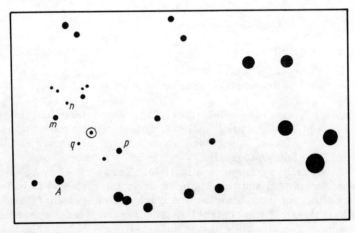

FIGURE 88. Comparison stars for VY Librae.

TABLE 183. Average light variation curve

φ	s	n	φ	s	n	φ	s	n
0^P.008	7.7	5	0^P.320	14.1	5	0^P.763	18.7	5
.031	8.6	5	.368	14.8	5	.810	18.8	5
.050	8.9	5	.424	15.5	5	.859	18.8	5
.089	10.5	5	.504	15.1	5	.889	16.8	5
.123	11.2	5	.597	18.4	5	.918	12.4	5
.170	12.4	5	.644	19.0	5	.935	11.9	3
.209	13.2	5	.674	19.0	5	.955	9.3	4
.238	14.1	5	.714	17.4	5	.974	8.5	4
.274	14.5	5						

TABLE 184. Visual observations

JD hel	s	JD hel	s	JD hel	s
243...		243...		243...	
0885.201	13.0	0896.216	14.1	0904.277	18.4
.207	13.0	0897.215	13.4	0905.192	14.1
.227	11.0	.226	14.1	.222	14.5
.231	10.5	.234	14.6	.263	15.2
.244	8.7	0898.234	13.1	0906.181	14.1
.251	10.5	.244	13.4	.209	12.5
.262	11.1	.250	13.7	.222	13.7
.267	9.8	.283	15.1	.237	14.0
.273	9.4	0899.184	10.0	.306	16.2
.287	8.6	.240	13.1	0907.252	13.7
.297	9.2	.298	15.7	.265	14.0
.308	10.4	0900.204	6.9	0908.210	6.3
.320	11.4	.218	6.6	.219	7.1
.335	11.8	.265	10.3	.232	8.1
.361	12.6	.274	12.6	.241	8.2
.383	14.2	.302	13.4	.270	10.8
.398	15.0	.324	14.3	0909.205	17.7
0886.250	15.7	.349	17.4	.213	18.0
.260	15.7	0901.184	18.7	.222	16.7
.269	11.4	.198	18.7	.234	12.0
.272	12.6	.206	17.7	.244	10.8
.277	11.4	.241	7.3	.250	7.5
.282	13.4	.253	7.3	.255	7.5
.291	11.4	.267	7.3	.301	8.9
0887.175	19.4	.275	7.6	0911.201	17.7
.195	19.7	.283	8.6	.231	18.4
.206	19.0	.291	10.0	.262	18.4
.210	18.7	.308	13.1	0914.186	11.8
.231	18.7	.319	11.6	.193	12.8
.245	19.0	.332	13.1	.201	13.4
.268	18.7	.349	13.0	.223	13.7
.289	18.7	0902.212	18.7	.248	14.1
.314	19.0	.226	18.7	.264	15.2
0889.316	19.4	.233	19.0	.285	16.7
.342	19.7	0903.217	19.0	0915.277	14.7
0894.206	19.7	.233	17.8	0925.176	18.7
.239	20.1	.251	15.7	0926.201	14.8
0895.187	18.7	.266	17.6	0928.189	18.7
.221	19.7	.314	18.7	.246	18.4
.242	19.6	0904.228	18.4		
0896.201	13.1	.261	17.9		

TABLE 185. Simeise observations

JD hel	s	JD hel	s	JD hel	s
242...		242...		242...	
0247.510	7.2	1373.368	21.5	3586.369	12.5
	5.1	3194.445	11.7	3943.332	22.2
0992.528	18.7		11.7		21.5
1373.368	21.1	3586.369	9.2	4268.474	19.6

161

TABLE 185 (continued)

JD hel	s	JD hel	s	JD hel	s
242...		242...		243...	
4268.474	19.3	5772.366	12.4	0161.379	16.9
4644.444	18.7	5798.362	9.6		13.4
.548	4.5		10.8	2674.458	9.7
.444	16.0	5800.358	18.0		8.6
.548	6.4		18.5	2699.369	19.1
4668.348	22.0	6093.511	20.4		20.1
5002.412	13.4		17.8	3035.468	3.5
	15.0	6095.480	16.0	3419.412	11.0
5005.417	16.0		15.5		11.8
	16.0	6145.387	16.0	3767.463	9.2
5008.494	19.5		16.0		5.5
	18.3	6483.363	18.7	3773.473	15.0
5027.436	8.0		18.7		15.1
	6.7	6500.376	18.0	3775.446	9.5
5028.336	20.3	6833.408	20.1		3.6
5035.365	8.0		19.5	3794.405	18.7
	6.6	.514	18.0	3829.331	18.7
5732.475	20.5		19.6		18.3
	18.7	.425	18.7	3832.322	22.8
5735.481	13.8	6857.397	16.0		22.8
	13.8		16.0	3834.327	16.0
5742.453	16.0	7563.41:	19.4		12.6
	15.0		20.9	3835.325	6.6
5743.390	6.8	7572.41:	21.5		6.3
	7.5		22.8	3838.358	22.3
.485	10.0	7963.448	20.7		22.8
	11.1	8696.393	20.5	4130.496	9.9
5768.340	21.1		20.3	4510.434	20.7
	19.4	243...			22.4
5772.366	16.0	0136.478	20 7:		

SX Aquarii

Studied by various astronomers since its discovery by Leavitt /335/.
A complete list of maxima is given in Table 186.

TABLE 186. List of maxima

Source	Max hel JD	E	O — A	O — B
Zinner /376/	2419776.25:	— 553	(+ $0^d.10$)	—
	20035.42:	— 69	(— .015:)	—
	0070.256	— 4	.000	—
	0072.400	0	+ .001	—
	0091.151:	+ 35	+ .002:	—
	20092.200	+ 37	(—0.020)	—
	0122.222:	+ 93	+ .002:	—
Haas	4429.93:	+ 8134	(+ .025:)	—
Tsesevich /371/	5087.224	+ 9361	— .004	—
	5088.295	+ 9363	— .004	—
	5089.377:	+ 9365	+ .007:	—
	5095.258	+ 9376	— .005	—
	5096.331	+ 9378	— .004	—
	5097.404	+ 9380	— .002	—
	5104.368	+ 9393	— .002	—
	5110.268	+ 9404	+ .005	—
	5112.407	+ 9408	+ .001	—
	5118.298	+ 9419	— .001	—
	5120.452	+ 9423	+ .010	—
	5125.260	+ 9432	— .003	—
	6194.545	+ 11428	— .006	—
Tsesevich /372/	6948.301	+ 12835	— .001	—
Dombrovskii /33/	7312.594	+ 13515	+ .006	—
Lange /58/	7338.838	+ 13564	.000	—
Tsesevich /372/	7660.267	+ 14164	.000	—
Alaniya /1/	33952.223	+ 25909	— .017	$+0^d.002$

162

TABLE 186 (continued)

Source	Max hel JD	E	O — A	O — B
Lange /65/	2437171.322:	+ 31918	$-0^d.030$:	$+0^d.011$:
	7172.377	+ 31920	— .046	— .006
	7193.267	+ 31959	— .049	— .009
	7195.412	+ 31963	— .047	— .006
Lange /68/	7518.446	+ 32566	— .049	— .006
	7524.338	+ 32577	— .050	— .007
	7525.419	+ 32579	— .041	+ .002
	7531.310	+ 32590	— .042	.000
	7533.450	+ 32594	— .045	— .002
Lange	7869.343	+ 33221	— .046	.000
	7878.457	+ 33238	— .039	+ .006
	7884.342	+ 33249	— .047	— .001
	7885.419	+ 33251	— .041	+ .004
	7900.417	+ 33279	— .043	+ .002
	7906.312	+ 33290	— .041	+ .004

The O — A residues were calculated from

$$\text{Max hel JD} = 2420072.399 + 0.53571505 \cdot E. \qquad (139)$$

The results were formed into normal groups:

$\bar{E}$	$\overline{O—A}$	n
31	$+ 0^d.001$	4
9390	.000	12
13101	.000	5
25909	— .017	1
31940	— .043	4
32581	— .045	5
33255	— .043	6

Despite the gap in observations between $E = 13101$ and $E = 25909$, the period seems to have jumped by $- 0^d.00000356$. The observations can be described by two equations: before $E = 20000$ by (139), after that epoch by the equation (O — B)

$$\text{Max hel JD} = 2433952.221 + 0.5357115 \ (E — 25909). \qquad (140)$$

The variation of O — A is shown in Figure 89.

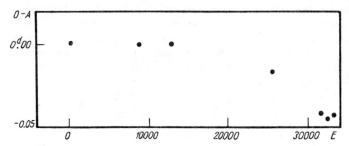

FIGURE 89. O—A residues for SX Aquarii.

V 347 Herculis

Discovered by Morgenroth. Initially regarded as a cepheid with a period of 5.8 days. Examination of Moscow and Odessa photographs definitely identified it as an RR Lyrae variable. This is also borne out by visual observations (Tsesevich). Seasonal light curves constructed using the tentative equation

$$\text{Max hel JD} = 2436104.26 + 0.537312 \cdot E \tag{141}$$

gave the maxima listed in Table 187.

TABLE 187. List of maxima

Source	Max hel JD	E	O — A	E'	O — B
Tsesevich (Mos.)	2415288.21	—38740	$-0\overset{d}{.}53$	0	$0\overset{d}{.}00$
	8596.21	—32584	—.23	+6156	.00
	29495.31	—12300	.00	—	—
Tsesevich (Odessa)	33922.24	— 4061	.00	—	—
	4621.28	— 2760	.00	—	—
	6076.330	— 52	.000	—	—
Tsesevich (visual)	6386.344	+ 525	—.016	—	—
Tsesevich (Odessa)	6408.374	+ 566	—.015	—	—
	6791.481	+ 1279	—.013	—	—
	7170.287	+ 1984	—.013	—	—
	7488.384	+ 2576	—.006	—	—
	7869.345	+ 3285	.000	—	—

The O — A residues were calculated from

$$\text{Max hel JD} = 2436104.27 + 0.5373136 \cdot E;\ P^{-1} = 1.86111053. \tag{142}$$

Table 187 shows that at the beginning of this century the O — A residues nearly reached the length of the period. The period apparently changed abruptly. Prior to 1910,

$$\text{Max hel JD} = 2415288.21 + 0.5373619 \cdot E';\ P^{-1} = 1.86094325. \tag{143}$$

The variation of the O — A residues in recent years shows that the period is still varying.

Comparison stars for the observations in Tables 189 and 190 (Figure 9 photographic scale $u = -15^s.8;\ m = -6^s.6;\ n = 0^s.0;\ a = 0^s.0;\ p = 4^s.9;$ $q = 8^s.7;$ visual scale $m = 0^s.0;\ n = 11^s.5;\ q = 19^s.8.$ The average light curve (Figure 91, 1) was obtained from the visual observations.

The average photographic light curve derived from the 1939—1957 observations is given in Table 188 (Figure 91, 2).

TABLE 188. Average photographic light curve

φ	s	n	φ	s	n	φ	s	n
$0^P.014$	—8.3	5	$0^P.257$	+4.0	5	$0^P.752$	+7.6	5
.058	—5.7	5	.327	+3.5	5	.834	+7.1	5
.119	—1.6	6	.419	+6.2	6	.886	+5.3	4
.162	—0.5	5	.549	+6.4	5	.944	—4.9	5
.205	—1.0	5	.638	+7.7	4	.968	—5.6	4

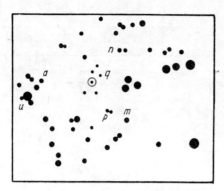

FIGURE 90. Comparison stars for V 347 Herculis.

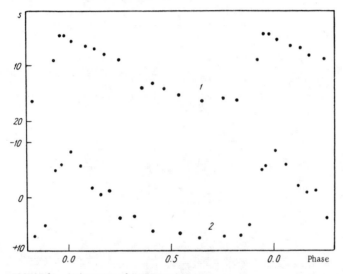

FIGURE 91. Light curves of V 347 Herculis.

TABLE 189. Visual observations

JD hel	s	JD hel	s	JD hel	s
243...		243...		243...	
3889.307	17.9	6347.362	13.5	6373.484	4.4
.323	19.8	.420	14.5	6375.475	14.3
3890.286	17.0	.462	17.0	6378.358	7.7
.308	17.0	6351.420	6.3	.381	6.7
.347	17.0	.430	5.8	.499	13.9
3892.297	16.5	.486	8.4	6381.441	17.4
.320	19.8	6352.503	3.5	6393.418	7.3
.347	17.4	6353.481	17.5	6395.327	16.2
3894.314	16.8	6371.365	6.6	.347	17.0
3897.279	16.2	6372.351	9.2	.370	17.5
.350	14.5	.370	5.4	.392	17.0
6344.415	4.2	.402	8.1	.419	16.2
.425	3.8	.453	7.7	.465	5.7
.442	2.3	6373.356	16.5	.473	5.3
.452	5.2	.424	9.2	.486	5.7
.462	5.0	.444	4.8	6396.344	17.4
.499	4.6	.455	4.7	.469	16.6

165

TABLE 189 (continued)

JD hel	s	JD hel	s	JD hel	s
243...		243...		243...	
6397.372	15.1	6405.362	14.3	6407.368	4.9
6399.322	7.4	.382	12.9	.377	5.5
.355	8.0	.395	14.3	.385	6.4
6400.407	6.7	.422	14.8	.394	8.0
.428	8.8	.458	16.0	.412	8.0
.474	9.8	6406.370	9.8	6453.339	18.5
6404.357	14.8	.381	9.8	6454.277	14.6
.383	14.3	6407.354	8.6	6455.275	11.5
		.362	6.8	.367	13.9

TABLE 190. Photographic observations

JD hel	s	JD hel	s	JD hel	s

Moscow

JD hel	s	JD hel	s	JD hel	s
241...		241...		242...	
4881.406	0.9	8927.333	2.6	9397.422	5.8
5143.443	5.8	9271.360	2.9	9488.404	−0.9
5255.338	6.5	9274.351	0.0	9495.334	−6.6
5288.264	−3.6:	242...		9495.393	1.1
6375.272	3.0:	9366.529	5.0	9496.378	−7.8
8230.240	6.5	9367.519	−2.4	9521.288	4.4
8596.222	−6.6	9376.503	6.2		

Odessa (seven-camera astrograph)

JD hel	s	JD hel	s	JD hel	s
243...		243...		243...	
6041.435	−5.8	6399.421	+1.6	6756.467	4.9
6043.475	7.4	6400.446	+2.8	6757.473	7.0
6047.403	−1.4	6401.448	0.0	6759.503	3.3
6047.426	−0.7	6406.435	4.9	6760.476	1.2
6049.435	−2.2	6407.436	5.5	6761.473	−3.3
6050.449	+4.9	6408.431	−5.1	6777.428	4.9
6053.415	6.2	6409.424	−4.0	6779.433	6.3
6067.326	+1.2	6410.421	6.8	6780.385	2.7
6069.362	−4.1	6419.333	3.7	6786.436	6.4
6070.362	+2.8	6423.380	5.7	6789.380	−4.9
6071.363	8.7	6424.380	7.0	6790.481	−2.9
6074.421	5.9	6426.359	7.1	6791.463	−7.1
6075.351	−1.5	6428.393	+1.4	6792.454	8.7
6076.349	−6.6	6429.422	0.0	6804.305	6.8
6079.325	6.6	6430.399	−4.5	6805.334	7.1
6080.346	6.8	6432.395	5.9	6806.317	7.4
6081.342	4.1	6451.318	2.3	6807.312	6.8
6082.340	+1.1	6453.307	3.9	6809.316	−2.5
6083.350	−6.6	6454.349	4.9	6813.377	6.4
6101.282	+4.2	6455.315	0.8	6814.363	5.9
6102.252	+1.6	6456.351	1.5	6815.332	0.0
6103.263	−1.7	6461.303	3.5	6817.292	2.8
6104.263	−6.6	6462.274	1.2	6834.247	4.9
6105.238	7.2	6463.287	−0.9	6838.277	−5.1
6126.203	+4.4	6465.285	3.1	6840.243	6.8
6344.520	−1.9	6478.214	−6.6	6869.202	7.1
6345.507	−7.6	6479.242	5.7	7135.487	1.0
6347.505	7.1	6481.221	5.7	7136.436	−5.8
6364.463	−0.8	6482.224	3.9	7137.454	7.8
6367.497	6.2	6483.235	1.0	7139.436	6.6
6371.463	+1.4	6484.227	+0.7	7140.442	6.2
6372.455	+2.0	6487.255	4.9	7142.415	−3.1
6373.461	−8.1	6488.260	6.8	7144.397	5.9
6375.484	6.4	6489.265	7.1	7145.431	7.0
6376.462	4.9	6728.496	5.7	7159.347	7.0
6379.478	0.0	6730.500	5.7	7161.370	2.9
6381.449	7.9	6734.504	−4.4	7162.352	−2.2
6395.450	−2.3	6749.474	6.4	7165.315	6.4
6396.424	7.0	6750.450	5.9	7167.343	3.9
6398.435	+3.5	6755.517	−3.5	7169.360	−0.7

166

TABLE 190 (continued)

JD hel	s	JD hel	s	JD hel	s
243...		243...		243...	
7170.337	—6.6	7466.478	1.0	7849.468	—5.5
7172.330	7.4	7470.450	7.3	7853.453	4.9
7174.352	4.5	7488.436	—2.2	7869.384	—3.5
7189.265	—0.7	7494.390	—1.8	7871.368	4.1
.289	4.9	7497.405	8.7	7872.362	6.9
7192.295	—3.5	7498.398	6.8	7878.354	5.9
7193.275	6.8	7512.331	7.0	7881.424	4.9
7195.261	4.9	7519.351	4.5	7882.355	0.0
7196.261	3.8	7520.346	5.9	7883.396	—0.8
7197.274	—1.2	7521.356	2.7	7900.311	6.8
7198.259	—6.6	7523.332	—2.6	7902.312	4.9
7199.300	—6.6	7524.337	0.0	7904.312	—2.9
7218.222	—1.2	7525.331	7.2	7906.311	6.6
7228.213	8.1	7526.328	4.2	7908.303	6.7
7458.469	2.8	7819.500	—0.8	7910.297	1.6
7463.482	7.0	7847.441	—0.7	7955.224	7.8
7464.499	7.1				

Odessa (three-camera astrograph)

JD hel	s	JD hel	s	JD hel	s
243...		243...		243...	
3775.416	6.8	3894.360	—2.2	4621.291	—8.6
3825.497	—4.4	3922.246	—9.6	4624.290	8.7
3836.367	1.4	4239.371	—3.6	4626.302	4.0:
3837.397	—5.6	4480.517	—8.9	4627.330	3.0:
3850.404	—1.6	4538.388	6.5	4636.300	—4.0
3859.350	—2.5	4539.370	5.0:	4649.288	—3.3
3860.374	8.7	4542.366	—1.5	4657.250	—3.8
.379	6.8:	4567.367	6.8	4662.227	—2.2
3861.386	8.7	4568.371	8.7	4677.226	—1.1
3862.400	7.8	4574.362	7.6	4685.198	—8.3
3868.345	7.0	4577.361	4.0	5398.189	4.4
3870.356	5.7:	4578.331	—8.6	5687.318	—6.6
3871.334	4.4	4592.335	0.0:	5691.373	6.8
3886.340	0.0	4601.377	—5.7	5692.383	6.8
3890.298	6.9	4606.390	6.0		

YZ Aquarii

Very few observations available. However, the marked deviation of the
last visual observations from the ephemeris has shown that the period
definitely changed. Its behavior was consequently investigated using the
Simeise photographs. Reduction of all observations gave the summary
list of maxima (Table 191).

TABLE 191. List of maxima

Source	Max hel JD	E	O — A	E_1	O — B
Tsesevich (Sim.)	2421075.461	— 2611	$\overset{d}{-}0.001$	—	—
	4726.474	+ 4004	+.003	—	—
Florya /126/	6604.130	+ 7406	—.002	—	—
Tsesevich (visual)	6983.319	+ 8093	+.012	— 2570	$\overset{d}{+}0.001$
	7386.221	+ 8823	+.006	— 1840	—.008
Solov'ev /108/	8401.782	+ 10663	+.018	0	—.005
Tsesevich (Sim.)	8403.446	+ 10666	+.026	+ 3	+.003
	33215.198	+ 19384	+.064	+ 8721	—.002
Alaniya /2/	5364.442	+ 23278	+.098	+ 12615	+.013
Tsesevich (visual)	7196.288	+ 26597	+.092	+ 15934	—.009

167

All the maxima can be fitted with two equations: before JD 2426930 (O−A)

$$\text{Max hel JD} = 2422516.548 + 0.5519287 \cdot E; \quad P^{-1} = 1.811828231; \tag{144}$$

after JD 2426930 (O−B)

$$\text{Max hel JD} = 2428401.787 + 0.5519336 \cdot E_1; \quad P^{-1} = 1.811812146. \tag{145}$$

The period jumped by $+0^d.0000049$. These equations were used to obtain the average light curves (Tables 192, 193). Comparison stars for the observations in Tables 194, 195 (Figure 92): photographic scale $k = 0^s.0$; $a = 9^s.1$; $b = 16^s.2$; $d = 21^s.2$; visual scale $k = 0^s.0$; $a = 6^s.7$; $b = 10^s.3$; $c = 16^s.2$; $d = 22^s.1$.

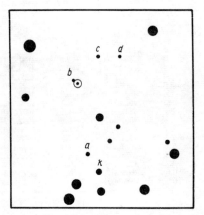

FIGURE 92. Comparison stars for YZ Aquarii.

TABLE 192. Average visual light curve

φ	s	n	φ	s	n	φ	s	n
0.011	5.9	10	0.514	17.0	10	0.920	13.7	10
.040	6.3	10	.628	16.7	10	.944	8.8	10
.089	8.0	10	.688	17.7	10	.967	5.8	10
.146	9.9	10	.761	17.2	10	.983	5.7	7
.214	11.4	10	.827	16.0	10	.994	5.3	7
.334	13.6	10	.882	18.5	10			

TABLE 193. Average photographic light curve

φ	s	n	φ	s	n	φ	s	n
0.019	7.0	5	0.322	16.0	5	0.736	19.2	4
.051	6.3	4	.349	17.2	6	.772	19.1	5
.074	7.0	6	.379	18.6	4	.809	18.3	6
.091	8.4	6	.422	17.7	4	.838	18.1	6
.118	9.4	6	.427	17.8	4	.844	19.0	5
.143	11.5	6	.452	18.7	4	.861	16.7	6
.169	11.6	5	.508	19.4	6	.905	11.8	3
.209	12.3	6	.571	18.9	5	.921	9.6	4
.234	16.0	5	.611	19.2	6	.953	10.0	3
.262	16.1	5	.655	19.2	6	.982	5.7	3
.290	14.9	6	.699	18.9	5			

TABLE 194. Visual observations

JD hel	s	JD hel	s	JD hel	s
242...		242...		243...	
6949.444	14.8:	7383.129	13.5	7172.335	19.2
.474	14.8	.134	13.5	.382	19.2
6950.348	15.3:	.162	13.7	.407	19.7
.372	12.2:	.203	13.6	.422	20.1
6952.273	14.1:	7384.167	14.0	.434	20.1
.310	15.6:	7385.076	12.7	7173.368	19.2
.407	3.9	.093	8.4	7175.347	8.3
.410	4.7	.097	6.0	.358	7.8
.420	5.4	.104	5.1	.390	13.7
.429	6.6	.110	4.9	7176.378	18.2
.476	11.8	.117	5.6	.427	8.3
.486	15.3	.123	6.0	.434	8.8
6959.417	12.2	.132	5.5	7193.323	13.7
6978.236	13.6	.140	5.9	.334	18.2
.247	14.3	.161	6.3	.371	20.6
6981.202	11.1	.173	6.8	.382	20.6
.208	10.3	.182	6.3	7196.257	14.8
.216	11.0	.208	7.2	.268	13.2
.225	11.7	.216	8.3	.282	5.7
.240	12.6	.223	9.0	.289	3.7
.248	12.4	7386.093	13.8	.298	5.9
.259	11.9	.100	14.0	.309	7.4
6983.263	15.4	.106	15.6	.327	10.3
.294	9.5	.127	14.0	.348	12.0
.299	7.9	.141	14.4	.363	12.0
.306	6.7	.151	14.6	.393	13.0
.310	5.9	.159	14.6	.414	12.8
.315	4.9	.164	14.4	7197.289	19.5
.323	3.9	.173	14.4	.310	24.1
.331	4.9	.197	9.6	.327	24.1
.346	5.5	.209	4.6	.339	24.1
.364	8.1	.212	4.9	.352	19.4
6984.205	13.6	.223	5.2	.335	13.5
.217	14.3	.245	5.3	.357	13.0
.229	14.4	7387.131	13.6	.361	12.6
.251	14.4	.148	14.6	.365	9.6
.265	14.4	7390.116	3.6	.367	9.1
.302	14.6	.138	5.5	.373	6.1
.319	15.0	.219	11.9	.380	7.2
.331	14.6	.235	13.2	.389	6.9
.336	14.4	7393.116	14.6	.394	7.2
.342	14.6	.153	14.2	.404	6.8
.357	14.6	.199	15.8	.409	6.9
.365	15.0	7415.121	14.8	.432	8.0
.376	12.4	.147	14.0	.453	9.3
.378	10.3	7422.078	9.1	7198.274	20.1
.383	10.3	.089	2.1	.313	23.1
.388	9.3	.094	2.8	.373	25.1
.392	7.2	.101	1.7	.433	24.1
.408	5.6	.113	3.6	7204.331	25.1
.414	8.1	.124	4.6	.357	25.1
6985.203	14.4	243...		.389	24.1
6987.257	8.2	7170.412	6.3	.402	23.1
.294	9.0	.424	5.3		
7383.113	13.2	.450	5.3		
.118	13.5	7172.321	15.4		

TABLE 195. Simeise observations

JD hel	s	JD hel	s	JD hel	s
242...		242...		242...	
0373.452	6.4	0743.350	17.2	1075.500	7.4
	8.0	0748.405	19.2	1102.419	18.2
0394.328	10.0		18.9		18.2
	11.1	0754.369	14.4	1108.449	19.2
0396.295	19.2		18.2		19.2
	19.2	0771.274	14.4	1112.429	5.7
0717.475	18.5		14.8		6.4
	19.2	0775.291	13.5	1134.330	19.8
0721.441	19.2		11.5		18.9
0726.507	19.2	1075.500	7.3	1466.380	14.2

TABLE 195 (continued)

JD hel	s	JD hel	s	JD hel	s
242...		242...		242...	
1466.380	16.2	5479.451	17.2	8461.301	20.0
1466.469	18.2	5482.466	19.2	8752.499	19.2
	18.2		19.2	8758.456	15.2
1485.334	19.2	5497.395	19.2		15.3
	19.2	5502.394	16.2	8779.373	8.0
1493.314	6.4		18.7		6.7
	6.8	5503.421	19.2	8786.40:	19.2
	7.8	5511.398	12.3		19.2
3647.406	11.5		9.1	9132.458	17.9
	7.8	5854.425	19.2		18.4
3649.409	19.5		19.2	9514.394	18.2
	19.2	6240.371	13.8		19.2
3664.287	20.0	6248.330	16.2	9518.403	6.1
	19.5		17.2	243...	
3673.307	17.9	6264.233	10.3	0236.443	7.0
	18.5		11.9		6.1
3674.300	19.2	6268.242	16.7	0251.311	5.0
3678.315	7.5		16.2	2769.441	18.2
	11.8	6566.431	20.2		18.7
3698.254	6.1		19.5	2775.476	16.2
3731.218	19.2	6984.351	16.2		14.2
	18.7		18.2	3122.475	7.3
3996.502	16.2	7302.472	14.6		7.1
	18.7		15.2	3152.373	12.1
4033.380	17.2	7329.408	6.1		11.5
	16.2	7343.411	16.2	3178.322	12.1
4379.491	17.4		17.6	3187.337	19.2
	16.2	7356.261	17.2		19.8
4385.490	8.6		18.2	3215.245	7.5
	9.1	7362.284	19.2		8.0
4389.416	16.2		19.2	3860.485	10.5
	16.2	7367.243	19.2		11.9
4732.493	14.2		19.2	3886.417	10.3
4736.426	6.1	7685.430	12.2		8.5
4740.471	17.2		7.8	4223.477	17.9
	19.2	8394.496	17.9		17.2
4744.409	18.5		17.4	4242.414	10.5
	19.2	8396.446	16.2		12.1
4760.320	15.2	8397.496	14.2	4246.373	15.2
	16.2	8403.459	7.5		13.2
5470.384	20.2		7.1	4596.464	18.2
5474.448	12.2	8460.273	8.1		18.5
	13.2		8.1		
5476.389	19.2	8461.301	19.2		
5479.451	16.2				

RR Ceti

Discovered in 1906 by Oppolzer. Repeatedly observed before 1925 by Oppolzer /327/, Ichinohe /277/, Pračka /339/, Luizet /311/, Martin and Plummer /319/, Jordan /290/, Stearns /355/, Florya /131/, Lange /63/, Solov'ev /108/, Selivanov /97/, and Gur'ev /27/. Florya /131/ systematized the old observations discovering a periodic variation of the period. During the last 30 years, however, only two small observation series were carried out by Alaniya /1/ and Tsarevskii. The complete history of the star therefore can be recovered only from sky photographs. Tsesevich estimated the magnitude from Simeise plates; at Tsesevich's request, Filatov estimated the magnitude from the Dushanbe photographs, and Migach from the Odessa photographs.

Seasonal light curves constructed using the equation

$$\text{Max hel JD} = 2417501.4421 + 0.5530253 \cdot E;\ P^{-1} = 1.8082355 \qquad (146)$$

gave the list of maxima (Table 196). It includes all the published maxima, except Alaniya's, as it greatly deviates from the ephemeris.

TABLE 196. List of maxima

Source	Max hel JD	E	O — A	O — B	O — C
Oppolzer /327/	2417501.445	0	$+0\overset{d}{.}003$	$+0\overset{d}{.}001$	—
Ichinohe /277/	8071.061	+ 1030	+.003	.000	—
Pračka /339/	8281.214	+ 1410	+.006	+.004	—
Tsesevich (Sim.)	20451.285	+ 5334	+.006	+.002	—
Luizet /311/	0618.306	+ 5636	+.013	+.009	—
Martin, Plummer /319/	1762.504	+ 7705	+.002	—.004	—
Jordan /290/	1801.772	+ 7776	+.005	.000	—
Stearns /355/	3194.822	+ 10295	—.016	—.022	—
Tsesevich (Sim.)	5100.579	+ 13741	+.017	+.008	—
Lange /63/	7484.116	+ 18051	+.014	+.004	—
Solov'ev /108/	7638.396	+ 18330	.000	—.010	—
Selivanov /97/	7683.204	+ 18411	+.013	+.003	—
Gur'ev /27/	8075.302	+ 19120	+.016	+.005	—
Tsesevich (Sim.)	8811.379	+ 20451	+.016	+.005	—
Filatov	33181.396	+ 28353	+.028	—	$-0\overset{d}{.}008$
Tsesevich (Sim.)	3207.394	+ 28400	+.033	—	—.002
Migach	4277.520	+ 30335	+.055	—	+.014
Filatov	5036.260	+ 31707	+.045	—	.000
Alaniya	5392.405	+ 32351	+.041	—	—.006
Migach	5403.463	+ 32371	+.039	—	—.008
	6163.324	+ 33745	+.043	—	—.008
Tsarevskii	6518.395	+ 34387	+.072	—	+.019
Migach	6518.376	+ 34387	+.053	—	.000
	6960.247	+ 35186	+.057	—	+.002
	7197.499	+ 35615	+.061	—	+.005
	7583.497	+ 36313	+.047	—	—.011
	7942.418	+ 36962	+.055	—	—.005

The variation of the O — A residues is plotted in Figure 93. We see that the period is subjected to irregular, aperiodic fluctuations. The average period has changed, and the observations can be fitted with two equations: from 2417501 to 2428811 (O — B)

$$\text{Max hel JD} = 2417501.444 + 0.55302577 \cdot E; \qquad (147)$$

from 2433181 to 2437942 (O — C)

$$\text{Max hel JD} = 2433181.404 + 0.55302814 \cdot E'. \qquad (148)$$

The period thus increased by $0^d.00000237$.

The observations are listed in Tables 197, 198. Comparison stars:

BD	s_1	s_2
+ 0°243	—	0.0
246	0.0	11.9
250	4.7	26.4
253	9.7	—
247	13.8	—
+ 1 270	—	38.4

171

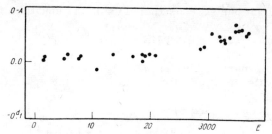

FIGURE 93. O—A residues for RR Ceti.

TABLE 197. Simeise observations (Tsesevich)

JD hel	s_1	JD hel	s_1	JD hel	s_1
241...		242...		242...	
9715.391	5.4	4416.468	2.6	8124.379	7.7
	6.1		3.1		8.7
9721.337	1.2	4419.463	7.5	8433.526	14.8
	2.4		9.0		8.5
9721.399	3.5	4820.297	2.9	8811.468	2.4
	2.4		1.9		2.9
242...		4823.415	12.2	9173.539	6.2
0045.455	2.8		12.8	9175.536	11.8
	3.1	4853.314	5.0		10.7
0047.483	11.1		7.2	9548.468	12.2
	11.8	5510.520	7.7		11.3
0050.482	3.1		7.2	9577.411	3.8
	3.5	6267.444	0.0		3.8
0067.445	3.8		0.0	243...	
	6.1	6269.432	10.7	2862.360	0.0
0073.387	9.7		11.5		0.0
	10.9	6270.317	1.9	2890.275	10.7
0451.297	—2.0		2.9		11.7
	—1.5	6296.328	2.9	3185.504	10.7
0774.487	7.7		2.9		10.7
	7.2	6624.338	4.7	3206.383	3.4
1170.439	8.0		4.7		2.8
	10.5	6629.383	8.7	3207.383	1.9
1513.479	9.7		10.9		1.9
	8.5	7339.495	10.7	3897.486	11.8
1515.384	2.4	7714.462	10.7		12.2
	3.5		9.7		
3701.369	3.1	8079.386	6.7		
	4.2		7.2		
4415.516	2.8	8080.395	3.1		
	3.8		2.7		

TABLE 198. Observations of G. S. Filatov

JD hel	s_2	JD hel	s_2	JD hel	s_2
242...		243...		243...	
9881.376	5.0	3181.418	4.3	4285.404	20.2
9903.324	22.3	3183.425	21.5	4353.160	26.4
9905.356	13.8	3186.348	7.9	4610.446	5.5
9911.334	7.6	3190.364	15.0	4639.414	22.5
9962.197	8.9	3238.244	30.4	4769.125	7.3
9967.186	6.9	3244.225	23.4	4994.414	20.2
9990.090	20.2	3262.148	4.0	5022.331	30.4
.112	21.3	3274.218	32.4	5036.291	11.9
243...		3545.416	15.7	5049.297	31.4
0284.311	21.6	3597.208	30.4	5480.112	26.4
0348.113	26.4	3895.328	11.9	5753.262	28.4
2829.313	6.8	3897.376	23.4	5755.348	18.6
3156.430	26.4	3898.388	22.5	5781.213	5.4
3158.471	24.4	3915.337	8.6	5812.168	3.3
3159.469	22.3	3918.316	22.4	5813.172	30.4
3161.436	26.4	4224.433	7.6	6074.429	17.4
3178.369	23.4	4279.352	18.1	6820.385	6.9
3179.410	21.5	4280.343	14.8		

V 734 Ophiuchi

Excellently visible on the photographs taken with the 16-in. astrograph at the Southern Section of the Shternberg Astronomical Institute. The Odessa photographs show this star only near the maximum, whereas in the Simeise photographs it is weak, so that the estimates are not quite reliable. Tsesevich nevertheless investigated its light variation. After a lengthy search for an improved period value, seasonal light curves were plotted and the maxima determined (Table 199; it also includes the maximum listed in the Catalogue of Variable Stars).

TABLE 199. List of maxima

Source	Max hel JD	E	$O - A$	E'	$O - B$
Tsesevich (Sim.)	2420636.497	0	0.000	—	—
	5770.313	+ 9127	−.021	—	—
Hughes—Boyce /275/	7959.55	+ 13019	+.009	—	—
Tsesevich (Sim.)	33803.41	+ 23408	+.170	−5920	0.000
Tsesevich (Odessa)	6756.401	+ 28658	+.094	− 670	+.003
	7133.252	+ 29328	+.076	0	−.003

The $O - A$ residues were calculated from

$$\text{Max hel JD} = 2420636.497 + 0.562489 \ E. \qquad (149)$$

This equation fits the observations between $E = 0$ and $E = 15000$. The last two maxima, which are highly reliable, do not fit this equation, however. The period apparently changed in a jump. The three last maxima are described by the equation

$$\text{Max hel JD} = 2437133.255 + 0.5624738 \cdot E'; \ P^{-1} = 1.77786059. \qquad (150)$$

The Moscow observations are the best of the lot; Odessa observations are somewhat poorer, and the Simeise observations are not particularly reliable. Figure 94 correspondingly gives the light curve plotted from Moscow and Odessa observations. Comparison stars (Figure 95): $a = 0^s.0$; $b = 5^s.1$; $c = 14^s.5$. The observations are listed in Tables 200−202.

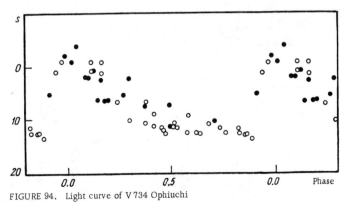

FIGURE 94. Light curve of V 734 Ophiuchi
(● Odessa observations, ○ Moscow observations).

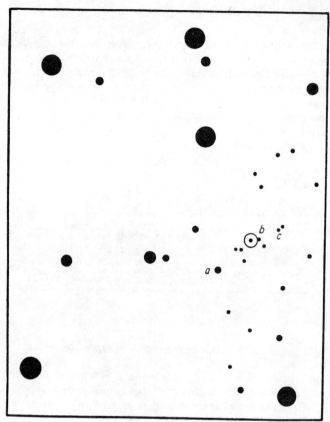

FIGURE 95. Comparison stars for V 734 Ophiuchi.

TABLE 200. Simeise observations

JD hed	s	JD hel	s	JD hel	s
242...		242...		243...	
0636.497	—1.0	5744.497	6.1	3034.505	5.0
1369.461	+3.8	5745.470	[5.1	3766.488	13.1
3193.376	[5.1	5764.404	[12.1	3802.362	6.1
3931.370	5.1:	5770.348	—0.5	3803.364	5.0
3942.463	9.1	5799.345	[5.1	3806.452	6.1
5382.442	12.1	5820.338	2.0:		
5385.465	10.1	5826.354	10.1		
5388.489	8.1	9071.384	8.1		

TABLE 201. Odessa observations

JD hel	s	JD hel	s	JD hel	s
243...		243...		243...	
6744.358	[5.1	6758.389	[5.1	7080.400	—4.0
6749.363	7.1	6769.389	2.0	.451	0.6
6755.424	5.1	7077.396	[5.1	7102.407	2.5
6756.344	5.1	.452	[5.1	7107.369	—2.0
.401	—1.0	7079.367	6.1	7111.370	2.2
6757.362	10.1	.418	2.0	.396	6.1
6758.363	11.1	.467	7.1	.424	6.1

174

TABLE 202. Moscow observations

JD hel	s	JD hel	s	JD hel	s
243...		243...		243...	
7052.547	11.4	7100.394	9.3	7137.327	6.6
7074.462	12.6	7102.381	0.9	7138.297	—1.0
7077.490	12.4	7103.344	11.7	7139.305	12.6
7078.482	12.4	7106.423	10.0	7140.300	11.7
7079.511	11.7	7109.360	10.7	7143.312	13.5
7080.472	—1.0	7112.360	12.4	7144.302	12.6
7087.457	12.4	7132.362	11.4	7145.299	8.6
7088.468	6.8	7133.341	1.0	7161.282	12.4
7089.441	—1.0	7135.330	10.7	7163.276	10.3
7099.355	11.7	7136.327	12.0	7165.281	1.0

TT Cancri

Discovered by Kulikovskii /53/ and independently by Hoffmeister /266/.
Lange /60/ and Jensch /288/ independently determined its period. It was
subsequently studied by McKnelly /315/, Payne —Gaposchkin /329/,
Alaniya /1/, and Tsesevich /160/. Variations of its period were established
by Tsesevich. Observed by Huth /276/ on Sonneberg plates. Magnitude
estimated by Tsesevich from Odessa and Simeise photographs; the seasonal
light curve was derived from Kulikovskii's observations based on old Moscow
photographs /53/.

Visually observed by Tsesevich in 1944. Comparison stars (Figure 96):
$a = 0^s.0$; $k = 7^s.0$; $b = 16^s.0$; $c = 24^s.8$.

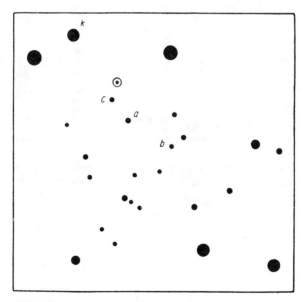

FIGURE 96. Comparison stars for TT Cancri.

The light curve (Table 203) was derived from Jensch's equation:

$$\text{Max hel JD} = 2425647.421 + 0.563457 \cdot E; \quad P^{-1} = 1.774758.$$

(151)

175

The maximum was found to depart from the linear ephemeris almost by as much as three hours!

TABLE 203. Average visual light curve

φ	s	n	φ	s	n	φ	s	n
$0^p.043$	17.7	5	$0^p.403$	22.3	5	$0^p.790$	12.2	5
.124	18.3	6	.437	21.9	5	.827	12.4	6
.185	20.1	5	.497	22.4	5	.859	12.7	5
.222	21.4	5	.564	21.6	5	.894	13.5	5
.259	20.5	5	.648	22.8	5	.939	14.2	5
.277	21.6	5	.681	22.3	5	.981	16.2	5
.318	21.6	5	.722	18.3	5			
.359	22.1	5	.758	14.8	5			

The maximum corresponds to the age of $0^p.809$. The average maximum is therefore 2431158.486. The summary list of maxima is given in Table 20

TABLE 204. List of maxima

Source	E	Max hel JD	O — A
Tsesevich (Sim.)	— 9059	2420543.300	$+0^d.236$
Kulikovskii /53/	— 8317	0961.31	+ .161
Tsesevich (Sim.)	— 1926	4562.293	+ .090
Jensch /288/	0	5647.422	+ .001
	+ 632	6003.528	+ .002
	+ 694	6038.462	+ .002
	+ 1361	6414.341	+ .055
	+ 1370	6419.368	+ .011
	+ 3216	7459.474	— .025
	+ 3340	7529.329	— .038
	+ 3349	7534.463	+ .025:
	+ 3351	7535.561	— .004
	+ 3358	7539.510	.000
	+ 3365	7543.459	+ .005
	+ 3413	7570.492	— .008
Lange /60/	+ 3943	7869.14	+ .003
McKnelly /315/	+ 9077	30761.818	— .102
Tsesevich (visual)	+ 9781	1158.486	— .108
McKnelly /315/	+ 9847	1195.678	— .104
	+ 10328	1466.710	— .095
	+ 10875	1774.914	— .102
	+ 11794	2292.718	— .115
Alaniya /1/	+ 15616	4446.220	— .146
Huth /276/	+ 18664	6163.630	— .152
Tsesevich (Odessa)	+ 20218	7039.230	— .165

The O — A residues show that Jensch's formula is in need of improvem
To simplify the calculations, closely spaced maxima are averaged (Table 2

TABLE 205. Average maxima

E	Max hel JD	O — A	O — B	O — C
—9059	2420543.300	$+0^d.236$	$+0^d.092$	—
—8317	0961.31	+ .161	+ .026	—
—1926	4562.293	+ .090	+ .035	—
+ 442	5896.471	+ .002	— .023	—
+1366	6417.136	+ .033	+ .019	—
+3302	7507.943	— .013	— .003	—
+3372	7547.396	— .002	+ .009	—
+3943	7869.140	+ .008	+ .026	—
+ 9077	30761.818	— .102	— .020	$-0^d.006$
+ 9781	1158.486	— .108	— .017	—.007

176

TABLE 205 (continued)

E	Max hel JD	O — A	O — B	O — C
+ 9847	1195.678	— .104	— .012	$-0\overset{d}{.}003$
+10328	1466.710	— .095	+ .003	+.010
+10875	1774.914	— .102	+ .003	+.007
+11794	2292.718	— .115	+ .002	.000
+15616	4446.220	— .146	+ .019	—.004
+18664	6163.630	— .152	+ .050	+.010
+20218	7039.230	— .165	+ .057	+.008

The least squares method gave the equation

$$\text{Max hel JD} = 2425647.452 + 0.5634445 \cdot E. \qquad (152)$$

The O — B residues are substantially less than the O — A residues, but both sets show systematic variation. This suggests that near $E = 7000$ the period jumped by $+0^d.00000560$. In this case the last maxima, starting with McKnelly's data, can be described by the equation

$$\text{Max hel JD} = 2430761.824 + 0.5634501 \cdot E. \qquad (153)$$

The O — C residues bear out this assumption. Tsesevich's observations are listed in Tables 206 — 208. Comparison stars in photographic observations: $k = - 8^s.2$; $a = 0^s.0$; $b = 9^s.4$.

TABLE 206. Visual observations

JD hel	s	JD hel	s	JD hel	s
243...		243...		243...	
1138.190	10.2	1150.164	20.4	1162.350	22.8
.217	11.8	.183	20.9	.409	21.3
1139.291	15.2	.211	20.9	1163.162	14.4
.317	13.3	.231	21.3	.183	18.2
.332	12.5	.270	22.5	.220	18.9
.344	12.2	.292	22.2	.225	20.9
.358	13.1	.304	22.8	.349	20.8
.375	13.7	.335	21.9	1164.205	15.5
1143.180	22.6	.377	21.9	.292	18.3
.189	23.2	.403	21.9	.321	17.1
.202	23.0	.438	22.8	.381	20.4
.216	21.9	1154.270	22.8	1165.289	13.0
.224	18.2	1157.121	23.0	1169.138	18.2
.226	18.2	.135	23.2	.172	13.8
.237	12.8	.151	22.8	.184	14.0
.241	13.9	.165	22.8	.243	14.0
.251	11.8	.187	22.8	.354	14.5
.269	9.9	.206	22.5	.382	19.8
.278	11.6	.216	21.3	1170.239	22.2
.287	12.6	.237	22.8	.295	14.4
.298	12.0	.288	22.8	.320	13.3
.314	12.0	1158.138	23.4	.345	12.0
.326	13.1	.156	23.4	1173.236	14.0
.340	12.6	.199	22.2	.287	14.0
1149.142	22.2	.223	22.8	1174.302	13.3
.161	21.3	.262	23.4	1175.172	21.9
.176	22.2	1162.135	21.6	1176.173	19.8
.199	22.2	.162	19.9	.210	20.4
1150.115	14.2	.195	21.3	.260	22.2
.144	18.9	.243	21.3	.287	21.5

TABLE 206 (continued)

JD hel	s	JD hel	s	JD hel	s
243...		243...		243...	
1176.366	22.2	1180.373	22.6	1197.247	19.3
1177.310	22.8	1181.199	18.2	1206.216	22.3
1178.255	14.6	1193.185	15.5	1207.221	21.0
.282	14.2	1194.215	16.0	1213.235	14.8
.307	13.3	.225	20.4	1225.163	19.8
.329	14.4	.242	18.2	1232.230	21.3
1179.224	22.6	.302	20.4	1235.220	18.9
.265	22.2	.325	21.9		

TABLE 207. Odessa observations

JD hel	s	JD hel	s	JD hel	s
243...		243...		243...	
6279.345	6.3	6628.383	10.4	7016.331	5.2
6281.343	2.8		7.3	7017.333	2.1
6286.329	6.6	6655.318	6.7	7020.354	6.6
6288.327	5.2	6656.360	6.3	7023.328	6.9
6306.305	4.9	6657.337	2.1	7039.296	−1.2
6526.600	0.0	6660.355	2.8	7317.450	3.8
6528.614	8.1	6663.327	4.7	7369.403	−3.5
6544.568	−2.5	6664.320	8.2	7377.361	−1.0
6546.533	6.6	6667.302	−2.0	7378.337	6.3
6555.509	5.2	6954.381	6.9	7396.307	7.2
6583.454	1.7	6959.474	7.4	7397.311	8.5
6607.434	7.3	6993.409	6.8	7398.301	2.9
6608.416	3.5	6996.435	−4.7	7400.307	5.8
6612.433	6.3	7015.351	7.9		

TABLE 208. Simeise observations

JD hel	s	JD hel	s	JD hel	s
242...		242...		242...	
0160.466	7.5	0894.531	7.7	4578.254	6.6
	6.3		6.6	4607.276	8.1
0177.3:	5.9	0918.287	8.4		8.5
	5.9		7.5	4611.281	2.1
0180.314	7.9	1250.502	0.0		−1.4
0186.3:	5.4		−0.4	5998.460	−0.9
	6.6	1284.273	−0.5		−3.4
0188.3:	−1.6		1.2	5999.351	9.4
	−0.9	4196.311	6.6		8.1
0190.3:	8.5		6.9	7843.492	7.4
	8.5	4559.289	7.5		8.2
0543.278	−0.9		7.3	243...	
	1.2	4562.300	−2.7	0024.433	5.2
0893.378	6.3		−1.2		4.2
	7.1	4578.254	7.3		

RX Ceti

Observed by Morgenroth /321/, Lange /56—58/, Solov'ev /108/, and Selivanov /97/ some 30 years ago. The old observations are adequately fitted with Morgenroth's formula

$$\text{Max hel JD} = 2425938.272 + 0.57371 \cdot E; \quad P^{-1} = 1.7430409. \tag{154}$$

The new observations indicate a variation of the period. Tsesevich estimated the magnitude from Harvard photographs, T. Nikulina from Dushanbe plates.

The maxima, except the erroneous determination of Alaniya /2/, are listed in Table 209. The O−A residues are calculated using (154).

TABLE 209. List of maxima

Source	Max hel JD	E	O − A
Tsesevich (Harv.)	2415860.538	− 17566	$+0^{d}.056$
	7437.656	− 14817	+ .045
	8962.589	− 12159	+ .057
	20630.918	− 9251	+ .037
	2448.423	− 6083	+ .029
	3894.746	− 3562	+ .029
	5149.438	− 1375	+ .017
Morgenroth /321/	5938.272	0	.000
Tsesevich (Harv.)	6986.437	+ 1827	− .003
Lange /56-58/	7336.405	+ 2437	+ .002
Tsesevich (visual)	7663.416	+ 3007	− .002
Selivanov /97/	7694.388	+ 3061	− .010
Tsesevich (Harv.)	8065.573	+ 3708	− .016
Gur'ev /27/	8228.522	+ 3992	.000
Solov'ev /108/	8417.276	+ 4321	+ .003
Gur'ev /27/	8426.456	+ 4337	+ .004
Tsesevich (Harv.)	8994.394	+ 5327	− .031
	9890.529	+ 6889	− .031
Nikulina	9961.092	+ 7012	− .035
Tsesevich (Harv.)	30974.835	+ 8779	− .037
	1874.943	+ 10348	− .080
	2618.451	+ 11644	− .100
	3528.917	+ 13231	− .112
Nikulina	3596.624	+ 13349	− .103
	5754.272	+ 17110	− .178
Tsesevich (visual)	7912.518	+ 20872	− .229

The period changed abruptly near JD 2428907. After that epoch, we have

$$\text{Max hel JD} = 2429961.099 + 0.5736957 \cdot E. \qquad (155)$$

Tsesevich's observations, reduced to one period using Morgenroth's equation, gave the average light curve (Table 210).

TABLE 210. Average visual light curve

φ	s	n	φ	s	n	φ	s	n
$0^{p}.043$	19.8	3	$0^{p}.510$	14.5	4	$0^{p}.698$	10.0	4
.146	21.8	3	.536	9.3	4	.795	15.0	4
.198	21.0	3	.565	9.2	5	.872	15.1	4
.369	18.9	3	.584	8.2	5	.966	16.1	3
.430	22.0	2	.617	7.4	5			
.471	18.4	3	.660	8.6	5			

Nikulina's observations were reduced to one period using (155) with $P^{-1} =$ = 1.743084356. The average light curve is given in Table 211.

TABLE 211. Average photographic light curve

φ	s	n	φ	s	n	φ	s	n
$0^{p}.033$	3.9	3	$0^{p}.392$	27.8	3	$0^{p}.867$	25.9	3
.086	5.6	2	.478	26.4	4	.925	4.4	1
.129	18.3	6	.511	27.2	3	.970	3.9	4
.200	21.3	3	.585	30.7	5	.987	0.0	3
.255	20.3	3	.719	27.0	5			
.334	25.5	4	.811	24.5	3			

179

Comparison stars used in the observations of Tables 212—214 (Figure 97):

		s_1	s_2
a	BD —16°89	0.0	—
k	BD —16 92	—	0.0
b	BD —16 93	7.3	—
c	BD —16 98	17.3	—
d	BD —16 91	26.3	11.0
f	—	—	20.2
e	—	34.8	26.4

TABLE 212. Visual observations (Tsesevich)

JD hel	s_2	JD hel	s_2	JD hel	s_2
243...		243...		243...	
7904.495	6.4	7911.372	7.8	7912.470	9.8
7908.425	16.6	.384	6.4	.475	8.4
.434	16.8	.389	8.0	.481	7.7
.454	16.0	.406	8.8	.490	8.0
.463	11.0	.415	8.9	.496	8.2
.472	10.2	.421	8.6	.503	8.8
.483	9.0	.442	12.8	.510	7.8
.489	8.6	.465	14.4	.525	8.0
.495	8.0	.480	15.6	.534	8.2
7910.342	14.4	.494	15.6	.548	8.0
.373	14.4	.514	16.1	.559	9.4
.398	15.1	.526	14.7	.565	9.4
.413	15.1	7912.362	19.2	.572	9.1
.434	16.1	.392	18.7	7913.369	24.3
.451	17.1	.401	18.7	.396	23.3
.461	17.1	.412	22.7	.405	19.1
.469	17.9	.428	21.2	.413	22.3
7911.352	12.0	.440	21.7	.424	19.1
.358	8.2	.458	16.1	.434	22.0
.362	8.2	.465	16.1	.446	22.0

TABLE 213. Old visual observations (Tsesevich)

JD hel	s_2	JD hel	s_2	JD hel	s_2
242...		242...		242...	
7663.379	9.1	7663.404	5.8	7663.443	7.0
.385	9.2	.410	3.8	.453	7.7
.389	7.3	.422	4.5		
.393	7.7	.430	6.1		

TABLE 214. Photographic observations (Nikulina)

JD hel	s_1	JD hel	s_1	JD hel	s_1
242...		243...		243...	
9879.339	24.3	3186.312	30.5	5036.261	26.3
9902.270	30.5	3236.204	25.8	5037.272	25.3
9903.272	19.5	3244.194	32.7	5038.262	26.3
.294	21.3	3596.237	25.4	5048.228	26.3
9906.315	30.3	3618.148	25.4	5049.225	13.3
9927.186	17.3	3886.367	0.0	5374.328	28.3
.208	4.4	3889.412	31.7	5376.344	7.3
9961.120	2.7	3895.421	30.9	5395.254	24.1
.181	7.0	3897.343	20.3	5401.250	29.3
243...		3898.354	0.0	5750.280	7.0
0258.339	22.3	3915.306	24.3	5752.269	24.3
0344.080	31.0	3916.338	26.3	5754.272	0.0
0383.242	17.3	3969.154	26.3	5758.281	0.0
0877.147	17.3	4255.407	23.9	5811.131	11.3
0882.153	26.3	4272.341	29.5	.140	21.3
0887.124	24.5	4279.325	2.1	5813.146	26.3
3155.435	0.0	4300.281	26.3	6046.438	22.3
3159.438	4.8	4301.296	26.3	6074.379	3.6
3177.315	17.3	4381.090	25.3	6844.373	17.3
3180.399	27.0	4721.129	20.3	7229.235	7.3
3183.373	31.0	4994.387	26.3	7255.218	17.3

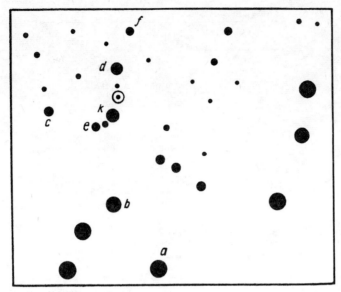

FIGURE 97. Comparison stars for RX Ceti.

IO Lyrae

Discovered and investigated by Hoffmeister /269, 270, 272/. Elements improved by Razgulyaeva /93/ from Moscow photographs. Later studied by Batyrev /17/. Odessa photographs analyzed by Tsesevich and Migach.

Tsesevich derived light curves from Razgulyaeva's published observations and found the maxima. Table 215 is a summary of all the maxima. The O−A residues were calculated from

$$\text{Max hel JD} = 2429374.435 + 0.5771203 \cdot E. \qquad (156)$$

The residue values show that the period corresponding to this formula should be somewhat increased. However, the exact change in the period is difficult to determine for the following reasons. First, the star probably reveals Blazhko effect; second, Hoffmeister did not construct seasonal light curves, and only reported the epochs when the star was near maximum. These data can be highly inaccurate. Tsesevich therefore averaged the data listed in the summary table (Table 216, n is the number of averaged maxima).

TABLE 215. List of maxima

Source	Max hel JD	E	O − A
Razgulyaeva /93/	2418560.31	− 18738	−0^d.045
Hoffmeister	29374.520	0	+.085
/269, 270/	9375.534	+ 2	−.054

181

TABLE 215 (continued)

Source	Max hel JD	E	O — A
Razgulyaeva /93/	242 9527.355	+ 265	−0^d.017
Hoffmeister	30209.531	+ 1447	+.003
/269, 270/	0386.676	+ 1754	−.028
	0430.615	+ 1830	+.050
	0497.508	+ 1946	−.003
	0549.425	+ 2036	−.027
	0590.423	+ 2107	−.004
	0793.620	+ 2459	+.046
	0848.419	+ 2554	+.019
	0852.464	+ 2561	+.024
	1001.332	+ 2819	−.005
	1204.474	+ 3171	−.009
	1222.398	+ 3202	+.024
	1589.422	+ 3838	−.001
	2615.547	+ 5616	+.004
	2638.643	+ 5656	+.016
	2641.527	+ 5661	+.014
	2648.445	+ 5673	+.007
	2652.475	+ 5680	−.003
Hoffmeister	2655.340	+ 5685	−.024
/269, 270/	2659.400	+ 5692	−.004
	2682.495	+ 5732	+.006
	2686.523	+ 5739	−.005
	2689.412	+ 5744	−.002
	2690.563	+ 5746	−.005
	2708.465	+ 5777	+.006
	2712.492	+ 5784	−.007
	2716.532	+ 5791	−.007
	2760.385	+ 5867	−.015
	2764.438	+ 5874	−.002
	2775.415	+ 5893	+.010
	2779.445	+ 5900	.000
	2780.602	+ 5902	+.003
	2794.456	+ 5926	+.006
	2795.602	+ 5928	−.002
	2797.350	+ 5931	+.015
Batyrev /17/	4267.240	+ 8478	−.021
	4282.258	+ 8504	−.008
	4289.195	+ 8516	+.004
	4304.189	+ 8542	−.008
	4600.259	+ 9055	.000
	4608.340	+ 9069	+.001
	4989.229	+ 9729	−.009
	5008.280	+ 9762	−.003
	5012.319	+ 9769	−.004
	5034.252	+ 9807	−.002
Tsesevich	6082.327	+ 11623	+.023
	6408.390	+ 12188	+.013
Migach	6815.257	+ 12893	+.010
	7518.218	+ 14111	+.038

TABLE 216. Averaged data of Table 215

E	O — A	n	(O—A)$_{calc}$	δ	Max hel JD	O — C
− 18738	−0^d.045	1	−0^d.018	−0^d.027	—	—
+ 265	− .017	1	−.003	−.014	—	—
+ 1853	− .002	6	− .002	.000	—	—
+ 2943	+ .014	7	−.001	+.015	—	—
+ 5688	+ .001	10	+.001	.000	—	—
+ 5860	.000	12	+.002	−.002	—	—
+ 8510	− .008	4	+.004	−.012	2434285.721	0^d.000
+ 9062	.000	2	+.004	−.004	4604.299	+.004
+ 9767	− .005	4	+.005	−.010	5011.164	−.006
+ 11623	+ .023	1	+.006	+.017	6082.327	+.009
+ 12188	+ .013	1	+.007	+.006	6408.390	−.005
+ 12893	+ .010	1	+.007	+.003	6815.257	−.013
+ 14111	+ .038	1	+.008	+.030	7518.218	+.008

5443

Assigning the same weight to all the data of Table 216, we obtain

$$O - A = -0^d.003 + 0^d.00000078 \cdot E,$$

whence

$$\text{Max hel JD} = 2429374.432 + 0.57712108 \cdot E. \tag{157}$$

The last maxima starting with $E = 8510$ are described by the following equation, from which the $O - C$ residues were calculated:

$$\text{Max hel JD} = 2434285.721 + 0.5771272 (E - 8510). \tag{158}$$

Comparison stars in the photographic observations of Table 217 (Figure 98)

	s	m	m'
k	0.0	—	11.43
b	6.9	11.72	11.77
c	11.6	12.07	12.01
e	18.9	12.32	12.37

The column m lists the stellar magnitudes according to Razgulyaeva. The power scale s is converted to stellar magnitudes using the equation $m = 11.43 + 0.0486 \cdot s$. Under m' the stellar magnitudes calculated by this relation from the power scale are given.

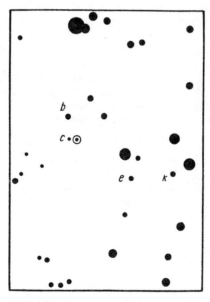

FIGURE 98. Comparison stars for IO Lyrae.

TABLE 217. Odessa observations

JD hel	s	JD hel	s	JD hel	s
243...		243...		243...	
6041.435	4.9	6344.520	10.3	6426.359	4.9
6043.475	14.7	6345.507	1.4	6428.393	15.0
6047.403	14.3	6347.505	11.6	6429.422	10.9
6047.426	11.6	6364.463	14.4	6430.399	8.5
6049.435	3.1	6367.497	5.0	6432.395	15.0
6050.449	14.7	6371.463	1.4	6434.363	—2.0
6053.415	9.6	6372.455	14.7	6451.318	9.7
6067.326	2.6	6373.461	13.7	6453.307	15.8
6069.362	14.0	6375.484	—1.0	6454.349	13.6
6070.362	9.7	6376.462	15.5	6455.315	9.7
6071.363	3.9	6379.478	5.9	6456.351	3.0
6074.421	13.3	6381.449	9.6	6461.303	15.2
6075.351	4.6	6395.450	14.5	6462.274	9.7
6076.349	13.7	6396.424	8.1	6463.287	8.1
6079.325	14.7	6398.435	13.1	6465.285	13.7
6080.347	12.6	6399.421	14.8	6478.214	—0.5
6081.342	9.5	6400.446	9.5	6479.242	10.4
6082.340	3.4	6401.448	1.9	6481.221	8.1
6083.350	15.7	6406.435	13.7	6482.224	4.9
6101.282	15.4	6407.436	9.7	6483.235	13.1
6102.252	13.8	6408.431	4.2	6484.227	11.1
6103.263	9.9	6409.424	15.0	6487.255	11.6
6104.263	—1.5	6423.380	3.1	6488.261	10.0
6105.238	13.4	6424.380	14.3	6489.265	4.6
6126.203	—2.0	6425.403	14.5		

RU Ceti

Lange /70/ used all the old and two new maxima to show that the period changed markedly. New data which became available since then make it possible to check Lange's results. A list of maxima is given in Table 218. The O−A residues were calculated from Lange's equation

$$\text{Max hel JD} = 2428722.920 + 0.5862758 \cdot E. \qquad (159)$$

The period clearly jumped by $0^d.0000245$. Tsesevich's main series of observations is published in /147/. During a repeated series (Table 219 the following comparison stars were used (Figure 99): $k = 0^s.0$; $a = 7^s.7$; $b = 18^s.3$; $c = 25^s.9$. Reduction to one period gave Max hel JD = 2437910.444. The maxima are probably of variable amplitude and the star has Blazhko effect.

TABLE 218. List of maxima

Source	Max hel JD	E	O − A
Tsesevich /147/	2426576.473	— 3661	—0^d.091
	6964.029	— 3000	— .064
Lange /70/	7332.213	— 2372	— .061
Tsesevich /147/	7405.502	— 2247	— .056
	7663.462	— 1807	— .058
Gur'ev /27/	8090.302	— 1079	— .026
Solov'ev /108/	8426.259	— 506	— .005
Gur'ev /27/	8427.428	— 504	— .009
Solov'ev /27/	33190.359	+ 7620	+ .017
	3274.217	+ 7763	+ .038
	3597.203	+ 8314	— .014
	3897.371	+ 8826	— .019
	5038.287	+ 10772	+ .004
	5812.155	+ 12092	— .012

TABLE 218 (continued)

Source	Max hel JD	E	O — A
Lange /70/	2436897.368	+ 13943	+0d.005
Tsesevich	7176.427	+ 14419	— .004
Lange	7910.444	+ 15671	— .004
	7910.442	+ 15671	— .006
	7913.385	+ 15676	+ .006
	8294.450	+ 16326	— .009

TABLE 219. New visual observations (Tsesevich)

JD hel	s	JD hel	s	JD hel	s
243...		243...		243...	
7904.474	16.4	7910.432	4.1	7912.411	15.6
.487	17.2	.443	2.6	.428	15.4
.508	18.3	.451	2.1	.439	16.6
7906.400	11.1	.460	2.2	.457	15.7
.404	12.6	.470	1.9	.462	17.1
.422	13.2	.477	3.8	.550	20.8
.453	13.8	7911.373	18.3	.571	22.9
.466	14.6	.405	16.8	7913.370	12.4
.482	14.4	.416	17.2	.384	10.1
7908.423	20.3	.443	20.8	.395	5.5
.433	21.3	.466	21.6	.404	5.1
.453	17.2	.482	22.7	.412	8.8
.462	20.3	.494	22.7	.419	10.9
.472	19.8	.515	20.8	.425	11.9
.487	21.3	.526	18.3	.433	11.6
7910.398	12.1	.531	20.8	.445	12.5
.413	9.4	7912.392	15.4	.455	13.0
.421	6.0	.401	15.4		

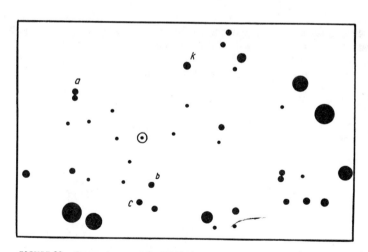

FIGURE 99. Comparison stars for RU Ceti.

185

Discovered by Hoffmeister /267/. Elements determined by Lange /62/ and Jacchia /283/. Grigor'eva /26/ estimated the magnitude from old Moscow photographs (maxima listed in Table 220). Jacchia /285/ examined 677 Harvard plates taken between 1898 and 1939 and obtained

$$\text{Max hel JD} = 2414815.759 + 0.5941295 \cdot E. \tag{160}$$

This equation was confirmed by Payne—Gaposchkin /330/ between JD 2414000 and JD 2429000. However, the new observations of Alaniya /1/, Batyrev /15/, Chuprina /164/, and Tsesevich indicate that near JD 2426000 the period jumped by $- 0^d.00000412$, and that now Batyrev's equation is valid:

$$\text{Max hel JD} = 2434244.333 + 0.59412538 \cdot E; P^{-1} = 1.68314641. \tag{161}$$

TABLE 220. List of maxima

Source	Max hel JD	E	O − B	O − A
Jacchia	2414815.759	− 32701	−0ᵈ.080	0ᵈ.000
Grigor'eva /26/	8230.221	− 26954	−.057	.000
Lange /62/	27980.468	− 10543	−.001	—
	8012.553	− 10489	+.001	—
	8014.341	− 10486	+.007	—
	8017.300	− 10481	−.005	—
	8023.247	− 10471	+.001	—
Jacchia /283/	8446.261	− 9759	−.002	—
Chuprina /164/	33540.301	− 1185	+.007	—
	3834.392	− 690	+.006	—
	3959.348	− 648	+.008	—
Batyrev /15/	4244.336	0	+.003	—
	4272.254	47	−.003	—
Alaniya /1/	4579.431	+ 564	+.011	—
Tsesevich (Odes.)	6397.449	+ 3624	+.006	—
Lange	6791.342	+ 4287	−.007	—
	6819.271	+ 4334	−.001	—

Comparison stars used in the evaluation of Odessa plates (Table 221, Figure 100): $k = - 9^s.5;$ $a = 0^s.0;$ $b = 10^s.1;$ $c = 15^s.1.$ Reduction to one period according to Batyrev's equation gave the average light curve (Table 222). The O−A residues were calculated from (160), O−B from (16

TABLE 221. Odessa observations (Tsesevich)

JD hel	s	JD hel	s	JD hel	s
243...		243...		243...	
6041.494	12.6	6081.377	−1.5	6381.475	1.0
6049.474	7.1	6082.371	12.4	6395.482	9.1
6050.483	−2.0	6083.377	10.1	6396.456	7.1
6051.464	10.1	6101.313	9.5	6397.447	−2.5
6053.449	0.0	6104.294	10.1	6398.464	9.0
6069.393	13.1	6105.270	6.1	6399.448	11.1
6070.402	10.1	6128.215	12.8	6400.473	0.5
6071.397	5.5	6129.229	13.6	6401.475	13.4
6075.378	9.0	6131.223	10.1	6402.473	11.1
6076.381	8.5	6138.211	12.6	6404.461	11.1
6078.377	0.0	6371.490	6.0	6406.464	0.0
6079.360	9.0	6372.482	0.0	6407.464	11.6
6080.415	7.6	6379.508	16.1	6408.359	11.6

TABLE 221 (continued)

JD hel	s	JD hel	s	JD hel	s
243...		243...		243...	
6410.448	12.6	6451.347	8.7	6481.250	0.0
6422.394	—4.5	6453.370	0.0	6482.248	8.5
6423.408	10.1	6454.385	13.2	6484.256	—1.0
6424.411	8.1	6455.396	11.6	6485.261	12.1
6425.430	0.0	6461.363	11.4	6487.282	5.5
6426.415	12.8	6462.347	5.6	6488.286	12.6
6428.421	1.6	6463.344	7.3	6490.263	4.0
6429.453	13.4	6465.346	6.7	6518.195	4.6
6432.431	13.1	6478.238	—1.0		
6434.394	4.6	6479.269	8.7		

TABLE 222. Average light curve

φ	s	n	φ	s	n	φ	s	n
0.010^p	—1.5	4	0.417^p	+ 8.9	3	0.855^p	+12.8	5
.096	—0.1	4	.552	+11.2	4	.895	+12.2	5
.154	+1.4	5	.599	+ 9.8	4	.918	+ 8.2	2
.222	+5.3	4	.686	+10.9	5	.976	0.0	2
.282	+6.0	5	.735	+ 9.5	5	.992	— 2.8	2
.366	+10.0	4	.805	+13.2	4			

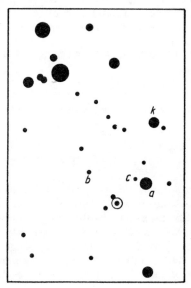

FIGURE 100. Comparison stars for BN Vulpeculae.

CX Lyrae

Observed by Zverev /38/ (visual) and Ahnert /177/ (photographic).
Magnitude estimated by Tsesevich from Odessa photographs taken in
1957 —1958; the observed maximum was found to precede the ephemeris
by a few hours. Moscow and Harvard photographs were then re-examined

and new visual observations made. Complete revision of the observation data showed that in 50 years the period changed substantially.

First the times of magnitude enhancement were picked out. Unfortunately in the old Moscow negatives the star is observed at varying distances from the maximum. The maxima could nevertheless be estimated with fair accuracy. The times of brightness enhancement and several exact maxima are listed in Table 223.

TABLE 223. List of enhancements and maxima

Source	Effect	Max hel JD	E	O — C
Tsesevich (Mosc.)	Enhancement	2415288.26	— 20121	—0.14^d
	Ditto	8927.33	— 14220	— .03
Zverev /38/	Maximum	27696.396	0	.000
Tsesevich (Mosc.)	Enhancement	8750.33	+ 1708	+ .05
	Ditto	9367.52	+ 2710	— .05
		9488.40	+ 2906	—. 04
Ahnert /177/	Maximum	9750.520	+ 3331	.000
	Ditto	9752.370	+ 3334	.000
		9784.440	+ 3386	+. 003
Tsesevich (Mosc.)	Enhancement	33775.41	+ 9858	— .11
	Ditto	4476.53	+ 10995	— .14
Tsesevich (Odes.)	Maximum	6489.28	+ 14259	— .20

The plot of O —C residues (Figure 101) shows that the period changed three times in a jump. The seasonal curves were derived using three different equations. Between JD 2414881 and JD 2420005

$$\text{Max hel JD} = 2415288.265 + 0.616689 \cdot E; \ P^{-1} = 1.6215629; \tag{162}$$

between 2427712 and 2430607

$$\text{Max hel JD} = 2427696.396 + 0.616669 \cdot E; \ P^{-1} = 1.6216155; \tag{163}$$

after 2432000

$$\text{Max hel JD} = 2429784.440 + 0.616650 \cdot E; \ P^{-1} = 1.6216655. \tag{164}$$

The seasonal curves cover very long time intervals, but the internal consistency of observations is satisfactory. Some average maxima were determined. The list of maxima (Table 224) includes the determinations of Makarenko from Zverev's observations; the data of Oliinyk /84/ obtained from the L'vov photographs; the maximum of Lange determined from new visual observations. The Blazhko effect is very prominent.

The light variation of the star is thus described by three linear relations before JD 2420005

$$\text{Max hel JD} = 2418596.111 + 0.616689 \cdot E_1; \tag{165}$$

from JD 2427712 to JD 2430607

$$\text{Max hel JD} = 2427696.396 + 0.616669 \cdot E_2; \tag{166}$$

after JD 2433069

$$\text{Max hel JD} = 2433069.365 + 0.61664495 \cdot E_3. \tag{167}$$

188

For the last equation $P^{-1} = 1.62167873$. It was used to calculate the average light curve from Odessa photographic observations (Table 225). The O−C residues were calculated from (166), O−D residues from (167).

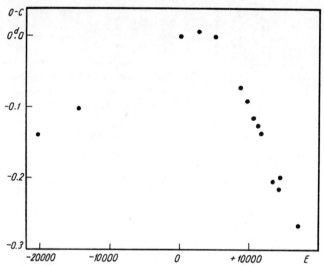

FIGURE 101. O−C residues for CX Lyrae.

TABLE 224. List of maxima

Source	Max hel JD	E	O − C	O − D
Tsesevich (Mosc.)	2415162.40:	− 20325	− 0.20:	—
	5288.26:	− 20121	− .14:	—
	5616.34:	− 19589	− .13:	—
	8566.30:	− 14805	− .31:	—
	8596.111	− 14757	− .101	—
	8888.44:	− 14283	− .07:	—
	8927.33:	− 14220	− .03:	—
Zverev	27551.490	− 235	+ .011	—
	7598.363	− 159	+ .017	—
	7662.465	− 55	− .014	—
	7677.266	− 31	− .013	—
	7678.512	− 29	− .001	—
	7696.393	0	− .003	—
	7916.517	+ 357	− .030	—
	7929.470	+ 378	− .027	—
	7934.405	+ 386	− .025	—
Tsesevich (Harv.)	8560.349	+ 1401	.000	—
Tsesevich (Mosc.)	9367.575	+ 2710	+ .006	—
Tsesevich (Harv.)	9684.525	+ 3224	− .012	—
Ahnert	9752.370	+ 3334	.000	—
Tsesevich (Harv.)	32891.759	+ 8425	− .073	—
Oliinyk	3069.359	+ 8713	− .074	−0.006
	3853.744	+ 9985	− .092	+ .007
Tsesevich (Mosc.)	3858.661:	+ 9993	− .108:	− .010:
Oliinyk	4232.978	+ 10600	− .111	+ .002
	4620.232	+ 11228	− .124	+ .005
Tsesevich (Mosc.)	4982.204	+ 11815	− .136	+ .006
Oliinyk	6095.841	+ 13621	− .203	− .017
Tsesevich (Odes.)	6489.280	+ 14259	− .199	+ .003
Tsesevich (visual)	6730.382	+ 14650	− .215	− .004
Lange (visual)	8281.255	+ 17165	− .264	+ .007

TABLE 225. Average photographic light curve

φ	s	n	φ	s	n	φ	s	n
p 0.035	1.9	5	p 0.439	11.4	5	p 0.856	11.4	5
.088	3.5	5	.520	11.9	6	.930	5.6	3
.163	5.3	5	.597	12.4	5	.943	2.8	4
.223	7.1	5	.665	11.8	4	.974	2.3	2
.275	8.6	4	.726	11.7	5			
.360	10.7	5	.772	11.8	5			

Comparison stars (Tables 226—228, Figure 102): Odessa photographic scale $k = -5^s.5$; $a = 0^s.0$; $b = 8^s.8$; $c = 14^s.8$; Moscow photographic scale $k = 3^s.5$; $a = 0^s.0$; $b = 8^s.6$; $c = 16^s.5$; visual scale $a = 0^s.0$; $b = 8^s.8$; $c = 18^s.1$.

TABLE 226. Visual observations (Tsesevich)

JD hel	s	JD hel	s	JD hel	s
243...		243...		243...	
6702.447	15.7	6720.496	5.5	6730.377	3.3
.458	16.0	.500	4.7	.387	3.5
.478	16.0	.506	—1.0	.410	3.9
.486	16.0	.510	—1.2	6732.350	11.1
.498	16.0	6722.338	7.0	.359	10.7
6703.376	12.6	.345	6.5	6736.344	16.0
.388	12.2	.352	5.3	6748.370	10.9
.397	12.9	.360	5.1	.386	12.3
.416	12.9	.370	2.9	6806.302	5.6
.435	14.3	.378	3.5	6814.306	6.2
.465	14.0	.399	4.6	6815.289	14.4
6720.414	16.4	.415	6.6	.304	14.0
.428	16.6	.425	7.2	.314	14.7
.451	16.2	.446	9.8	6819.293	9.8
.471	14.7	.459	11.0	.301	7.7
.476	14.0	6729.327	9.8	.322	10.9
.482	12.5	6730.348	7.6	.366	11.8
.486	10.3	.359	6.8		

TABLE 227. Moscow observations (Tsesevich)

JD hel	s	JD hel	s	JD hel	s
Series S					
241...		241...		241...	
4881.409	12.0	6375.274	14.8	8927.335	4.8
5143.444	12.0	8230.240	15.5	9727.354	10.6
5288.265	4.8	8564.323	13.0	242...	
5290.266	11.7	8596.222	8.6	0005.335	6.4
6349.345	15.4	8923.320	16.5		
Series T					
242...		242...		243...	
7712.246	13.1	9495.337	8.6	4978.289	14.8
7750.178	5.7	9496.381	8.6	4980.302	6.5
7927.410	13.9	9521.290	5.4	4982.287	5.7
8043.306	13.9	243...		4982.393	12.6
8081.217	—1.8	0607.278	15.4	4983.281	13.1
8665.436	12.6	3775.413	1.9	4989.280	12.0
8750.338	4.9	3776.469	8.6	5011.259	5.7
8751.356	14.5	4121.494	8.6	5011.285	9.7
8776.296	10.9	4333.150	9.6	5012.264	9.7
8785.280	13.1	4480.401	7.8	5012.317	12.6
8789.301	8.6	4518.390	12.6	6071.388	13.1
9366.528	11.6	4610.362	4.9:	6071.415	8.6
9367.518	2.9	4628.254	—1.2	6072.334	12.6
9376.503	14.3	4630.329	15.4	6084.358	7.8
9488.407	4.3	4869.409	7.8		
Series A					
243...		243...		243...	
2034.552	13.1	4476.531	3.8	4480.436	10.4
4471.472	13.9	4477.460	13.6		
4472.544	14.9	4479.542	13.6		

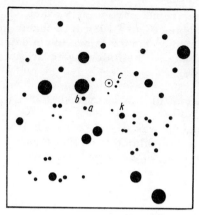

FIGURE 102. Comparison stars for CX Lyrae.

TABLE 228. Odessa observations (Tsesevich)

JD hel	s	JD hel	s	JD hel	s
243...		243...		243...	
6041.435	11.8	6345.506	8.7	6424.380	12.1
6043.475	2.4	6347.504	5.5	6425.403	12.6
6049.435	10.5	6364.462	12.7	6426.359	3.1
6050.449	11.2	6367.497	10.4	6428.393	7.5
6053.415	7.6	6371.462	0.8	6429.422	8.4
6067.326	12.0	6372.454	11.8	6430.399	12.4
6069.361	1.8	6373.460	5.7	6432.395	10.4
6070.362	10.7	6375.483	11.6	6434.363	6.6
6071.363	10.0	6376.462	3.9	6451.318	12.3
6074.421	9.4	6379.478	1.4	6453.307	12.4
6075.351	11.6	6381.449	4.9	6454.349	8.4
6076.349	10.6	6395.449	13.0	6455.315	3.9
6079.326	5.0	6396.424	9.2	6461.304	12.5
6080.347	11.4	6397.419	1.0	6462.275	4.8
6081.342	11.2	6398.435	11.6	6463.288	11.6
6082.342	1.1	6399.421	7.5	6465.286	4.4
6083.350	12.6	6400.446	2.4	6478.215	2.9
6101.283	13.0	6401.448	12.0	6479.243	11.4
6102.253	11.6	6406.436	12.6	6481.221	4.4
6103.264	0.5	6407.436	9.4	6482.225	12.1
6104.264	12.7	6408.431	12.4	6484.227	12.1
6105.239	8.6	6409.424	12.6	6488.261	11.5
6126.204	2.9	6410.421	3.4	6489.266	1.5
6344.519	9.6	6423.381	3.4	6490.214	12.1

AF Herculis

Discovered and first studied by Furujelm /236/. Parenago improved the elements using magnitudes from old Moscow photographs /86/. Tsesevich's visual observations and magnitudes from Odessa photographs give more detailed information on the behavior of this star. The maxima (Table 229) were determined from the seasonal curves.

The O −A residues were calculated from

$$\text{Max hel JD} = 2427663.330 + 0.630336 \cdot E; \ P^{-1} = 1.58645548. \tag{168}$$

191

Parenago's maxima are not particularly exact. It seems, however, that the period is variable (Figure 103); it probably increased in an abrupt jump. This is confirmed by the exact maximum determined by Furujelm.

The last seven maxima fit the following equation, which is applicable after JD 2426950 (O — B):

$$\text{Max hel JD} = 2427663.3095 + 0.63034554 \cdot E. \tag{169}$$

Up to that epoch the period was $0^d.630336$, i.e., it increased by $0^d.00000954$.

No light curves were calculated from Tsesevich's photographic and old visual observations. The 1962 observations gave a good average curve (Table 230).

TABLE 229. List of maxima

Source	Max hel JD	E	O — A	O — B
Parenago /86/	2417826.304	— 15606	-0.002^d	—
	8535.349	— 14481	— .085	—
	8888.366	— 13921	— .057	—
Furujelm /236/	22935.150	— 7501	— .030	—
Tsesevich (visual)	6952.281	— 1128	— .030	$+0^d.001$
	7663.310:	0	— .020:	.000
Tsesevich (Odes.)	36373.415	+ 13818	+ .102	— .009
	7077.517	+ 14935	+ .119	— .003
	7790.444	+ 16066	+ .136	+ .003
Tsesevich (visual)	7903.276	+ 16245	+ .138	+ .003
Tsesevich (Odes.)	8203.321	+ 16721	+ .143	+ .004

The maximum observed on different evenings shows definite differences. The shape of the light curve is apparently variable. It is impossible to decide whether this phenomenon is periodic or not.

TABLE 230. Average light curve

φ	s_2	n	φ	s_2	n	φ	s_2	n
$0^P.038$	25.6	5	$0^P.241$	8.1	5	$0^P.546$	20.0	5
.100	26.4	5	.272	9.9	4	.606	21.0	6
.124	23.8	5	.318	11.6	5	.682	20.8	5
.153	15.4	5	.364	11.9	5	.733	21.2	5
.171	12.0	5	.407	14.9	5	.786	22.3	5
.187	10.1	5	.448	17.9	5	.826	22.3	5
.208	8.8	6	.475	18.2	5	.892	22.3	5
.227	6.8	5	.509	18.1	5	.952	24.3	4

Comparison stars (Figure 104):

	s_1	s_2	s_3	m
u	0.0	—	0.0	11.94
k	4.4	0.0	14.5	12.39
a	14.2	8.4	22.7	12.88
b	22.6	11.8	25.9	13.04
c	31.7	16.9	—	—

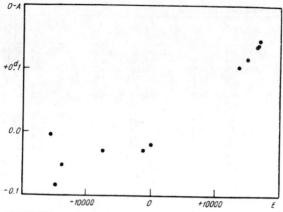

FIGURE 103. O—A residues for AF Herculis.

The last column gives Furujelm's stellar magnitudes. The power scale can be converted to photographic stellar magnitudes using the relation $m = 11.89 + 0.0426 \cdot s_3$.

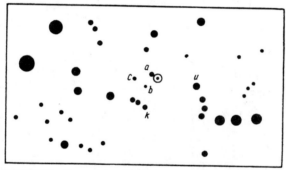

FIGURE 104. Comparison stars for AF Herculis.

Tables 231—233 list the observations of Tsesevich.

TABLE 231. Old visual observations (Tsesevich)

JD hel	s_2	JD hel	s_2	JD hel	s_2
242...		242...		242...	
6947.244	6.1	6949.324	10.7	6952.332	4.2
6948.232	14.8	.361	10.9	.342	5.0
.256	15.9	.371	11.8	.349	5.0
.262	14.9	6950.233	10.9	.362	5.0
.277	14.9	.259	12.8	.370	5.3
.288	14.9	.266	12.8	.377	6.2
.296	14.9	.282	13.8	6953.223	13.8
.305	14.9	.293	14.8	.260	14.9
.313	15.2	.305	14.9	6957.345	1.3
.324	15.2	.323	14.9	.353	1.4
.333	14.9	6952.228	11.8	6959.220	0.9
.342	13.8:	.246	9.4	6972.201	9.5
.360	13.8	.263	3.4	.256	10.9
6949.238	9.2	.271	0.9	.267	11.8
.242	10.7	.283	1.0	6973.212	9.5
.259	9.5	.290	1.5	.242	10.3
.273	10.7	.306	3.0	.262	11.8
.296	10.7	.311	3.4	.289	11.8
.306	10.7	.321	3.7	6977.214	13.8

193

TABLE 231 (continued)

JD hel	s₂	JD hel	s₂	JD hel	s₂
2426978.189	4.8	2427385.110	13.8	2427663.235	14.9
6980.216	11.8	7386.079	9.5	.258	14.6
6981.176	14.8	.087	9.5	.268	8.2
.187	14.9	.111	8.4	.272	6.5
.200	15.5	.121	6.7	.276	6.1
.221	14.9	7390.103	9.5	.281	5.8
.229	14.9	.108	9.5	.287	5.4
.248	10.1	.115	12.1	.296	5.9
.256	5.6	7393.087	9.5	.307	5.9
.261	4.4	.098	8.9	.327	5.8
.266	3.2	7424.523	10.3	.338	5.9
.270	2.5	.535	7.1	.351	6.0
.290	2.8	7425.533	12.8	7664.214	10.9
.296	4.7	7595.361	6.0	.221	10.9
6983.248	5.3	7654.249	14.9	.239	12.3
.258	5.3	.315	14.9	7869.370	13.8
.264	5.9	.364	13.3	.374	10.7
6984.196	14.9	7660.223	4.5	.377	9.8
.202	14.9	.232	4.8	.388	8.4
.209	14.9	.255	6.0	.393	7.5
.217	14.9	.289	7.4	.402	6.5
.225	15.1	7662.188	7.4	.415	3.4
.247	14.9	.196	7.5	.424	3.5
6985.213	9.2	.215	9.8	.435	3.2
7384.122	6.3	.232	10.7	.442	3.9
.128	6.3	.300	10.7	.463	4.6
7385.068	13.8	.351	12.3	.472	4.5
.078	14.9	7663.192	15.3	.492	5.0
.087	10.7	.198	14.9	.518	5.4
.097	11.8	.218	15.2		

TABLE 232. Visual observations, 1962 (Tsesevich)

JD hel	s₁	JD hel	s₁	JD hel	s₁
243...		243...		243...	
7854.438	23.6	7873.435	20.5	7908.280	12.6
7847.353	10.1	7899.295	20.1	.291	10.4
.360	10.6	.314	24.6	.299	10.4
.391	12.5	7900.278	17.9	.310	10.3
.412	17.8	.284	17.4	.320	9.7
.424	19.0	.301	16.1	.326	10.1
.465	17.0	.345	19.0	.330	10.0
7868.338	21.8	7901.269	25.6	.339	10.4
.344	20.1	.316	22.6	7910.263	12.6
.362	24.6	.357	10.9	.287	11.7
.375	21.7	.370	8.8	.302	13.6
7869.358	18.9	7902.274	19.4	.325	14.2
.367	17.7	.286	19.1	.346	17.9
.374	17.9	.299	19.8	.356	17.9
.393	18.9	.325	23.6	.401	19.4
.400	19.2	.357	24.6	7911.272	24.6
.420	19.4	.382	24.6	.279	25.6
.433	20.5	7903.268	9.9	.287	25.6
7871.354	20.7	.276	2.6	.302	25.6
.366	20.4	.281	2.2	.338	25.6
.389	19.8	.287	2.2	7912.270	16.6
.421	20.9	.294	8.1	.276	17.8
.440	20.9	.304	10.5	.285	18.0
7872.328	24.1	.321	11.0	.301	19.2
.357	12.6	.337	12.3	.314	24.1
.362	12.3	.354	12.6	.346	24.6
.380	10.0	.374	12.6	7913.274	26.6
.393	10.1	7904.268	20.2	.287	28.7
.400	9.7	.295	24.6	.292	28.7
.417	10.1	7906.288	21.3	.298	28.4
.443	10.3	.302	25.6	.314	20.5
.459	11.3	.320	25.6	.318	13.6
.490	11.3	.340	25.6	.322	12.9
7873.326	20.7	.361	25.6	.329	12.9
.334	20.4	.391	17.4	.337	10.6
.350	20.1	.399	11.3	.344	9.6
.358	21.1	.412	10.4	.356	10.3
.374	20.4	7908.245	25.6	.370	2.0
.395	20.7	.261	21.5	.395	7.9
.424	20.9	.269	19.4		

TABLE 233. Odessa observations (Tsesevich)

JD hel	s_3	JD hel	s_3	JD hel	s_3
243...		243...		243...	
6286.559	12.9	6756.399	22.7	7789.477	21.9
6313.535	10.0	6757.360	11.6	7790.400	17.8
6314.555	20.3	6758.361	24.5	7790.450	11.2
6322.545	17.5	6758.387	24.5	7790.498	12.0
6336.536	20.9	6759.355	20.6	7793.422	25.9
6339.472	14.5	6759.383	18.8	7793.477	26.9
6344.472	12.7	6761.377	19.6	7793.523	26.9
6345.437	20.3	6781.329	13.4	7794.442	20.6
6347.438	22.7	7073.434	24.3	7810.405	24.8
6351.487	15.5	7075.461	23.8	7810.451	21.9
6362.409	24.3	7077.393	24.3	7812.360	24.8
6364.407	24.6	7077.449	21.5	7812.410	25.9
6367.421	24.6	7078.385	19.4	7813.386	20.3
6371.403	26.9	7078.433	23.7	7813.434	24.6
6372.395	16.8	7079.364	17.2	7814.441	12.9
6373.402	7.7	7079.414	8.3	7817.408	24.8
6376.361	21.7	7079.464	9.7	7817.458	24.3
6391.393	21.8	7080.397	23.8	7818.392	21.2
6392.363	14.5	7080.448	24.1	7818.440	21.8
6395.386	23.7	7087.451	24.5	7821.363	11.6
6396.363	26.9	7099.358	20.6	7821.420	14.0
6397.364	10.5	7101.397	22.7	7839.379	24.7
6398.374	24.8	7102.404	22.7	7840.380	14.5
6399.366	14.5	7102.453	21.2	7840.412	20.0:
6400.357	27.9	7105.454	24.3	7844.415	22.7
6401.350	22.7	7107.367	19.5	7847.391	21.7
6404.353	10.5	7107.420	21.1	7849.413	24.8
6405.377	24.6	7111.367	20.3	7850.368	11.6
6406.349	16.5	7111.393	23.7	7850.423	12.7
6407.339	26.4	7111.421	24.5	7851.360	22.7
6408.349	21.8	7111.449	25.9	7881.361	15.8
6420.326	24.3	7128.351	26.9	7884.388	11.6
6422.328	24.7	7137.364	13.4	8143.508	14.5
6424.302	25.9	7137.393	8.7	8144.512	24.8
6428.306	13.1	7373.625	24.6	8164.431	20.3
6429.319	24.8	7378.595	21.9	8165.432	24.8
6432.284	21.2	7402.536	24.8	8170.444	27.9
6453.271	25.9	7404.549	25.1	8173.462	24.8
6454.283	20.0	7405.544	20.3	8194.391	24.8
6455.278	24.5	7427.475	14.0	8195.371	20.3
6661.551	12.8	7454.395	21.1	8199.400	26.9
6662.571	24.8	7457.409	21.9	8200.418	20.3
6667.539	21.2	7458.390	18.0	8203.397	11.6
6668.559	17.9	7462.422	26.9	8204.417	24.3
6744.355	24.5	7464.420	25.9	8223.347	24.3
6749.361	18.6	7406.552	9.4	8225.370	10.9
6749.403	24.7	7488.359	24.3	8114.506	14.5
6755.422	10.9	7764.551	23.5	8114.596	17.6
6756.342	24.3	7789.424	19.5		

AG Herculis

Discovered by Furujelm /236/, who also determined its elements. Parenago /86/ subsequently improved Furujelm's equation using Moscow photographs and reported 4 maxima. Observed visually by Tsesevich for several seasons and using Odessa photographs. Comparison stars (Figure 105):

	s_1	s_2	s_3	m
m	0.0	0.0	—	—
p	—	—	0.0	11.86
q	—	—	5.6	—
s	—	13.0	17.0	12.87
t	10.0	17.5	12.6	12.86
w	15.2	28.4	—	—

195

The last column gives Furujelm's photographic stellar magnitudes. The power scale, however, does not correlate with these magnitudes, and the main reduction was therefore carried out in powers. All the observations were combined to give seasonal brightness curves, from which the maxima were determined (Table 234). The general summary also includes the maxima from other sources.

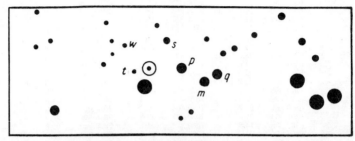

FIGURE 105. Comparison stars for AG Herculis.

TABLE 234. List of maxima

Source	Max hel JD	E	O — A	O — B	O — C
Parenago /86/	2418210.274	— 14554	$+0^d.010$	$0^d.000$	—
Furujelm /236/	22935.584	— 7278	$+.007$	.000	—
Tsesevich (visual)	6952.361	— 1093	$+ .009$	---	$-0^d.003$
	7662.215	0	$+ .027$	---	$+ .006$
Tsesevich (photogr.)	36367.381	+ 13404	$+ .123$	—	$— .009$
	6756.408	+ 14003	$+ .136$	---	.000
	7378.576	+ 14961	$+ .143$	—	$— .002$
	7810.462	+ 15626	$+ .153$	—	$+ .002$
Tsesevich (visual)	7903.326	+ 15769	$+ .147$	—	$— .005$
Tsesevich (photo.)	8164.422	+ 16171	$+ .169$	—	$+ .014$

The O — A residues were calculated from

$$\text{Max hel JD} = 2427662.188 + 0.6494382 \cdot E; P^{-1} = 1.53979239. \qquad (170)$$

The variation of O — A residues (Figure 106) shows that the period increased abruptly, jumping by $0^d.0000087$. Before JD 2426952, Parenago's formula applied:

$$\text{Max hel JD} = 2422935.584 + 0.6494378 \cdot E. \qquad (171)$$

After JD 2426952,

$$\text{Max hel JD} = 2427662.209 + 0.6494465 \cdot E. \qquad (172)$$

The visual observations of 1962 were reduced to one period using (170). They were used to derive the average light curve (Table 235), which gives a fairly faithful representation of light variation. Tsesevich's observations are listed in Tables 236 — 238. The O — B and O — C residues were calculated from (171) and (172), respectively.

196

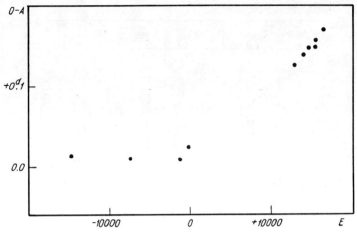

FIGURE 106. O—A residues for AG Herculis.

TABLE 235. Average light-variation curve.

φ	s_2	n	φ	s_2	n	φ	s_2	n
$0^P.012$	24.8	5	$0^P.264$	7.6	4	$0^P.641$	23.7	5
.034	25.5	5	.377	10.7	4	.679	24.0	5
.050	24.9	5	.462	15.5	5	.742	24.5	5
.085	26.0	5	.512	18.3	5	.821	25.3	5
.124	25.2	4	.545	20.1	5	.881	25.6	5
.146	17.0	4	.569	21.0	5	.925	23.5	5
.183	8.7	5	.599	23.0	5	.953	24.2	4
.212	7.2	4	.619	23.3	5	.989	23.5	4

TABLE 236. Old visual observations (Tsesevich)

JD hel	s_1	JD hel	s_1	JD hel	s_1
242...		242...		242...	
6947.244	8.4	6950.293	13.1	6973.242	9.0
6948.232	8.9	.305	14.2	.330	11.0
.257	13.1	.323	14.7	6977.215	12.1
.263	13.1	6952.227	14.0	6980.216	14.3
.277	13.1	.247	14.2	6981.176	15.2
.288	13.1	.263	14.3	.191	13.6
.295	13.8	.284	13.1	.200	13.5
.304	13.5	.290	13.3	.220	13.1
.313	13.8	.311	13.1	.228	13.1
.324	13.8	.320	10.0	.248	13.1
.333	13.8	.333	8.9	.256	13.8
.360	14.2	.341	6.8	.268	14.2
6949.238	13.5	.348	6.1	.296	13.5
.245	10.0	.362	4.0	6983.249	13.8
.259	13.1	.369	4.5	.264	13.8
.274	11.4	.377	5.6	6984.196	6.4
.296	12.8	6953.223	10.0	.202	7.3
.306	12.7	.247	11.0	.209	7.5
.324	12.8	.261	11.9	.218	8.2
.360	13.3	6957.345	13.8	.225	7.4
.371	13.1	.353	13.8	.248	8.0
6950.233	14.7	6972.202	13.1	6985.213	14.2
.259	14.2	.256	13.1	7384.121	13.0
.266	13.0	.267	13.5	.128	13.0
.282	14.2	6973.212	8.0	7385.068	9.0

197

TABLE 236 (continued)

JD hel	s_1	JD hel	s_1	JD hel	s_1
242...		242...		242...	
7385.078	9.5	7660.224	7.4	7663.328	13.1
.087	9.5	.233	7.1	7664.215	9.5
.097	10.0	.265	7.8	.222	8.2
.110	10.0	.289	8.3	.229	8.3
7386.079	11.0	7662.188	5.7	.240	9.5
.087	12.1	.196	6.0	7869.374	3.0
.112	14.3	.215	6.8	.377	3.0
.121	13.1	.232	8.0	.388	3.8
7390.104	6.0	.264	8.4	.393	3.6
.108	5.7	.300	10.0	.403	4.3
.116	6.0	.351	10.0	.416	5.5
7393.087	12.3	7663.193	13.6	.424	5.5
.099	13.1	.198	13.1	.435	5.8
7424.525	5.5	.218	13.4	.443	5.5
.535	3.8	.235	13.5	.463	7.4
.544	3.2	.258	13.5	.473	7.4
.552	3.9	.281	13.4	.492	7.8
7595.363	7.0				

TABLE 237. Visual observations, 1962 (Tsesevich)

JD hel	s_2	JD hel	s_2	JD hel	s_2
243...		243...		243...	
7854.439	22.5	7873.424	10.2	7908.243	25.2
7857.354	19.5	.435	6.9	.261	25.6
.361	20.5	.445	7.3	.270	24.9
.391	19.0	.463	7.6	.280	25.2
.412	18.5	7899.295	23.8	.299	25.2
.425	22.5	.313	24.2	.311	25.8
.465	21.5	7900.278	23.8	.320	25.5
7868.337	10.6	.284	23.8	7910.263	25.2
.343	9.8	.301	23.8	.287	24.6
.362	11.2	.345	24.2	.302	25.5
.375	11.2	7901.269	24.9	.325	25.6
7869.357	20.5	.317	23.8	.346	24.9
.367	21.5	.357	8.7	.356	25.2
.373	22.5	.370	7.0	.401	25.2
.393	22.5	7902.274	24.2	7911.272	10.4
.399	22.8	.286	24.2	.279	14.5
.419	24.6	.299	23.8	.287	12.4
.432	25.2	.325	24.9	.302	15.2
7871.354	24.2	.357	25.2	.338	21.0
.366	24.9	.382	24.9	7912.270	24.6
.388	25.2	7903.268	16.4	.276	25.2
.421	26.4	.277	15.2	.285	25.2
.440	26.6	.281	11.0	.301	25.2
7872.328	14.8	.287	8.0	.314	25.6
.357	15.2	.294	8.7	.346	25.2
.362	15.7	.304	7.8	7913.274	23.5
.380	19.8	.321	7.4	.287	19.6
.393	21.4	.337	7.4	.293	20.7
.400	23.0	.354	7.8	.298	23.0
.417	23.0	.374	7.8	.314	23.3
.443	22.3	7904.268	24.2	.318	23.4
.459	23.8	.295	23.8	.322	23.8
7873.326	26.0	7906.288	24.9	.329	23.7
.334	26.0	.302	25.4	.337	23.8
.350	25.3	.320	25.5	.344	23.8
.358	26.2	.340	26.0	.356	24.2
.374	26.4	.361	25.8	.370	24.2
.395	25.6	.391	26.2		

TABLE 238. Odessa observations (Tsesevich)

JD hel	s	JD hel	s	JD hel	s
243...		243...		243...	
6286.559	15.9	6756.399	−2.0	7789.476	15.2
6313.535	3.7	6757.360	12.6	7790.400	7.0
6314.555	15.9	6758.361	2.8	7790.450	7.3
6322.545	2.5	6758.387	1.9	7790.498	8.6
6336.536	10.3	6759.355	12.6	7793.422	15.5
6339.472	−1.5	6759.383	14.8	7793.477	15.2
6344.472	14.1	6761.377	11.7	7793.523	11.6
6345.437	8.6	6781.329	12.4	7794.442	10.5
6347.438	12.6	7073.434	8.7	7810.405	9.9
6351.487	14.8	7075.461	9.3	7810.451	−3.5
6362.409	10.0	7077.393	9.8	7812.360	8.8
6364.407	12.6	7077.449	9.3	7812.410	0.0
6367.421	−1.5	7078.385	15.2	7813.386	14.1
6371.403	8.8	7078.434	13.6	7813.434	15.5
6372.395	15.2	7079.364	9.7	7814.441	5.6
6373.402	10.3	7079.414	11.8	7817.408	12.6
6376.361	14.8	7079.464	14.4	7817.458	14.6
6391.393	−2.0	7080.397	15.5	7818.392	7.2
6392.363	13.7	7080.448	8.9	7818.440	9.7
6395.386	6.5	7087.451	15.6	7821.363	17.0
6396.363	15.1	7099.358	4.5	7821.420	14.6
6397.364	4.5	7101.397	7.6	7839.379	11.2
6398.374	15.9	7102.404	14.8	7840.380	3.2
6399.366	9.1	7102.453	15.2	7840.412	3.7
6400.357	12.2	7105.454	11.5	7844.415	7.9
6401.350	9.6	7107.367	12.6	7847.391	17.0
6404.353	7.0	7107.420	14.4	7849.413	1.4
6405.377	15.1	7111.367	15.5	7850.368	12.6
6406.349	2.8	7111.393	14.4	7850.423	12.6
6407.339	12.6	7111.421	15.2	7851.360	1.5
6408.349	3.4	7111.449	12.6	7881.361	4.9
6420.326	15.2	7128.351	12.1	7884.388	12.6
6422.328	15.2	7137.364	11.4	8114.505	9.1
6424.302	14.4	7137.393	11.8	8114.596	9.1
6428.306	15.5	7373.625	7.3	8143.508	9.1
6429.319	12.1	7378.595	−2.0	8144.512	10.3
6432.284	7.2	7402.536	15.9	8164.431	−2.5
6453.271	7.9	7404.549	−5.0	8165.432	11.4
6454.283	15.6	7405.544	12.6	8170.444	8.6
6455.278	10.7	7406.552	2.5	8173.462	5.6
6661.551	3.4	7427.475	7.9	8194.391	3.7
6662.571	14.1	7454.394	14.4	8195.371	9.6
6667.539	6.6	7457.409	8.7	8199.400	15.5
6668.559	14.3	7458.390	14.4	8200.418	10.3
6744.355	14.1	7462.422	4.1	8203.396	−0.5
6749.361	7.3	7464.420	2.6	8204.417	14.8
6749.403	6.8	7488.359	0.0	8223.347	12.6
6755.422	15.2	7764.551	8.1	8225.370	15.5
6756.342	10.9	7789.424	8.4		

X Arietis

Discovered by Leavitt /333, 334/. Studied by Hoffmeister /256/,
Nijland /325/, Robinson /346/, and others. The old observations reveal
remarkable constancy of the period. However, the relatively few observations of Latyshev made in 1957 point to considerable variations in the
period. Tsesevich collected all the available data on this star. The magnitudes were estimated from Simeise planetary photographs (where the
star's image is unfortunately overexposed). After that, the 1930−1931
visual observations were reduced. At Tsesevich's request, V. Karamysh
estimated the magnitude of this variable using photographs obtained with
the seven-camera astrographs. The seasonal light curves were then
derived from these data. The table of maxima was compiled using the
seasonal curves and all the published data (Table 239).

TABLE 239. List of maxima

Source	Max hel JD	E	O — A
Tsesevich (Sim.)	2420461.388	— 498	$+0^d.013$
Robinson /346/	0727.065	— 90	+ .031
Hoffmeister /256/	2420785.620	0	— .015
Payne—Gaposchkin,			
Gaposchkin /329/	0785.662	0	$+0^d.027$
Tsesevich (Sim.)	4825.200:	+ 6204	— .013:
Nijland /325/	6023.276	+ 8044	— .007
Tsesevich (visual)	6250.516	+ 8393	— .009
Lange /66/	6331.256	+ 8517	— .009
Tsesevich (visual)	6590.400	+ 8915	— .013
Tsesevich (Sim.)	6678.331	+ 9050	+ .017
Florya /125/	6991.505	+ 9531	.000
Gur'ev /27/	8068.416	+ 11185	.000
	8427.893	+ 11737	+ .006
Tsesevich (Sim.)	33210.503:	+ 19082	+ .105
Latyshev	6195.327	+ 23666	+ .172
Karamysh	7583.568	+ 25798	+ .235

The O — A residues were calculated from the equation

$$\text{Max hel JD} = 2420785.635 + 0.6511248 \cdot E. \qquad (173)$$

The last two maxima fit the equation

$$\text{Max hel JD} = 2437583.568 + 0.651139 \cdot E'. \qquad (174)$$

The period thus increased by $0^d.0000142$. Comparison stars:

		s_1	s_2
k	BD + 9°407	—	0.0
a	BD + 9 401	0.0	7.2
b	BD + 9 402	11.6	16.6
c	BD + 9 404	13.8	—

The average light curve was derived from visual observations (Table 240). The visual observations of Tsesevich are given in Table 241, of Latyshev in Table 242. Simeise photographic observations are given in Table 243.

TABLE 240. Average light curve

φ	s_1	n	φ	s_1	n	φ	s_1	n
$0^P.009$	1.8	5	$0^P.411$	9.3	5	$0^P.933$	4.6	5
.026	1.6	5	.448	10.7	4	.948	4.6	5
.051	2.4	5	.583	9.7	5	.960	2.4	5
.078	2.2	5	.698	10.4	5	.970	1.2	5
.115	4.5	5	.788	10.8	5	.984	1.2	5
.188	6.4	4	.895	10.4	4	.994	1.9	4
.368	9.6	5						

TABLE 241. Visual observations (Tsesevich)

JD hel	s_1	JD hel	s_1	JD hel	s_1
242...		242...		242...	
6011.271	8.3	6234.503	10.4	6239.471	1.7
.293	8.1	.520	7.6	.488	2.0
.319	8.5	.532	10.8	.518	3.9
6025.252	1.3	6239.422	4.2	.538	5.8
.272	1.7	.436	2.8	.552	6.2
.287	2.6	.448	1.9	6250.387	10.0
6216.414	10.6	.459	1.5	.413	10.8

200

TABLE 241 (continued)

JD hel	s_1	JD hel	s_1	JD hel	s_1
242...		242...		242...	
6250.454	8.4	6322.276	6.1	6590.367	5.4
.481	2.5	.313	8.1	.375	4.2
.494	1.5	.372	9.1	.391	1.6
.502	0.7	.416	9.2	.399	0.8
.519	0.5	.440	10.0	.408	2.0
.533	1.4	6324.399	12.0	.447	2.8
.545	1.4	.454	13.1	.463	3.1
.568	1.4	6325.193	11.0	.493	3.4
6254.396	4.2	.198	10.4	6592.328	6.1
.408	3.2	.214	10.4	.335	5.4
.424	1.4	.251	9.9	.340	1.2
.438	1.9	.279	11.3	.346	0.9
6282.357	10.6	6326.313	7.6	.351	0.0
.379	4.8	6336.298	10.0	.357	0.3
.411	0.7	6541.419	12.1	.369	0.9
.427	2.1	6588.401	11.8	.388	1.3
.442	2.1	.422	7.7	.408	2.4
.490	2.1	.429	3.8	.419	1.8
6298.274	10.7	.434	2.9	.433	3.4
.328	12.3	.446	1.5	.468	5.3
6305.475	7.8	.457	3.0	6595.425	10.0
.490	10.0	.469	3.1	6598.473	9.9
.515	10.1	.482	3.0		
6322.233	6.1	6590.342	11.0		

TABLE 242. Visual observations (Latyshev)

JD hel	m	JD hel	m	JD hel	m
243...		243...		243...	
6113.544	9.70	6135.337	9.56	6246.175	8.81
6114.417	9.32	6137.474	8.87	.219	9.04
.459	9.50	.552	9.28	6247.132	9.30:
6115.399	9.35	6138.247	9.28:	.173	9.40
.496	9.30	.485	9.75:	.215	9.76
6128.316	8.92	6139.474	9.34	6248.129	8.70
.332	8.81	6168.439	9.40	.171	8.92:
.334	8.85	6184.228	8.92:	6249.223	9.30:
.352	8.92:	6188.220	8.80	6251.150	9.28
6132.386	9.67:	6195.312	8.65	6254.132	9.30
6133.292	9.28:	6197.264	8.56	6276.158	8.87
6134.241	9.24	6242.220	8.87	6279.155	9.28:
.298	9.16	6246.132	8.65	6280.154	9.04

TABLE 243. Simeise observations

JD hel	s_2	JD hel	s_2	JD hel	s_2
241...		242...		242...	
8986.493	12.4	4449.459	5.1	6329.322	3.9
	9.7	4822.289	12.1	6625.529	8.7
9259.437	4.3		13.5		7.2
242...		4823.509	13.9	6655.416	14.2
0447.291	12.8		12.1	6655.509	10.3
	8.7	4825.317	7.2	6655.462	10.3
0461.385	—1.0		7.2	6678.408	4.5
	4.1	4826.281	15.3	7418.426	12.4
0462.270	12.6		14.5		18.6:
	11.6	4828.341	14.1	8482.467	8.4
0811.376	14.7		14.7		11.6
	12.4	4845.239	14.1	9228.302	11.5
1170.535	8.4		13.8		10.6
	7.2	4854.334	13.1	9935.473	11.6
1518.567	14.0		10.0		6.2
	13.5	5186.456	13.8	243...	
3736.275	9.1	5918.482	5.8	3210.503	2.4
	9.3		5.1		5.8
3737.514	8.4	6269.538	8.8	3238.350	15.4
	11.9		14.0		13.5
4439.506	11.5	6307.454	13.1	4687.462	10.6
	14.2	6329.322	4.2		9.3
4449.459	4.5				

AW Draconis

Studied by Tsesevich in 1957 using Odessa and Moscow photographs. The elements were tentatively determined from

$$\text{Max hel JD} = 2436079.32 + 0.6871977 \cdot E; \ P^{-1} = 1.45518531, \qquad (175)$$

and then seasonal light curves were constructed, from which highly reliable maxima were determined. These maxima gave the equation

$$\text{Max hel JD} = 2436075.224 + 0.6871941 \cdot E; \ P^{-1} = 1.45519294. \qquad (176)$$

New photographs have become available during the six years since then. The new seasonal curves gave additional maxima. A complete list of maxima is listed in Table 244 (which includes the maxima obtained from Tsesevich's Harvard observations).

Equation (176) fits the new observations, but fails to describe the earlier epochs. This is due to secular variation of the period. The average light curve for all the observations between 1937 and 1957 was derived from these observations (Table 245, old observations omitted). Another average curve was derived from observations made with the seven-camera astrograph (Table 246).

TABLE 244. List of maxima

Source	Max hel JD	E	O — B
Tsesevich (Mosc.)	2415251.344:	—30302	—0^{d}.522:
	6734.397:	—28144	—. 436:
	7554.181	—26951	—. 475
	7793.367:	—26603	—. 432:
Tsesevich (Harv.)	9160.784:	—24613	—. 531:
	21132.728:	—21744	—. 147
	7458.494	—12539	—. 003
	9142.799	—10088	—. 011
Tsesevich (Mosc.)	30580.416	— 7996	—. 004
Tsesevich (Odes.)	4331.143	—2538	+. 018
	5363.281	—1036	—. 010
Tsesevich (Odes.)	6075.221	0	+. 003
Tsesevich (visual)	6424.325	+ 508	+. 006
Tsesevich (Odes.)	6806.400	+1064	+. 001
	7173.367	+1598	+. 007
	7496.348	+2068	+. 007
	7878.428	+2624	+. 007
	8234.402	+3142	+. 014

TABLE 245. Average light curve, 1937—1957 observations (Figure 107, 1)

φ	s	n	φ	s	n	φ	s	n
0^{p}.017	2.8	5	0^{p}.367	11.6	5	0^{p}.805	16.8	5
.040	3.2	5	.400	12.3	5	.837	15.4	5
.083	5.0	5	.454	11.9	5	.872	15.1	3
.159	6.9	5	.493	14.5	5	.912	7.5	4
.211	9.7	5	.570	14.9	5	.943	5.4	3
.267	9.5	5	.669	14.9	5	.984	3.0	3
.336	11.7	5	.727	14.9	5			

TABLE 246. Average light curve, 1957—1962 observations (Figure 107, 2)

φ	s	n	φ	s	n	φ	s	n
0^P.012	1.5	5	0^P.361	10.6	10	0^P.836	15.3	10
.059	2.4	5	.398	11.6	10	.859	14.4	10
.082	2.8	5	.442	12.6	10	.882	13.1	5
.105	4.0	5	.501	12.7	10	.898	14.5	5
.148	5.6	5	.577	13.1	10	.921	9.5	5
.182	6.3	5	.625	13.2	10	.931	9.2	5
.221	8.1	10	.685	14.2	10	.952	7.2	5
.257	8.9	10	.728	14.4	10	.968	5.3	5
.292	10.5	10	.769	15.0	10	.987	2.4	5
.320	10.5	10	.804	14.5	10			

The visual observations fit (176). Comparison stars (Figure 108): photographic scale $a = 0^s.0$; $b = 5^s.1$; $d = 12^s.0$; $c = 16^s.1$; $e = 19^s.0$; visual scale $k = 0^s.0$; $e = 11^s.6$; $c = 21^s.3$.

All observations are listed in Tables 247—250.

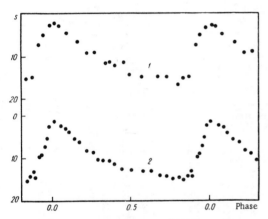

FIGURE 107. Light-variation curves of AW Draconis.

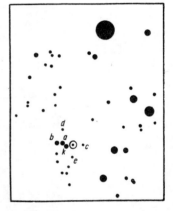

FIGURE 108. Comparison stars for AW Draconis.

TABLE 247. Moscow observations

JD hel	s	JD hel	s	JD hel	s
241...		242...		243. .	
4926.281	11.1	9907.291	17.1	3951.227	8.1
5251.344	9.1	9908.238	17.1	4121.467	9.1
6734.397	7.1	243...		4128.427	0.0
7124.339	15.1	0580.355	5.1	4223.381	9.9
7554.170	4.2	0583.317	10.9	4224.445	16.1
7793.367	8.1	0583.356	11.5	4250.401	16.1
7852.249	14.3	0585.293	5.1	4330.156	13.9
8567.269	14.7	0587.290	4.1	4331.166	3.1
242...		0588.272	14.1	4480.436	11.1
9402.405	17.1	0588.355	15.6	4678.210	3.1
9487.435	14.1	0589.314	8.1	4681.167	11.8
9488.431	5.1	0590.293	13.1	4683.202	8.4
9490.425	14.5	0664.251	3.6	4684.177	13.9
9497.417	5.1	3775.480	8.1	4871.433	13.5

TABLE 247 (continued)

JD hel	s	JD hel	s	JD hel	s
243...		243...		243...	
4978.338	17.1	5336.495	2.6	5363.298	2.0
4982.339	14.1	5337.307	6.9	5365.247	14.7
5006.329	15.0	5341.429	7.8	5366.281	14.3
5012.397	15.0	5342.463	14.1	5366.376	14.7
5333.322	9.5	5344.341	13.7	5369.249	14.3
5334.446	3.1	5347.367	14.7	5718.361	15.5
5335.306	10.1	5348.413	12.8	5721.372	5.1
5336.379	15.2	5361.304	6.1		

TABLE 248. Odessa observations (old series)

JD hel	s	JD hel	s	JD hel	s
243...		243...		243...	
3824.439	17.0	4594.337	0.0	5454.222	13.0
3951.339	14.0:	4599.415	5.1	5458.193	7.4
4538.474	15.0	4627.344	5.9	5717.268	8.6
4552.419	4.1	4654.368	9.0	5727.413	16.0
4566.496	12.0	4678.259	8.6	5746.375	15.0
4576.466	4.1	4954.458	3.1	5749.278	14.0

TABLE 249. Odessa observations (seven-camera astrograph)

JD hel	s	JD hel	s	JD hel	s
243...		243...		243.	
6041.431	14.7	6401.445	14.3	6791.470	9.9
6049.431	10.0	6404.432	3.2	6792.479	14.7
6050.445	3.1	6406.432	3.4	6806.344	8.8
6053.412	8.1	6407.433	11.0	6807.361	10.5
6067.323	14.8	6408.428	12.0	6809.379	12.0
6069.358	13.0	6410.418	13.6	6817.325	14.3
6070.359	8.1	6423.378	9.0	6834.283	12.8
6071.360	10.8	6424.378	3.9	6837.325	2.7
6072.356	16.1	6425.401	11.1	6840.272	11.6
6074.341	15.1	6426.357	8.1:	6862.240	10.1
6075.348	3.8	6428.391	11.3	6863.234	14.7
6076.346	13.8	6429.420	11.0	6868.242	1.3
6079.323	1.3	6430.397	14.3	7135.456	15.2
6080.344	13.6	6432.393	14.6	.511	12.0
6081.340	14.7	6451.316	9.7	7136.432	11.0
6082.340	7.6	6453.305	6.3	.494	10.4
6083.348	17.1	6454.347	13.9	7137.451	15.4
6101.281	7.6	6455.313	2.8	.485	12.0
6102.251	10.2	6456.349	11.3	7139.464	13.7
6103.262	18.1	6461.302	14.5	7140.475	5.8
6104.260	6.5	6462.273	6.6	7142.457	2.0
6105.237	14.9	6463.286	10.8	7144.423	15.1
6126.204	7.4	6478.214	9.2	.449	7.9
6344.518	13.8	6479.242	11.0	7145.454	10.9
6345.505	9.7	6481.221	10.8	7162.414	2.7
6347.503	8.1	6482.224	9.0	7165.395	12.5
6364.460	13.5	6483.234	11.0	7166.382	14.3
6367.494	9.7	6484.227	3.7	7167.388	10.9
6371.460	3.4	6488.260	1.9	7169.387	8.4
6372.452	10.8	6489.265	10.5	.418	9.5
6373.458	2.2	6490.213	10.0	7170.363	14.0
6375.481	8.1	6756.484	12.0	.389	14.6
6376.459	11.0	6757.498	14.6	7172.353	14.3
6379.475	13.6	6760.505	8.7	.401	14.3
6381.446	14.3	6765.500	13.4	7173.355	1.7
6395.447	1.7	6766.509	6.0	.407	2.0
6397.416	14.7	6780.411	10.4	7174.379	12.0
6398.432	9.7	6781.447	14.3	7175.376	8.7
6399.418	15.8	6789.421	12.0	7176.373	10.9
6400.443	11.0	6790.511	13.9	.401	13.2

TABLE 249 (continued)

JD hel	s	JD hel	s	JD hel	s
243...		243...		243...	
7189.315	8.1	7521.406	14.7	7900.382	8.9
.338	6.0	7522.343	17.1	7901.342	9.0
7192.320	14.7	7523.360	11.1	7902.336	15.0
.359	12.8	.408	11.2	.360	15.1
7193.298	1.7	7524.362	14.7	7903.321	8.6
7195.286	14.7	.420	14.3	.372	11.0
.333	5.1	7525.361	9.2	7904.340	16.1
7196.285	9.4	.408	11.0	7906.337	14.7
.308	12.0	7526.355	14.9	7908.330	13.0
7197.300	15.3	.408	13.6	7913.364	15.0
.325	14.9	7544.296	17.1	7939.293	12.0
7198.283	9.3	.348	14.0	7959.216	13.9
.331	10.1	7545.291	7.1	7962.229	8.9
7199.324	14.3	.346	10.3	7963.249	12.7
7204.338	3.1	7547.285	5.8	7964.229	14.3
7228.239	14.7	7549.313	1.0	8198.470	15.2
7461.497	9.9	7555.292	14.3	8210.454	5.9
7463.504	8.9	7557.292	14.6	8225.459	2.0
7470.472	14.0	7844.471	13.9	8228.446	10.6
7472.505	10.8	7847.463	9.1	.473	15.0
7473.472	14.3	7848.490	13.0	8230.471	10.6
7486.436	15.2	7854.457	6.2	8231.482	17.0
7488.457	13.8	7855.458	13.1	8233.466	14.0
7493.454	15.2	7871.396	14.7	8234.462	6.2
7496.374	2.0	7872.386	6.5	8236.428	9.2
.397	2.6	7873.418	14.7	8254.386	3.4
.448	4.3	.473	14.2	8259.396	11.0
7497.430	12.0	7878.375	10.2	8260.392	15.4
.483	15.1	.424	—1.0	.416	14.0
7501.413	11.1	7880.382	14.5	8262.357	14.7
.463	12.6	7881.448	10.5	8263.346	6.0
7518.392	1.3	7882.378	14.7	.401	9.0
7519.374	14.4	.423	14.0	8268.365	14.0
.420	12.0	7883.418	9.7		
7520.371	4.5	.441	9.6		
7521.382	13.5	7900.335	15.6		

TABLE 250. Visual observations

JD hel	s	JD hel	s	JD hel	s
243...		243...		243...	
6339.449	24.7	6340.455	14.6	6730.331	15.7
.464	24.3	.466	8.9	.388	17.6
.488	24.9	.471	9.7	6732.331	8.5
.501	24.9	.479	9.1	.342	10.1
6340.349	25.3	6344.336	24.4	.358	9.7
.386	25.7	.384	25.3	6736.342	10.3
.419	24.4	.421	24.9	6819.375	23.4
.425	22.2	.458	25.3	.406	15.4
.431	19.9	.503	26.1	.413	18.3
.438	18.6	6722.440	26.7		
.449	17.5	6730.321	15.7		

BO Aquarii

Observed visually by Tsesevich and Lange. Parenago reduced the Simeise photographs. The observation dates used in his work were only accurate to 0.01 days. This is much too crude for this star. The Simeise photographs were therefore revised; by that time a much higher number was available.

The seasonal curves derived from this work gave the reliable maxima listed in Table 251 (which also includes the maxima derived by Tsesevich from Harvard photographs). It follows from the table that the period jumped several times, recovering its original value. The O−A residues were calculated from

$$\text{Max hel JD} = 2426589.421 + 0.6940238 \cdot E. \tag{177}$$

Recent maxima fit the equation

$$\text{Max hel JD} = 2430618.207 + 0.6940179 \cdot E'. \tag{178}$$

TABLE 251. List of maxima

Source	Max hel JD	E	O − A	O − B
Tsesevich (Harv.)	2415638.452	−15779	$+0^{d}.033$	—
	7108.410	−13661	+ .048	—
	8364.591	−11851	+ .046	—
Tsesevich (Harv.)	9877.551	− 9671	+.034	—
	21299.597	− 7622	+.025	
Tsesevich (Sim.)	2554.369	− 5814	+.002	—
Tsesevich (Harv.)	3670.364	− 4206	+.007	—
	4548.292	− 2941	−.005	—
	5488.730:	− 1586	+.031:	—
Tsesevich (Sim.)	5530.340	− 1526	−.001	—
Tsesevich (visual)	6573.464	− 23	+.006	—
	6963.487	+ 539	−.013	—
Tsesevich (Sim.)	8776.296	+ 3151	+.006	—
Tsesevich (visual)	30618.210	+ 5805	−.019	$+0^{d}.003$
Tsesevich (Sim.)	3215.215	+ 9547	−.051	−.007
Tsesevich (visual)	6490.308	+14266	−.057	+.016
Lange (visual)	7195.407	+15282	−.086	−.008
	7501.479	+15723	−.078	+.002
	7517.451	+15746	−.069	+.012
	7519.511	+15749	−.091	−.010
	7524.381	+15756	−.070	+.002
	7526.462	+15759	−.080	+.001
	7549.355	+15792	−.090	−.009

Comparison stars used in the observations of Tables 252, 253 (Figure 109):

	s_1	s_2
m	0.0	0.0
n	10.5	11.9
p	15.1	—
q	20.3	—
r	—	24.3

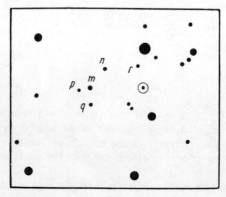

FIGURE 109. Comparison stars for BO Aquarii.

TABLE 252. Simeise observations

JD hel	s_1	JD hel	s_1	JD hel	s_1
242...		242...		242...	
0394.425	19.0	5510.477	9.9	8403.459	17.1
	18.1	5530.334	−1.0	8433.324	17.2
0396.295	13.1		6.1		17.7
0717.474	8.1	5855.441	12.0	8434.393	12.8
0748.405	13.0		13.0		12.5
	13.1	5856.437	14.3	8776.368	5.7
0754.468	17.4		13.6		5.2
	15.1	6189.493	15.1	8779.373	13.3
0773.365	14.2		17.2		12.8
0777.282	12.0	6209.431	13.1	8780.373	14.2
	12.8		13.3	8786.37:	12.5
0786.341	12.5	6216.388	13.3		12.3
1101.478	13.3		13.3	9514.394	14.0
	18.2	6222.483	10.5		14.0
1105.450	9.6		10.5	9518.403	10.5
1108.449	15.1	6237.428	18.1	243...	
	13.4	6240.371	4.8	2771.475	14.1
1112.429	13.4	6241.367	14.0		13.6
	11.6		13.1	2775.475	10.5
1464.502	18.0	6242.377	3.3		11.6
	13.4	6249.334	2.8	3151.454	5.7
1466.469	10.5		4.3		6.0
	13.0	6264.331	12.8	3157.422	13.1
2554.423	2.5		13.3		14.2
3647.406	7.3	6566.431	8.2	3187.337	21.3
	8.4		14.0		15.1
3649.409	17.7	6947.449	14.0	3215.246	0.0
	18.6		13.6		1.0
3666.322	7.6	6948.431	11.6	3867.506	16.0
	6.3		9.5		15.8
4033.381	0.0	6972.40:	15.1	3892.402	17.2
	−2.0		16.1		15.1
4036.451	9.5	6984.351	2.3	3915.340	18.6
4385.490	10.5	7310.401	15.1	4242.413	1.0
4389.504	7.0	.516	4.2		1.9
	9.5	7310.401	15.1	4253.427	7.7
4449.311	15.1	.516	5.7		8.6
4736.425	4.4	7329.408	9.0	4255.442	18.2
5124.414	4.8	7343.412	10.5		18.6
	8.4		12.8	4269.320	15.1
5129.430	8.2	7685.430	7.6	.410	4.2
.455	10.5		6.7	4601.470	13.9
5129.430	9.5	8080.301	10.5		12.5
.455	10.5		10.5	4653.306	5.7
5485.469	13.3	8394.496	12.8		6.2
	10.5	8403.459	15.1		

TABLE 253. Visual observations

JD hel	s_2	JD hel	s_2	JD hel	s_2
243...		243...		243...	
6454.411	14.2	6455.334	17.6	6482.269	10.6
.431	16.0	.403	17.6	.285	13.8
.443	15.7	.421	18.1	.298	17.1
.449	14.2	6461.330	10.2	6489.303	19.5
.471	15.4	.351	13.0	6490.235	13.0
.483	16.6	.378	15.3	.307	8.9
.495	16.9	6462.299	19.0	.328	10.4
6455.286	19.5:	.436	17.5	.375	15.0
.319	16.3	6466.433	19.1	.396	13.1

207

CONCLUSIONS

The jump ΔP is often comparable with the period P. Some observers established definite correlation between the two quantities. The currently available data, unfortunately, are insufficient for conclusively resolving this problem.

Table 254 lists the variables with their period P, the period increment ΔP, the length of the observation series ΔE, the spectral type at the minimum from calcium lines (Preston's SpCaII), the difference between hydrogen and calcium spectral types ΔS, the radial velocity v_r (km/sec), and the galactic coordinates l and b.

TABLE 254. Stars with regular variations of the period

Star	P	$\Delta P \cdot 10^8$	ΔE	Sp Ca II	ΔS	v_r	l	b	$\frac{\Delta P}{P} \cdot 10^8$
IV Cyg	0.334	+1024	34500	—	—	—	37°	+ 3°	+ 3066
AX Aqr	0.388	+2000	33900	—	—	—	8	−57	+ 5155
BN Vir	0.391	+ 865	44045	—	—	—	263	+65	+ 2212
V 696 Oph	0.393	− 393	32500	—	—	—	334	+26	− 1000
DM Cyg	0.420	(+ 150)	(53800)	(F6)	(0)	(−50)	(47)	(−13)	(+ 357)
VZ Her	0.440	+ 237	35235	F0	4	−120	27	+33	+ 539
RV Cap	0.448	− 973	30435	A8	6	−110	1	−37	− 2172
CP Aqr	0.463	+ 180	39400	F2	3	—	17	−33	+ 389
BN Aqr	0.470	+1338	35710	—	(4)	—	25	−52	+ 2847
BW Vir	0.471	− 410	34900	—	—	—	272	+61	− 870
CVS 194	0.472	+3360	37002	—	—	—	114	−37	+ 7119
UU Vir	0.476	(+ 34)	(45264)	(F3)	(2)	(−15)	(251)	(+61)	(+ 71)
VV Lib	0.478	+1170	26000	—	—	—	316	+26	+ 2448
BR Aqr	0.482	− 750	35600	F2	3	—	46	−66	− 1556
VV Peg	0.488	+ 542	37200	A5	9	+10	47	−31	+ 1111
BV Vir	0.497	+7570	34600	—	—	—	271	+65	+15231
RZ Cet	0.511	−1990	22300	F0	4	0	147	−59	− 3894
BO Vir	0.520	− 799	33755	—	—	—	264°	+63°	−1536
EZ Lyr	0.525	+ 650	41000	A 7	7	−75	33	+15	+1238
BH Aqr	0.526	−2710	25600	—	—	—	8	−60	−5152
CS Ser	0.527	+ 690	30500	—	(6)	—	335	+44	+1309
RY Psc	0.530	+1130	34500	A 8	7	+25	71	−63	+2132
TW Boo	0.532	− 180	31000	—	—	—	37	+62	− 338
CVS 1410	0.533	−1750	—	—	—	—	189	+35	−3283
VY Lib	0.534	− 530	32800	—	—	—	322	+27	− 993
SX Aqr	0.536	− 356	33800	A 7	9	−220	26	−35	− 664
V 347 Her	0.537	+4830	42000	—	—	—	21	+10	+8994
YZ Aqr	0.552	+ 490	29200	—	—	—	18	−51	+ 888
RR Ceti	0.553	+ 237	37000	F 0	5	−95	113	−59	+ 429
V 734 Oph	0.562	−1520	29300	—	—	—	341	+26	−2705
TT Cnc	0.563	+ 560	29300	A 7	7	+55	180	+30	+ 995
RX Cet	0.574	−1430	20872	—	—	—	77	−78	−2491
IO Lyr	0.577	+ 612	32850	F 3	3	—	28	+19	+1061
RU Cet	0.586	−2450	20000	—	(9)	—	108	−78	−4181
BN Vul	0.594	− 412	37035	A 8	6	−235	26	+02	− 694
CX Lyr	0.617	−2000	37300	—	(7):	—	27	+12	−3241
CX Lyr	0.617	−2405	37300	—	—	—	27	+12	−3898
AF Her	0.630	+ 954	32300	—	—	—	32	+41	+1514
AG Her	0.649	+ 870	30725	—	—	—	31	+40	+1341
X Ari	0.651	+1420	26200	A 4	10	−40	137	−39	+2181
AW Dra	0.687	—	30000	—	—	—	48	+18	—
BO Aqr	0.694	− 590	21600	—	(6)	−55	25	−60	− 850
XX And	0.723	−1050	—	A 7	9	−15	97	−23	−1452

Almost all the data in Table 254 are completely reliable. The only doubtful cases are the two stars BV Vir and V 347 Her. These stars probably show continuous, and not jumplike, variation of the period.

Graphical plot of $\Delta P \cdot 10^8$ and P leads to the following conclusions.

1. There is no correlation between the two sets of data.

2. Increase and decrease of period are equiprobable. In Table 254, there are 23 positive and 19 negative ΔP.

208

3. Most of the objects studied show only one jump in 20,000—50,000 cycles. Change of period is thus an infrequent phenomenon. CX Lyr is the only star whose period changed twice in 37,300 cycles.

4. A certain feature which emerges from these data requires further verification. Dropping the two doubtful cases of BV Vir and V 347 Her, we calculate the average increment for stars with (+) and (−) separately: $10^8 \cdot \Delta P_+ = +882$; $10^8 \cdot \Delta P_- = -1195$. It seems that the average decrease in periods is greater than the average increase.

5. No correlation was observed between $\frac{\Delta P}{P} \cdot 10^8$ and P.

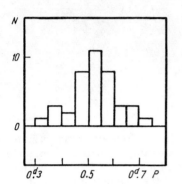

FIGURE 110. Frequency of periods for stars with suddenly changing period.

The spectral properties of the stars of this group are highly remarkable. DM Cyg and UU Vir are exceptions. They have very low values of $\Delta P \cdot 10^8$, low radial velocities, and small ΔS which is characteristic of group 1 stars (see Chapter 2). They should therefore be transferred from Table 254 to Table 109. This is the reason why the corresponding data are listed in parentheses. The spectral data of the other stars lead to the following average results:

δP	$\overline{P}$	n	$\overline{s_p}$	ΔS	$\overline{\lvert v_r \rvert}$
$0^d.40—0^d.50$	$0^d.464$	5	A 9.4	5.0	80
0.50—0.60	0.549	8	A 8.8	6.0	101
> 0.60	0.687	2	A 5.5	9.5	37

We see that longer periods correspond to earlier spectral types, whereas ΔS markedly increases. This is probably characteristic of stars with variable periods.

The distribution of stars of this group over the period P (Figure 110) is represented by a smooth curve with a maximum between $0.50 — 0^d.55$:

δP	n	δP	n	δP	n
$0.30^d — 0.35^d$	1	$0.45^d — 0.50^d$	8	$0.60^d — 0.65^d$	3
.35 — .40	3	.50 — .55	11	.65 — .70	3
.40 — .45	2	.55 — .60	8	.70 — .75	1

Chapter 4

IRREGULAR VARIATION OF PERIODS

In the previous chapter we have noted that the periods of some variables are subject to irregular fluctuations. The best examples are RV Coronae Borealis and RR Geminorum, which clearly demonstrate a range of effects characteristic of numerous RR Lyrae stars. In the present chapter it is shown that many other stars of this type also reveal abrupt and very considerable irregular fluctuations in their period.

SMALL FLUCTUATIONS

We will first establish that some of the well-known stars reveal completely irregular, though small, fluctuations of period. There are, however relatively few stars under this category.

SW Aquarii

Discovered by Leavitt in 1908 /335/. Observed by numerous authors. Table 255 lists all the published maxima.

TABLE 255. List of maxima

Source	Max hel JD	E	O — A
Zinner /375, 378/	2419674.398	— 11807	+0.011
	9685.419	— 11783	+ .009
	9686.338	— 11781	+ .009
	9687.257	— 11779	+ .009
	20024.383	— 11045	+ .007
	0037.243	—11017	+ .007
	0039.080	— 11013	+ .006
	0042.296	— 11006	+ .007
	0054.238	— 10980	+ .007
	0076.284	— 10932	+ .007
	0122.214	— 10832	+ .006
Ivanov /280/	4373.522	— 1576	+ .006
	4379.492	— 1563	+ .005
	4415.316	— 1485	+ .004
Tsesevich /229/	5069.353	— 61	— .007
	5086.350	— 24	— .004
	5087.272	— 22	.000
	5092.322	— 11	— .003
	5093.238	— 9	— .003
	5096.455	— 2	— .003

TABLE 255 (continued)

Source	Max hel JD	E	O — A
Tsesevich /229/	2425097.374	0	—0^d.003
	5098.293	+ 2	— .003
	5102.425	+ 11	— .004
	5104.264	+ 15	— .003
	5108.396	+ 24	— .004
	5120.338	+ 50	— .004
	5121.257	+ 52	— .004
	5125.391	+ 61	— .004
	5127.228	+ 65	— .004
Eropkin /229/	5127.229	+ 65	— .003
Tsesevich /149-151/	6577.251	+ 3222	.000
	6949.287	+ 4032	.000
Florya /149-151/	6971.331	+ 4080	— .002
Lange /58/	7325.458	+ 4851	.000
Tsesevich /150/	7662.584	+ 5585	.000
Lange /150/	7769.137	+ 5817	— .005
Alaniya /1/	33860.424	+ 19079	+ .006
Born /203/	3895.323	+ 19155	— .002
Sofronievitsch /204/	3895.322	+ 19155	— .003
Born /203/	3898.540	+ 19162	.000
Sofronievitsch /204/	3898.536	+ 19162	— .004
Born /203/	3900.380	+ 19166	+ .003
Sofronievitsch /204/	3900.380	+ 19166	+ .003
Born /203/	3917.372	+ 19203	.000
Sofronievitsch /204/	3917.369	+ 19203	— .003
Born /203/	3918.295	+ 19205	+ .005
Sofronievitsch /204/	3918.292	+ 19205	+ .002
Born /203/	3929.316	+ 19229	+ .002
	4272.416	+ 19976	+ .003
Lange /65/	7139.387	+ 26218	+ .005
	7144.440	+ 26229	+ .006
Lange /68/	7522.445	+ 27052	+ .004
	7523.358	+ 27054	— .001
	7528.419	+ 27065	+ .008
	7533.474	+ 27076	+ .010
Lange	7908.254	+ 27892	— .001
	7925.248	+ 27929	— .001

The O — A residues were calculated from the equation

$$\text{Max hel JD} = 2425097.3771 + 0.45930295 \cdot E. \qquad (179)$$

The period of this star is nearly constant. However, initially the residues were +0^d.007, near $E \sim 0$ they changed to —0^d.003, and then they again became positive, though close to zero. The period apparently slightly fluctuated in an irregular fashion.

RV Ursae Majoris

Discovered by Tserasskaya [Ceraski] /220/ and observed by numerous authors. Since we are interested in the slow variation of its period, we will only consider the average maxima. The old observations were reduced twice by Hertzsprung's method. The results are summarized in Table 256. The residues were calculated from

$$\text{Max hel JD} = 2417861.434 + 0.468062 \cdot E. \qquad (180)$$

211

TABLE 256. List of maxima

Source	Max hel JD	E	O — A
Luizet /310/	2418132.915	+ 580	+0.005 ᵈ
Blazhko /310/	8270.985	+ 875	— .003
Nijland /324/	20121.243	+ 4828	+ .006
Belyavskii /22/	0308.464	+ 5228	+ .002
Jordan /290/	0856.092	+ 6398	— .003
Blazhko /199/	3307.801	+ 11636	— .002
Subbotin /357/	3913.004	+ 12929	— .004
Tsesevich /165/	5095.336	+ 15455	+ .004
Sharonov /165/	5226.389	+ 15735	— .001
Dombrovskii /34/	6889.902	+ 19289	+ .020
	6923.126	+ 19360	+ .012
Dombrovskii, Zagrebin /34/	7090.221	+ 19717	+ .009
Dombrovskii /34/	7300.392	+ 20166	+ .020
Solov'ev /108/	7581.692	+ 20767	+ .014
Dombrovskii /34/	7613.521	+ 20835	+ .015
Solov'ev /108/	7860.663	+ 21363	+ .020
Rabkin	7895.769	+ 21438	+ .022
Ishchenko /41/	9056.578	+ 23918	+ .037
Yudkina /168/	33098.276	+ 32533	+ .020
	3486.336	+ 33382	+ .056
	3814.435	+ 34083	+ .044
Alaniya /1/	4099.460	+ 34692	+ .019
Yudkina /169/	4166.418	+ 34835	+ .044
Geyer /239/	5872.4960	+ 38480	+ .0362
	5899.6464	+ 38538	+ .039
Lange	7369.355	+ 41678	+ .033
	7378.258	+ 41697	+ .043
	7400.244	+ 41744	+ .030
	7405.383	+ 41755	+ .020
	7406.330	+ 41757	+ .031
Sakharov	8235.272	+ 43528	+ .035

The plot of O — A shows that the period undergoes irregular fluctuations; however, over the last 20,000 cycles the period remained constant (Figure 111).

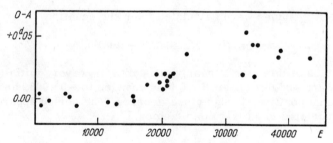

FIGURE 111. O — A residues for RV Ursae Majoris.

BB Virginis

Discovered by Hoffmeister in 1935 /267/. In the same year Jensch determined its period /289/. Later observed by Tsesevich /157/ and Alaniya /1/. Tsesevich estimated the magnitude from Odessa and Simeise photographs and using the equation

$$\text{Max hel JD} = 2431247.096 + 0.4710958 \cdot E \tag{181}$$

obtained seasonal light curves, from which the maxima were determined. The complete list of maxima is given in Table 257.

TABLE 257.   List of maxima

Source	Max hel JD	E	O — A
			d
Tsesevich (Sim.)	2419526.291	— 24880	+0.059
Jensch /289/	25707.480	— 11759	.000
Tsesevich (Sim.)	6118.329	— 10887	+ .053
	8258.450	— 6344	— .014
Tsesevich (visual)	31247.096	0	.000
Tsesevich (Sim.)	3736.387	+ 5284	+ .021
Alaniya /1/	4509.439	+ 6925	+ .005
Tsesevich (Odessa)	6362.315	+ 10858	+ .061

The O — A plot reveals irregular fluctuations of the period.   The star probably has Blazhko effect.   Comparison stars (Figure 112): Simeise scale (s_1), $B = 0^s.0$; $a = 10^s.0$; $b = 21^s.0$; Odessa scale (s_2), $B = 0^s.0$; $a = 8^s.6$; $b = 18^s.4$.

The observations are listed in Tables 258, 259.

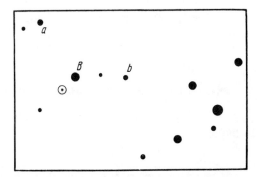

FIGURE 112.   Comparison stars for BB Virginis.

TABLE 258.   Simeise observations (Tsesevich)

JD hel	s_1	JD hel	s_1	JD hel	s_1
241...		242...		243...	
9526.290	9.0	5732.315	15.7	2675.327	17.0
	9.0		16.0		16.4
9886.439	14.4	6060.569	15.0	2998.518	16.0
	13.3	6095.348	13.7		15.0
9892.429	15.1		13.0	3437.41:	12.2
	15.0	6096.361	14.4	3443.346	7.8
9898.41±	15.1		14.4	3447.41:	16.0
	15.1	6118.343	8.0	3451.41:	6.7
9946.41±	17.6		8.4	3736.425	5.5
	15.9	6420.582	17.7		6.7
9927.41±	17.0		15.8	4120.393	14.4
	16.8	6806.555	8.7	4122.357	15.2
242...			6.2		12.4
0582.542	13.7	8258.518	8.5	4510.340	13.7
3173.457	14.0		7.0		12.2
	15.5	8631.527	9.0		
4621.371	14.4		7.6		
	15.7				

TABLE 259. Odessa observations (Tsesevich)

JD hel	s_2	JD hel	s_2	JD hel	s_2
243...		243...		243...	
6286. 485	11.0	6344. 383	2. 9	6607. 602	13. 1
6288. 517	13.0	6345. 379	4. 5	6608. 663	10. 0
6292. 536	6.3	6347. 383	8. 6	6613. 626	13. 0
6304. 430	9.5	6362. 352	3. 7	6646. 525	11. 5
6313.476	12.2	6364. 344	15. 1	6660. 513	4. 3
6314.508	10.4	6371. 338	6. 1	6661. 493	9. 7
6335. 441	10. 4	6372. 336	10. 8	6662. 502	7. 6
6338. 382	10. 8	6373. 336	10. 1	6663. 519	10. 6
6339. 411	14. 0	6376. 338	10. 8	6667. 486	7. 1
6340. 393	9. 6	6590. 697	13. 0	6668. 508	5. 4

DX Cephei

Discovered by Geyer, Kippenhan, and Stromeyer /240/, who designated it BV 71. Beyer /194/ established its elements:

$$\text{Max hel JD} = 2435202.3596 + 0.526178 \cdot E. \tag{182}$$

Tsesevich carried out visual observations and used Moscow photographs. At that time Stromeyer, Knigge, and Ott /356/ published their Bamberg photographic observations of maxima and showed that equation (182) require radical modification:

$$\text{Max hel JD} = 2426930.360 + 0.5260422 \cdot E. \tag{183}$$

A complete list of maxima and O−A residues calculated from (183) is given in Table 260.

TABLE 260. List of maxima

Source	Max hel JD	E	O − A
Tsesevich (Mosc.)	2414423.365	− 23776	$+0^d.184$
Stromeyer et al. /356/	26930.413	0	+.053
	6931.415	+ 2	+.003
	6931.440	+ 2	+.028
	6980.343	+ 95	+.009
	7473.282	+ 1032	+.046
Tsesevich (Mosc.)	7898.311	+ 1840	+.033
	8081.398	+ 2188	+.058
Stromeyer et al. /356/	8336.481	+ 2673	+.010
	8396.467	+ 2787	+.027
	8407.489	+ 2808	+.003
	8425.408	+ 2842	+.036
	9143.465	+ 4207	+. 045
Tsesevich (Mosc.)	9192.416	+ 4300	+. 075
	9310.210	+ 4524	+. 035
Stromeyer et al. /356/	9376.444	+ 4650	− .012
	9407.481	+ 4709	− .012
Tsesevich (Mosc.)	9527.474	+ 4937	+.044
	9547.409	+ 4975	− .011
Beyer /194/	35202.366	+ 15725	− .008
	5205.522	+ 15731	− .008
	5212.360	+ 15744	− .008
	5213.411:	+ 15746	− .009
	5214.467:	+ 15748	− .006:
	6189.218:	+ 17601	− .011:
	6232.353	+ 17683	− .011

TABLE 260 (continued)

Source	Max hel JD	E	O — A
Beyer /194/	2436233.407	+ 17685	−0^d.009
	6252.344	+ 17721	— .010
	6261.286	+ 17738	— .011
	6262.339	+ 17740	— .010
	6263.391:	+ 17742	— .011
Stromeyer et al. /356/	6436.468	+ 18069	+ .052
Tsesevich	7204.483	+ 19531	— .007
Stromeyer et al. /356/	7577.434	+ 20240	— .020
	7615.290	+ 20312	— .039
	7615.340	+ 20312	+ .011
Tsesevich (visual)	7911.469	+ 20875	— .022

Since the maxima determined from visual observations are much more accurate, they were averaged and compiled in Table 261. Only the old maximum obtained from the seasonal light curve using the Moscow photographs was retained individually. All the photographic data from $E = 0$ to $E = +4975$ were averaged. The last 4 photographic maxima were omitted.

TABLE 261. Averaged maxima

Max hel JD	E	O — A	E′	O — B
2414423.365	— 23776	+0^d.184	— 39501	+0^d.096
28349.648	+ 2698	+ .026	— 13027	+ .001
35209.730	+ 15739	— .008	+ 14	— .002
6241.823	+ 17701	— .010	+ 1976	+ .001
7204.483	+ 19531	— .007	+ 3806	+ .009
7911.469	+ 20875	— .022	+ 5150	— .003

The last maxima give the equation

$$\text{Max hel JD} = 2435202.367 + 0.5260398 \cdot E'; \; P^{-1} = 1.900996845. \tag{184}$$

The deviation of the old maximum by $+0^d.096$ indicates that the period has changed. However, the exact nature of this change cannot be established from the available observations. We only know that the period became longer. Comparison stars (Figure 113): photographic scale $k = 0^s.0$; $a = 11^s.6$; $b = 20^s.6$; $c = 25^s.0$; visual scale $k = 0^s.0$; $a = 15^s.1$; $c = 26^s.4$; $b = 35^s.7$.

TABLE 262. Average visual light curve

φ	s	n	φ	s	n	φ	s	n
0^P.005	8.6	5	0^P.552	28.4	5	0^P.900	26.3	5
.021	9.6	5	.646	28.7	5	.914	26.7	5
.039	9.3	5	.702	29.8	5	.925	24.9	5
.056	10.3	6	.743	28.5	5	.936	24.1	5
.077	11.8	4	.783	24.1	5	.950	21.2	6
.090	13.4	3	.810	25.2	5	.960	16.6	5
.172	17.1	3	.841	25.0	5	.972	12.6	5
.301	22.8	4	.869	27.4	5	.982	12.0	4
.451	26.1	5	.886	27.7	5	.990	9.4	4

The visual observations were used with equation (184) to obtain the average light curve (Figure 114, Table 262).

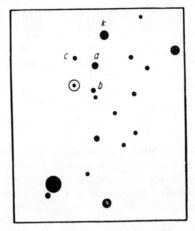

FIGURE 113. Comparison stars for DX Cephei.

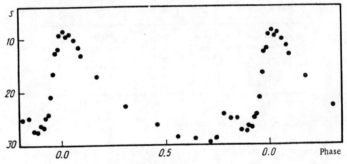

FIGURE 114. Light-variation curve of DX Cephei.

The magnitude enhancement at $0^P.8$ is not significant in my opinion. However, it also emerges from the average light curve of Stromeyer, Knigge, and Ott, and also from Beyer's light curve. Moscow photographic and visual observations are listed in Tables 263, 264.

TABLE 263. Moscow observations (Tsesevich)

JD hel	s	JD hel	s	JD hel	s
241...		242...		242...	
4194.40:	9.9	7898.311	6.3	9310.210	5.8
4421.390	18.3	8081.398	7.2	9367.398	21.9
4423.401	7.2	9146.482	22.1	9521.465	21.7
4782.395	26.0	9147.528	18.3	9527.474	7.3
5522.393	22.8	9166.484	18.8	9547.409	6.7
8295.158	22.4	9172.490	13.9	243...	
242...		9192.416	7.7	4061.486	20.6
0150.200	18.6	9284.182	21.6	4477.384	20.6
7813.259	17.9	9287.176	15.2	4480.518	21.1

216

TABLE 264. Visual observations (Tsesevich)

JD hel	s	JD hel	s	JD hel	s
243...		243...		243...	
7196.587	6.0	7904.500	29.9	7911.434	24.4
.595	7.6	7906.310	18.6	.435	24.4
7204.273	29.4	.346	21.4	.440	24.8
.293	29.4	.381	23.0	.442	23.9
.303	29.4	.422	24.1	.445	21.5
.310	27.9	.439	24.3	.446	21.6
.321	29.4	.470	25.9	.448	19.7
.331	29.4	.482	25.7	.450	16.9
.335	29.4	7908.306	12.6	.452	16.2
.344	28.9	.341	10.1	.456	13.2
.360	24.4	.350	11.2	.458	12.4
.363	24.9	.356	12.3	.461	10.1
.367	26.2	.361	12.7	.465	8.0
.374	24.5	.398	15.1	.468	8.0
.378	25.4	.407	17.6	.473	8.5
.384	25.0	.475	23.3	.478	8.5
.388	24.9	.485	23.6	.484	9.4
.394	25.0	7910.268	32.0	.491	8.8
.400	27.4	.290	23.9	.498	9.4
.410	28.9	.298	23.9	.506	11.6
.421	27.9	.305	23.5	.512	12.9
.425	27.9	.312	23.8	.517	13.9
.431	28.9	.318	25.0	.523	13.7
.435	25.4	.327	25.1	7912.375	29.8
.437	24.5	.352	25.9	.456	29.8
.441	22.6	.362	24.9	.462	29.5
.444	22.6	.386	25.9	.467	30.4
.447	21.4	.437	9.1	.474	28.6
.451	18.5	.441	8.0	.480	25.8
.455	12.1	.447	9.2	.483	25.9
.459	11.0	.457	10.5	.500	19.2
.464	12.5	7911.276	23.9	.504	19.4
.469	10.1	.386	25.3	.508	14.0
.477	10.1	.396	24.9	.511	12.5
.482	9.6	.403	24.8	.516	12.6
.490	10.1	.408	25.6	.519	11.5
.495	10.6	.412	28.0	.525	9.7
7904.368	31.7	.417	24.9	.533	11.4
.380	30.5	.419	24.9	.548	9.8
.414	32.2	.421	25.4	.557	10.4
.434	30.5	.426	25.9	.569	11.7
.451	30.5	.429	24.8	7913.281	24.4
.470	29.9	.431	24.4	.328	24.4

CVS 1389 = 78.1933 = SVS 540 Cancri

Discovered by Kulikovskii /53/ and independently by Morgenroth. Kulikovskii erroneously regarded the star as an eclipsing variable. Kordylewski carried out an extensive and long series of visual observations /289/, but did not reduce the results. Tsesevich studied Kordylewski's observations in order to determine the period. The seasonal maxima (Table 265) were derived from the observations of Kulikovskii and Kordylewski and from Tsesevich's estimates based on Moscow, Odessa, and Simeise photographs.(Table 265).

The O—A residues were calculated from the equation

$$\text{Max hel JD} = 2428247.323 + 0.54316 \cdot E, \qquad (186)$$

which was subsequently improved by the least squares method:

$$\text{Max hel JD} = 2428247 \cdot 3236 + 0.54315827 \cdot E; \; P^{-1} = 1.841084. \qquad (185)$$

217

TABLE 265. Seasonal maxima

Source	Max hel JD	E	O — A	O — B
Tsesevich (Sim.)	2418720.317	— 17540	$+0^d.020$	$0^d.011$
Kulikovskii /53/	9858.255	— 15445	+ .038	+ .011
Tsesevich (Sim.)	24225.244	— 7405	+ .021	+ .007
Kordylewski /298/	8247.323	0	.000	— .001
	8592.231	+ 635	+ .001	+ .002
	8938.236	+ 1272	+ .013	+ .015
	9318.450	+ 1972	+ .015	+ .018
	30468.287	+ 4089	— .017	— .011
	0796.335	+ 4693	— .038	— .030
	1909.280	+ 6742	— .028	— .017
	2243.339	+ 7357	— .012	.000
	2941.296	+ 8642	— .016	— .001
	3005.387	+ 8760	— .018	— .003
	3358.440	+ 9410	— .019	— .003
Tsesevich (Mosc.)	5219.328	+ 12836	+ .003	+ .025
Tsesevich (Odessa)	6544.601	+ 15276	— .034	— .008
Tsesevich (Mosc.)	6660.308	+ 15489	— .020	+ .006
Tsesevich (Odessa)	7378.361	+ 16811	— .025	+ .004

The O — B residues were calculated from this equation and found to be fairly large in some cases. They show a gradual variation (Figure 115). The period is apparently variable, fluctuating irregularly. The star possibly shows Blazhko effect. There are only few photographic observations, and no average photographic curve was derived. Kordylewski's observations cannot be reduced to a single average curve. The average curve is "fuzzed" due to period fluctuations. We therefore constructed an average visual curve from that part of Kordylewski's observations where the O — B residues were small, i.e., between JD 2432203 and JD 2433385 (Table 266, Figure 116).

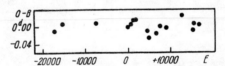

FIGURE 115. O — B residues for SVS 540 Cancri.

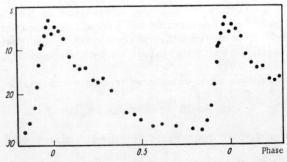

FIGURE 116. Light curve of SVS 540 Cancri, from K. Kordylewski's observations.

218

TABLE 266. Average visual light curve

φ	s_4	n	φ	s_4	n	φ	s_4	n
$0^P.015$	4.5	10	$0^P.326$	19.9	10	$0^P.890$	23.4	10
.038	5.7	10	.413	24.6	10	.910	18.3	10
.064	7.6	10	.452	25.2	10	.924	13.3	10
.092	11.7	10	.498	26.1	10	.933	9.6	10
.122	13.6	10	.561	27.7	10	.943	8.9	10
.150	14.4	10	.613	27.1	10	.953	6.4	10
.187	14.2	10	.696	28.0	10	.966	4.6	10
.225	17.1	10	.789	28.5	10	.977	3.0	7
.251	17.9	10	.832	28.8	10	.992	6.1	7
.285	16.6	10	.862	26.6	10			

Comparison stars (Figure 117):

	s_1	s_2	s_3	s_4
m	—	—	0.0	—
u	—15.5	0.0	—	0.0
k	0.0	10.8	9.1	—
c	6.1	—	—	—
b	11.6	—	—	—
a	16.3	18.9	19.3	27.3
d	24.9	26.0	—	—
A	—	—	23.9	14.3

Tsesevich's observations are listed in Tables 267—269.

TABLE 267. Simeise observations (Tsesevich)

JD hel	s_1	JD hel	s_1	JD hel	s_1
241...		242...		242...	
8708.328	17.3	0929.289	3.7	4225.262	2.0
	19.2	1284.274	15.0	4226.415	7.5
8720.318	1.5		15.1	4448.541	3.8
	2.4	3847.287	3.8	4559.289	—0.5
242...			4.1		3.0
0158.498	20.2	4177.371	16.3	4562.300	20.6
	21.0	4178.403	19.3		19.2
0180.314	1.2		20.3	4578.254	7.2
0894.531	11.2	4196.311	17.7		6.1
	14.4		20.1	4607.276	11.6
0918.287	17.4	4210.252	15.1		14.4
	18.0		14.7	4611.281	16.0
0929.289	4.9	4225.262	0.0		19.2

TABLE 268. Moscow observations (Tsesevich)

JD hel	s_2	JD hel	s_2	JD hel	s_2
243...		243...		243...	
4059.367	20.5	6639.322	15.3	7293.491	21.9
4062.253	20.5	.373	17.8	.512	20.9
4076.314	20.4	.422	17.9	7312.387	21.3
4390.448	8.8	6640.291	12.8	.409	21.3
4420.403	12.0	.347	13.2	.431	21.7
4421.311	21.9	.384	14.8	.452	21.7
4454.322	19.9	.431	16.2	.474	21.9
4455.283	15.9	.479	18.9	.496	21.3
5187.389	16.9	6656.303	18.0	7638.456	18.9
5219.327	—2.0	.349	18.0	.478	17.3
6626.246	16.5	6660.301	—1.8	.495	10.8
6632.369	16.6	.345	10.8	7699.283	17.9
.425	17.0	6661.260	18.9	.305	18.1
6635.322	—2.8	.305	22.4	.328	13.2
.375	10.8	7209.416	14.0		
.426	10.8	7293.469	20.7		

TABLE 269. Odessa observations (Tsesevich)

JD hel	s_3	JD hel	s_3	JD hel	s_3
243...		243...		243...	
6279.345	21.7	6612.434	20.4	7015.352	20.6
6281.343	20.5	6655.318	22.1	7016.332	20.6
6286.330	20.6	6656.360	20.6	7017.337	20.5
6288.327	7.9	6657.337	18.6	7020.354	22.4
6306.306	12.9	6660.356	6.1	7023.328	[15.1
6526.599	24.9	6661.310	20.9	7039.297	19.5
6528.613	24.9	6663.328	21.7	7317.450	[19.5
6544.568	8.6	6664.320	20.4	7377.362	9.1
6546.533	22.8	6667.303	17.3	7378.337	8.2
6555.509	6.8	6954.381	22.8	7396.307	7.4
6576.466	21.7	6959.474	24.9	7397.311	18.6
6583.454	20.4	6993.409	15.0	7398.302	[15.1
6607.435	21.7	6996.435	22.8	7400.307	19.5
6608.417	18.3				

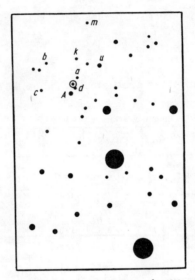

FIGURE 117. Comparison stars for SVS 540 Cancri.

TU Ursae Majoris

Discovered and investigated by Guthnick and Prager /247, 248/. Observed by numerous authors. Solov'ev /108/, in particular, prepared a list of old maxima and Silva /351/ derived the equation

$$\text{Max hel JD} = 2425760.441 + 0.557665 \cdot E - 0.403 \cdot 10^{-9} \cdot E^2. \qquad (187)$$

Silva's formula can be checked using the new observations added since that time. The maxima are listed in Table 270. The O—B residues were calculated from

$$\text{Max hel JD} = 2425760.441 + 0.557665 \cdot E. \qquad (188)$$

220

TABLE 270. List of maxima

Source	Max hel JD	E	O − B	O − C
Robinson /346/	2420160.447	− 10042	+0d.078	+0d.010
Prager /247, 248/	5006.480	− 1352	+ .002	− .012
Kukarkin /52/	5732.544	− 50	− .014	− .020
Jacchia /282/	5760.450	0	+ .009	+ .003
Tsesevich (visual)	6066.595	+ 549	− .004	− .007
Mustel /78/	7248.291	+ 2668	.000	+ .010
Solov'ev /108/	7593.480	+ 3287	− .006	+ .008
Dombrovskii /35/	7620.810	+ 3336	− .001	+ .004
Solov'ev /108/	7894.613	+ 3827	− .012	+ .006
Payne-Gaposchkin /329/	9040.610	+ 5882	− .017	+ .014
Silva /351/	32101.038	+ 11370	− .054	+ .010
Born /203/	4082.382	+ 14923	− .094	− .008
	4091.311	+ 14939	− .087	− .001
	4121.420	+ 14993	− .092	− .006
Sofronievitsch /204/	4451.557	+ 15585	− .093	− .003
Born /203/	4451.554	+ 15585	− .096	− .006
Sofronievitsch /204/	4455.457	+ 15592	− .097	− .007
Born /203/	4455.461	+ 15592	− .093	− .003
Sofronievitsch /204/	4478.321	+ 15633	− .097	− .007
Born /203/	4478.323	+ 15633	− .095	− .005
Alaniya /1/	4478.345	+ 15633	− .073	+ .017
Born /203/	4484.442	+ 15644	− .110	− .020
Sofronievitsch /204/	4484.459	+ 15644	− .093	− .003
Born /203/	4498.400	+ 15669	− .094	− .003
Geyer /239/	6611.3818	+ 19458	− .105	+ .009
Ahnert /172/	6659.357	+ 19544	− .089	+ .026

This equation was improved by the least squares method:

$$\text{Max hel } JD = 2425760.447 + 0.55765884 \cdot E. \qquad (189)$$

The O − C residues point to irregular fluctuations of the period. The quadratic term is rejected. Tsesevich observed the star in 1930 (Table 271). The chart of comparison stars was unfortunately lost, but their scale remained: $p = 0^s.0$; $q = 6^s.8$; $s = 13^s.3$. The 1930 maximum is also included in Table 270.

TABLE 271. The 1930 observations (Tsesevich)

JD hel	s	JD hel	s	JD hel	s
242...		242...		242...	
6011.345	6.8	6068.432	4.2	6074.514	5.8
.359	5.8	.442	4.5	6090.456	8.1
.382	2.3	.466	5.2	.502	9.0
.391	1.2	.482	5.1	6091.410	3.9
.411	2.3	.505	5.7	.497	7.5
.448	3.9	.518	5.8	6092.393	4.5
6065.488	1.9	.549	6.3	.410	4.2
.504	2.9	6071.408	7.3	6092.435	5.5
.572	4.2	.559	5.2	.450	5.7
6066.440	6.8	6072.453	6.3	6093.337	3.9
.506	6.6	.490	8.3	.341	2.7
.542	7.2	6073.321	2.1	.353	1.7
.567	4.5	.398	4.9	.360	1.9
6067.417	6.3	.453	5.8	.375	1.4
.438	7.4	6074.399	3.9	.385	2.4
.468	7.6	.455	4.4	.395	1.8
.517	7.9	.495	5.8	.418	2.9
.562	7.6	.505	6.3	.430	3.6

TZ Aquarii

Repeatedly observed by Tsesevich /145/, and later by Solov'ev /108/ and Lange. The period was determined by Tsesevich from the observations of Solov'ev, Lange, his own visual observations, and his measurements of Simeise photographs. Table 272 lists all the maxima.

TABLE 272. List of maxima

Source	Max hel JD		E	O — A	O — B
Tsesevich (photogr.)	2420775.242	—	6928	$+0^d.040$	$+0^d.018$
	4379.466	—	618	+ .023	+ .010
Tsesevich (visual)	4756.430	+	42	— .002	— .014
	5098.580	+	641	+ .002	— .009
	6573.408	+	3223	+ .004	— .003
	6965.244	+	3909	.000	— .006
	7426.192	+	4716	— .007	— .011
Solov'ev /108/	8398.378	+	6418	+ .005	+ .003
Tsesevich (photogr.)	8752.514	+	7038	.000	— .001
Tsesevich (visual)	30592.333	+	10259	— .001	+ .003
	7150.214	+	21740	— .012	+ .010
Lange (visual)	7878.482	+	23015	— .018	+ .006
	7885.333	+	23027	— .021	+ .011
	7906.454	+	23064	— .034	— .011

The O — A residues were calculated from

$$\text{Max hel JD} = 2424732.442 + 0.5711952 \cdot E. \qquad (190)$$

The entire set of O — A values yielded the improved equation

$$\text{Max hel JD} = 2424732.454 + 0.57119366 \cdot E; \; P^{-1} = 1.75071971. \qquad (191)$$

The variation of the O — B residues shows that the period is subject to small irregular fluctuations. The Simeise photographs were used to plot the average light curve (Table 273).

TABLE 273. Average light curve

φ	s	n	φ	s	n	φ	s	n
$0^P.024$	2.6	5	$0^P.435$	9.1	5	$0^P.838$	9.2	6
.058	0.4	6	.492	8.9	5	.862	6.5	5
.084	—0.2	5	.538	9.0	5	.886	4.2	4
.120	3.3	6	.555	9.0	6	.901	4.8	4
.164	3.3	5	.588	9.7	5	.916	3.0	6
.194	6.2	5	.621	9.0	6	.947	0.6	6
.252	7.0	5	.667	8.9	6	.972	0.6	3
.321	8.3	5	.714	8.6	5	.982	—0.2	4
.364	7.4	6	.744	9.1	6			
.404	8.1	6	.796	10.7	6			

Comparison stars (Figure 118): photographic scale $a = 0^s.0$; $b = 10^s.1$; $c = 13^s.7$; visual scale $a = -10^s.4$; $b = 0^s.0$; $c = 10^s.1$; $d = 17^s.8$. Tables 274, 275 give the photographic and the new visual observations of Tsesevich. The powers have been converted to visual stellar magnitudes of Solov'ev's system using the relation $m = 12.11 + 0.0546s$.

222

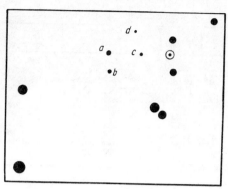

FIGURE 118. Comparison stars for TZ Aquarii.

TABLE 274. Simeise observations (Tsesevich)

JD hel	s	JD hel	s	JD hel	s
241...		242...		242...	
9629.314	8.3	4389.416	8.5	6984.351	8.6
	9.0		9.2		9.2
9987.508	2.2	4412.445	9.6	7329.408	10.1
9991.470	5.0		9.6	7343.411	2.0
	5.5	4414.368	1.4		1.0
242...			1.0	7356.261	10.1
0006.355	2.8	4771.365	9.1		7.8
	3.4		7.1	7663.453	7.1
0385.277	8.3	4787.275	0.0		4.0
	9.2		1.0	8394.496	8.1
0388.320	8.3	5474.441	—1.0		5.5
	7.9	5476.389	10.1	8396.446	10.8
0717.475	3.5	5478.487	0.0		10.1
	4.8		—1.0	8397.496	11.3
0743.350	8.7	5479.451	12.3	8399.428	9.2
0754.369	8.7		10.1		10.1
	7.1	5481.392	1.6	8401.465	7.4
0771.274	0.0	5482.466	1.6		7.7
	0.0		—1.0	8403.459	6.4
0775.291	2.4	5499.309	9.2		6.1
	—1.0		6.4	8752.499	0.5
0786.242	8.3	5502.394	—2.0		1.4
1075.500	11.5		—1.5	8753.453	9.1
	10.1	5508.333	8.6		9.1
1099.392	10.1		7.3	8758.456	7.1
	9.1	5511.397	7.6		8.3
1102.419	10.1		9.0	8779.373	5.0
	8.7	5525.286	—1.0	9132.458	2.0
1106.438	7.1	5527.327	8.5		3.7
	6.4		9.6	9495.410	8.3
1108.449	9.1	5528.262	8.1		8.6
	8.1	5534.247	10.8	9498.440	4.2
1112.429	7.9	5823.470	0.0		4.6
	7.9	5854.425	5.6	9512.296	3.7
1466.469	7.1		4.0		5.5
1483.359	8.5	5865.474	9.2	9514.394	9.1
	10.1		10.1		9.2
2552.353	9.1	5890.347	1.6	9518.403	8.5
	8.3	5892.354	8.5	243...	
3996.502	8.7		5.3	0251.311	0.0
	9.6	6208.362	3.0	2769.441	8.8
4051.357	10.1		3.7		8.6
	9.1	6240.371	1.0	2795.399	2.2
4379.491	—1.0	6248.330	3.0		2.2
	0.0		5.6	3152.373	0.0
4381.399	8.8	6264.233	8.8		0.0
	9.2		7.0	3215.245	—1.0
4385.490	12.2	6268.242	8.5		—1.0
	8.7		8.3	4246.273	5.6
4388.344	9.0	6566.431	11.3		4.5
	9.1		11.9		

223

TABLE 275. New visual observations (Tsesevich)

JD hel	m	JD hel	m	JD hel	m
243...		243...		243...	
7196.272	12.91	7197.341	12.89	7198.433	12.59
.312	12.85	.397	12.89	7204.333	12.90
.348	12.84	7198.314	12.42	.356	12.84
.362	12.84	.322	12.40	.369	12.69
.414	12.57	.327	12.44	.380	12.66
.434	12.40	.341	12.51	.389	12.84
.452	12.04	.350	12.53	.412	12.85
.471	11.88	.374	12.44	.426	12.51
.485	11.85	.403	12.59	.438	12.36
7197.290	12.90	.420	12.55	.454	11.98

RU Canum Venaticorum

Discovered by Tserasskaya /220/. Observed by Blazhko /25, 198/. Hoffmeister /255/, Jordan /290/, Parenago /328/, Gur'ev /27/, Payne-Gaposchkin /329/, Alaniya /1/, Strelkova /115/, Satanova /94/, and Tsesevich /374/. A list of maxima was compiled by Satanova. This list, however, was incomplete, since Blazhko had not published a substantial part of his observations. Tsesevich proceeded with a thorough revision of maxima, using Blazhko's archives. Two additional maxima were determined by Tsesevich from his unpublished observations. Hoffmeister's data were omitted.

Table 276 lists all the available maxima. The O—A residues were calculated from

$$\text{Max hel JD} = 2420227.340 + 0.5732492 \cdot E \qquad (192)$$

and are plotted in Figure 119. The period changes irregularly and the ephemeris can be calculated from Satanova's equation

$$\text{Max hel JD} = 2420227.398 + 0.57324647 \cdot E.$$

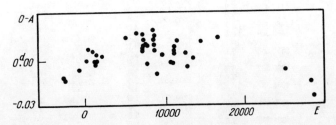

FIGURE 119. O—A residues for RU Canum Venaticorum.

TABLE 276. List of maxima

Source	Max hel JD	E	O — A
Blazhko	2418979 364	— 2177	— 0^d.012
	9673.574	— 966	— .007
	20277.340	0	.000
	0243.398	+ 28	+ .007
	0626.327	+ 696	+ .006
	0638.356	+ 717	— .004
	0743.263	+ 900	— .001
	0959.382	+ 1277	+ .003
Jordan /290/	1015.556	+ 1375	— .002
Blazhko	1138.232	+ 1589	— .001
	1374.413	+ 2001	+ .001
	3260.417	+ 5291	+ .015
Jordan /290/	3911.631	+ 6427	+ .018
Blazhko	4269.336	+ 7051	+ .016
	4277.358	+ 7065	+ .008
	4285.377	+ 7079	+ .006
	4293.402	+ 7093	+ .005
Jordan /290/	4314.619	+ 7130	+ .012
Parenago /238/	4609.266	+ 7644	+ .009
Blazhko	4653.406	+ 7721	+ .009
	4657.406	+ 7728	— .004
Tsesevich /374/	4936.595	+ 8215	+ .013
	4963.545	+ 8262	+ .020
	4966.405	+ 8267	+ .014
	4967.547	+ 8269	+ .009
	4986.458	+ 8302	+ .003
Blazhko	4997.351	+ 8321	+ .004
	5365.363	+ 8363	— .010
	5745.439	+ 9626	+ .002
Tsesevich /374/	6066.455	+ 10186	— .001
Blazhko	6073.333	+ 10198	— .002
Tsesevich /374/	6074.491	+ 10200	+ .009
Blazhko	6120.349	+ 10280	+ .007
Tsesevich /374/	6770.417	+ 11414	+ .011
	6771.557	+ 11416	+ .004
Florya	7243.341	+ 12239	+ .004
Blazhko	7580.400	+ 12827	— .007
	7812.256	+ 13371	+ .001
Gur'ev /27/	8365.197	+ 14196	+ .011
Payne-Gaposchkin /329/	9436.601	+ 16065	+ .013
Alaniya /1/	34483.465	+ 24869	— .009
Strelkova /115/	6406.711	+ 28224	— .014
Satanova /94/	6665.234	+ 28675	— .027

ST Bootis

Observed by numerous authors. List of maxima given in Table 277. The O — A residues were calculated from

$$\text{Max hel JD} = 2426068.381 + 0.62229075 \cdot E. \qquad (193)$$

TABLE 277. List of maxima

Source	Max hel JD	E	O — A
Tsesevich, Parenago /373/	2419181.51:	— 11067	— 0^d.021
Zinner /376/	20234.423:	— 9375	+ .018
	0239.379	— 9367	— .004
	0240.626	— 9365	— .002
	0242.489	— 9362	— .006
Zinner /376/	0249.343	— 9351	+ .003
	0346.402	— 9195	— .016

225

TABLE 277 (continued)

Source	Max hel JD	E	O — A
Nijland /325/	2423630.285	— 3918	+ 0^d.039
	3836.886	— 3586	+ .040
	3988.714	— 3342	+ .039
	4289.288	— 2859	+ .036
	4657.058	— 2268	+ .032
	4957.012	— 1786	+ .042
	5344.076	— 1164	+ .041
	5516.441	— 887	+ .032
Tsesevich /373/	6058.427:	— 16	+ .003:
	6066.508	— 3	— .006
	6068.373	0	— .008
	6071.49:	+ 5	— .002:
	6091.409	37	+ .003
Radlova /92/	7313.620	+ 2001	+ .035
Selivanov /95/	7681.347	+ 2592	— .012
Solov'ev /108/	7905.400	+ 2952	+ .017
Payne-Gaposchkin /329/	7967.612	+ 3052	.000
Yudkina /167/	34129.526	+ 12954	— .009
	4159.397	+ 13002	— .008
Geyer /239/	5549.614	+ 15236	+ .011
Geyer (photoel.) /239/	6279.530	+ 16409	— .020
Popov /91/	7520.402	+ 18403	+ .004
Lange	8143.315	+ 19404	+ .004

"On the average", this equation fits the observations. However, such an experienced observer as A. Nijland could not have made a biased mistake of 1 hour in the determination of the maximum. We are therefore led to the conclusion that the period is subject to irregular fluctuations.

LARGE IRREGULAR FLUCTUATIONS

Some objects were found to show considerable irregular fluctuations of periods.

ST Virginis

Observed repeatedly, but its peculiar properties were established only recently. Observed by Hartwig /251, 252/, Guthnick /245/, Nijland /325/, Solov'ev /108/, Florya /122/, Lause /304/, Tsesevich /144/, and Alaniya /2/. Florya's observations, unfortunately, were never published. When new observations revealed marked contraction of the period, the visual observations of Tsesevich and Lange were supplemented by magnitude estimates from Simeise, Odessa, and Dushanbe (G. S. Filatov) photographs (Tables 278 — 280).

The photographic magnitudes of the comparison stars (Figure 120) were determined using the BSD standard:

	m	s_F	s_S	s_O	m_F	m_O	m_S
k	9.09	—	0.0	—8.0	—	9.51	9.12
m	9.95	0.0	10.0	0.0	9.75	10.06	9.88
s	10.47	—	17.0	2.8	—	10.25	10.42
p	10.66	—	22.9	10.7	—	10.79	10.87
c	11.21	38.2	—	—	11.56	—	—
q	11.77	—	33.7	24.6	—	11.74	11.69
d	12.51	51.9	—	—	12.20	—	—
a	10.29	14.0	—	—	10.41	—	—
b	10.89	24.5	—	—	10.91	—	—

The estimates gave three power scales, Filatov's s_F, Simeise s_s and Odessa s_o. They are related to stellar magnitudes by $m_F = 9.75 + 0.0473 s_F$; $m_s = 9.12 + 0.07623 s_s$; $m_o = 10.06 + 0.0682 s_o$. From calculated m_F and m_s we determined all the stellar magnitudes listed in Tables 278—280.

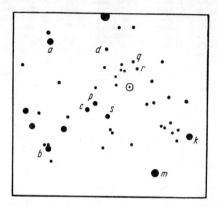

FIGURE 120. Comparison stars for ST Virginis.

TABLE 278. Simeise observations

JD hel	m	JD hel	m	JD hel	m
241...		242...		242...	
9530.307	11.32	4620.475	11.55	9077.358	11.61
	11.31		11.54		11.69
9530.392	11.48	4650.367	11.54	243...	
	11.54		11.51	0113.360	11.39
9531.309	11.48	5384.387	10.64		11.34
	11.48		10.53	0116.369	11.55
9886.439	10.57	5732.316	10.35	0130.343	11.54
	10.64		10.28		11.57
9887.380	11.39	6095.348	11.57	0132.302	11.53
	11.39		11.42		11.46
9888.403	10.99	6096.361	10.15	0136.377	11.51
	10.96		9.81		11.49
9900.432	10.72	6477.331	11.44	0138.433	11.51
242...			11.47		11.48
0245.542	10.80	6478.394	11.07	0146.386	11.50
	10.77	7211.350	10.72		11.49
0253.384	11.00		10.95	2647.496	11.32
	11.03	7543.478	11.38		11.44
0582.542	11.39		11.34	2655.485	10.72
0597.502	11.59	8631.527	11.60		10.33
0634.362	11.54		11.46	2675.327	11.03
	11.42	8633.456	11.17	3001.499	10.42
0948.528	11.09		11.11		10.79
	10.87	8637.503	10.53	3034.416	11.36
0983.463	11.30		10.53		11.09
0986.526	11.62	9018.477	11.33	3411.369	11.32
1369.353	11.48		11.47		11.42
	11.53	9025.434	11.42	4096.504	11.47
3173.457	11.47		11.39		11.52
	11.36	9069.352	11.00	4131.405	11.44
3938.434	11.51		11.00	4510.342	10.87
	11.58	9075.339	11.54		10.98
4268.356	11.58		11.59		
	11.53				

TABLE 279. Observations of G. S. Filatov

JD hel	m	JD hel	m	JD hel	m
243...		243...		243...	
0065.085	11.70	4181.233	11.91	5628.266	10.54
3381.405	10.22	4485.378	11.09	5638.247	11.56
3395.357	11.86	4537.248	10.71	5976.262	10.68
3408.240	11.28	4538.228	10.66	5981.299	11.21
3416.266	11.93	5577.275	9.51	6339.250	11.28
3439.246	11.88	5598.388	11.42	6368.241	11.21
3449.238	10.06	5602.383	10.15	6399.377	11.56:
4102.356	11.88	5610.304	11.23	6748.200	10.41
4152.294	11.74	5626.280	11.37	7079.283	10.91
4157.284	11.56:				

TABLE 280. Odessa observations

JD hel	m	JD hel	m	JD hel	m
243...		243...		243...	
6699.433	11.52	7373.583	11.46	7763.490	11.42
6701.418	10.94	7377.592	10.64	.513	11.55
6702.430	11.60	7378.524	11.05	7764.506	10.06
6703.440	10.98	.550	11.36	7781.463	11.34
6715.389	11.36	7400.505	9.94	7783.430	10.12
6716.407	11.53	.529	10.06	7786.452	11.54
6722.380	11.08	7402.446	11.34	7808.390	11.50
6726.354	10.06	.498	11.62	7810.362	11.55
6729.347	11.42	7405.473	10.40	.384	11.54
6730.352	11.48	.498	10.74	7813.362	10.00
6971.634	10.39	7406.529	11.52	7817.359	11.55
7002.610	11.42	7426.442	10.55	7818.345	10.40
7015.554	10.25	7427.382	11.26	.371	10.79
7016.486	10.96	.428	11.52	8141.482	11.49
7019.504	11.49	7429.414	11.05	.569	11.42
7020.532	10.16	7432.389	11.42	8143.435	11.37
7028.524	11.40	7464.369	11.34	.460	11.47
7029.531	10.16	7729.603	10.39	8144.445	11.23
7044.407	11.11	7734.564	10.86	.468	10.17
7046.474	11.17	.588	10.69	8162.372	11.55
7052.438	11.80	7758.517	11.52	8163.404	10.16
7071.424	10.06	.542	11.40	8165.374	11.53
7072.412	11.32	7759.504	11.33	8170.369	10.06
7073.403	11.53	.532	10.06	8172.401	10.14
7075.387	11.66	7761.515	11.50		
7373.557	10.96	.539	11.61		

All the observations were reduced to give seasonal light curves. The old observations were reduced using the equation

$$\text{Max hel JD} = 2425325.590 + 0.41084567 \cdot E; \; P^{-1} = 2.43400399, \tag{194}$$

and the new observations, after 2435000, using the equation

$$\text{Max hel JD} = 2436703.326 + 0.41081 \cdot E; \; P^{-1} = 2.4342153. \tag{195}$$

Near JD 2434000 — JD 2435000 the period changed abruptly. Filatov and Tsesevich made several observations in this epoch (from Simeise photographs). All attempts to construct a seasonal light curve for this interval of time failed! The maxima are listed in Table 281.

TABLE 281. List of maxima

Source	Max hel JD	E	O — A	O — B
Hartwig /251, 252/	2417797.260	—18324	$+0^d.006$	$—0^d.108$
Guthnick /245/	8093.487	—17603	$+.013$	$— .095$
Tsesevich (Sim.)	22053.281	—12346	$— .009$	$— .079$
Nijland /325/	3514.563	— 4408	$— .019$	$— .033$
	3913.499	— 3437	$—.015$	$— .021$
	4284.491	— 2534	$—.016$	$—.016$
	4618.516	— 1721	$—.009$	$—.003$
	5026.485	— 728	$—.010$	$+.004$
Tsesevich (visual)	5325.583	0	$—.007$	$+.011$
	5330.505	+ 12	$—.015$	$+.003$
Nijland /325/	5362.564	+ 90	$—.002$	$+.017$
Tsesevich (Sim.)	5384.337	+ .143	$—.004$	$+.015$
Nijland /325/	5759.441	+ 1056	$—.002$	$+.024$
Lause /304/	5780.389	+ 1107	$—.007$	$+.019$
Tsesevich (visual)	6067.572	+ 1806	$—.005$	$+.026$
	6091.398	+ 1864	$—.007$	$+.023$
Lause /304/	6440.625	+ 2714	$.000$	$+.038$
Tsesevich (visual)	6455.414	+ 2750	$—.002$	$+.037$
Solov'ev /108/	7618.531	+ 5581	$+.011$	$+.070$
Tsesevich (visual)	7842.443	+ 6126	$+.012$	$+.075$
Tsesevich (Sim.)	9069.253	+ 9112	$+.037$	$+.121$
Tsesevich (Sim.)	33001.416	+18683	$—.004$	$+.149$
Alaniya /2/	4478.396	+22278	$—.014$	$+.165$
Filatov	5577.275	+24953	$—.147$	$+.051$
Tsesevich (visual)	6702.507	+27692	$—.221$	$—.003$
	6703.327	+27694	$—.223$	$—.005$
Tsesevich (Odessa)	7020.500	+28466	$—.223$	$+.001$
	7400.503	+29391	$—.252$	$—.022$
Lange (visual)	7781.300	+30318	$—.309$	$—.072$
Tsesevich (Odessa)	7783.376	+30323	$—.287$	$—.050$
Lange (visual)	7808.437	+30384	$—.288$	$—.050$
	7813.353	+30396	$—.302$	$—.065$
	8152.286	+31221	$—.317$	$—.073$
Tsesevich (Odessa)	8163.378	+31248	$—.318$	$—.074$
Lange (visual)	8163.379	+31248	$—.317$	$—.073$

The O — A residues were calculated from equation (194), derived by N. F. Florya /122/.

Table 281 leads to the following conclusions. The period is variable. Deviations from the initial linear relation have almost reached 3/4 of the period by now. Although the last observations are described by Lange's equation

$$\text{Max hel JD} = 2436703.327 + 0.410817 \cdot E, \qquad (196)$$

the least squares method gives a common formula for all the maxima in the table:

$$\text{Max hel JD} = 2425325.5717 + 0.41083846 \cdot E, \qquad (197)$$

from which the O — B residues were calculated (Figure 121).

Lange is of the opinion that the star shows Blazhko effect.

Tsesevich's visual observations are listed in Tables 282, 283. Comparison stars (Figure 120): $m = -8^s.3$; $s = 0^s.0$; $p = 10^s.5$; $q = 21^s.0$; $r = 34^s8$.

229

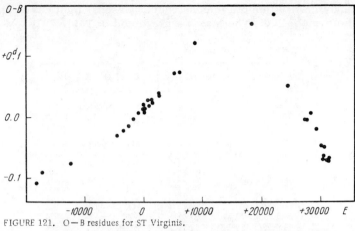

FIGURE 121. O—B residues for ST Virginis.

TABLE 282. Old visual observations (Tsesevich)

JD hel	s	JD hel	s	JD hel	s
242...		242...		242...	
5325.502	23.8	5332.467	23.1	6091.434	4.8
.516	25.1	.495	22.4	.494	14.0
.543	24.5	.526	23.8	.515	16.5
.563	14.7	5345.374	8.4	6092.445	17.5
.571	1.5	.378	9.0	.476	22.3
.574	—0.2	.396	10.9	6093.406	22.2
.580	—0.8	.415	16.8	.415	23.1
.587	—1.0	.439	19.2	.429	19.1
.598	—0.4	.466	20.2	.436	14.3
.607	0.0	5347.477	18.4	.443	7.2
5327.492	21.7	.503	19.5	.450	—0.3
.508	21.7	.529	20.0	.454	—1.6
5330.426	22.4	6067.502	22.8	.463	—1.4
.437	23.1	.509	22.8	6455.401	5.6
.480	18.9	.525	22.8	.420	—3.8
.485	16.8	.535	19.2	.426	—2.3
.493	2.6	.537	19.2	.435	2.1
.498	0.0	.542	22.2	.446	6.0
.503	—1.7	.554	17.1	.451	6.5
.516	—1.1	.571	—1.8	.460	7.9
.526	—0.4	.576	—1.8	.477	8.8
.539	1.0	.583	0.8	7842.424	12.6
.553	4.7	.587	3.1	.432	5.2
.575	7.4	6090.489	23.1	.438	—0.2
.581	7.6	.499	22.2	.441	—1.8
5331.395	7.9	.515	22.2	.447	—1.3
.412	9.8	6091.384	12.1	.454	—0.7
.440	12.6	.386	9.7	.461	0.0
.530	18.9	.390	4.8	.466	3.0
.549	21.0	.397	—1.5	.480	2.1
.580	21.3	.401	—1.6	.496	8.1
5332.424	21.0	.404	—1.2	.502	8.1
.431	21.7	.408	—1.0		
.445	22.4	.414	—0.8		

TABLE 283. New visual observations (Tsesevich)

JD hel	s	JD hel	s	JD hel	s
243...		243...		243...	
6702.419	27.4	6702.457	27.9	6702.486	14.0
.431	27.4	.464	28.1	.487	14.0
.438	27.0	.468	27.1	.488	13.1
.446	27.4	.475	27.9	.491	8.8
.454	27.9	.485	16.2	.494	7.3

TABLE 283 (continued)

JD hel	s	JD hel	s	JD hel	s
243...		243...		243...	
6702.499	6.3	6703.396	9.0	6720.431	25.2
.500	4.2	.405	8.6	.446	25.6
.504	4.2	.419	12.8	6722.328	9.3
.508	3.5	.440	15.8	.342	13.7
.516	5.5	.448	16.2	.360	16.0
6703.302	23.8	.463	16.2	.369	18.3
.303	23.1	.476	16.5	.400	24.9
.304	18.7	.512	19.7	.422	24.9
.305	16.3	.529	25.6	6730.321	26.5
.307	16.0	6717.371	15.4	.332	25.6
.312	9.3	.378	9.0	.348	25.9
.315	5.7	.382	14.0	.357	26.5
.318	5.0	.407	15.8	.367	28.4
.332	2.9	.420	16.6	.381	27.9
.335	3.8	.425	16.2	.410	27.9
.342	4.8	.430	16.6	6732.323	27.9
.348	5.7	.442	17.9	.349	27.9
.355	5.7	6720.369	17.2	.365	27.9
.362	6.4	.374	18.9	.380	27.9
.370	6.4	.398	23.0	.400	27.1
.383	8.0	.412	25.2	6734.420	27.0
.390	7.3	.422	24.2		

The new observations were reduced according to equation (194) to give the average light curve (Table 284).

TABLE 284. Average visual light curve

φ	s	n	φ	s	n	φ	s	n
$0^P.048$	25.1	5	$0^P.406$	16.2	5	$0^P.636$	9.9	5
.131	27.0	5	.414	13.6	2	.703	13.5	5
.211	26.8	5	.425	7.8	4	.756	15.7	4
.279	27.4	5	.444	4.9	4	.785	16.3	3
.336	27.8	4	.473	3.9	4	.813	17.6	3
.375	27.6	3	.509	5.4	3	.916	21.7	4
.398	23.4	2	.566	6.9	3	.950	22.2	2

EL Cephei

Discovered by Hoffmeister. Romano /347/ studied the star and assigned it to RR Lyrae variables. He also derived the equation

$$\text{Max hel JD} = 2436079.348 + 0.416671 \cdot E; \; P^{-1} = 2.399975. \tag{198}$$

Tsesevich made visual observations in 1962 and estimated the magnitude from old Moscow photographs. Visual observations were reduced using Romano's equation to give a seasonal light curve. The maxima deviated by half the period from the ephemeris calculated using (198). Romano's equation was thus improved by Tsesevich.

$$\text{Max hel JD} = 2436079.348 + 0.4166275 \cdot E; \; P^{-1} = 2.40022562, \tag{199}$$

and this equation was used to construct the seasonal curves from Moscow observations. The maxima are listed in Table 285.

TABLE 285. List of maxima

Source	Max hel JD	E	O — B
Tsesevich (Mosc.)	2416349.440	— 47356	— 0.$^{\rm d}$096
	29165.419	— 16595	+ .004
	30614.445	— 13177	.000
	3012.540	— 7361	— .013
	3382.510	— 6473	— .008
	5724.381	— 852	.000
Romano /347/	6079.348	0	.000
Tsesevich (visual)	7912.473	+ 4400	— .036

A plot of O — B residues shows that the period varies irregularly.
Comparison stars (Figure 122): visual observations $a = 0^s.0$; $b = 6^s.5$; $c = 12^s.8$; $d = 19^s.0$; $e = 23^s.8$; photographic observations $k = -9^s.4$; $a = 0^s.0$; $b = 8^s.4$; $c = 12^s.6$. The average visual light curve is given in Figure 123 and Table 286. Tsesevich's observations are listed in Tables 287 and 288.

TABLE 286. Average visual light curve

φ	s	n	φ	s	n	φ	s	n
0$^{\rm p}$·060	19.0	4	0$^{\rm p}$.351	18.8	5	0$^{\rm p}$.529	6.2	5
·144	17.3	5	.385	13.6	5	.565	6.7	5
·196	16.2	5	.400	9.6	5	.608	8.9	4
.243	18.1	5	.422	3.6	5	.655	10.4	3
.280	17.6	5	.451	2.2	5	.815	14.6	4
.318	18.4	5	.495	3.9	5	.962	17.5	3

TABLE 287. Visual observations (Tsesevich)

JD hel	s	JD hel	s	JD hel	s
243...		243...		243...	
7904.431	18.8	7911.303	10.7	7912.440	16.3
.464	18.8	.390	11.5	.443	14.8
.476	19.8	.519	16.5	.444	14.3
.499	18.8	7912.276	17.8	.446	12.8
7906.348	15.8	.300	18.0	.447	10.0
.384	17.4	.313	20.0	.448	10.0
.424	17.4	.323	20.0	.450	9.4
.465	17.8	.335	16.8	.452	8.8
.437	17.4	.338	17.8	.455	4.9
7908.338	9.3	.345	16.8	.459	4.8
.345	10.0	.355	17.4	.463	3.2
.364	9.6	.366	16.8	.465	2.5
.380	11.0	.370	20.7	.470	1.6
.397	11.2	.378	22.6	.476	2.3
.454	13.8	.383	16.8	.483	3.2
7910.274	11.9	.390	18.0	.488	3.5
.295	19.0	.395	16.8	.494	3.5
.309	18.1	.399	18.0	.500	3.2
.334	21.0	.405	16.8	.507	4.0
.364	16.3	.409	16.8	.512	3.9
.374	2.8	.413	16.8	.522	5.2
.377	2.9	.416	17.8	.532	8.6
.385	2.4	.419	17.8	.553	9.2
.396	2.4	.422	18.8	7913.291	4.3
.404	3.7	.426	18.8	.330	5.4
.416	4.3	.429	18.8	.344	10.3
.436	4.6	.434	18.8	.367	10.7
.452	5.4	.438	18.8		

TABLE 288. Photographic observations (Tsesevich)

JD hel	s	JD hel	s	JD hel	s
241...		243...		243...	
4784.395	[8.4	0594.325	15.6	3382.418	16.6
4989.199	[8.4	0594.349	[12.6	3382.481	4.7
5254.445	[8.4	0614.389	15.6	3389.402	11.2
5612.448	6.5	0614.426	2.0:	3408.427	10.5
5613.436	[12.6	0674.241	14.1	3412.386	14.6
5642.336	10.5	0675.324	5.6	3512.480	0.0
6349.429	7.0	2977.603	5.3	3708.524	7.5
6708.356	[8.4	3005.491	4.4	3711.505	[12.6
6736.382	[8.4	3006.478	12.6	3718.553	10.5
8235.244	7.4	3007.532	2.5	3761.420	15.6
8238.395	5.9	3011.507	10.9	3766.397	16.6
8239.390	10.1	3012.537	—1.4	3949.422	14.6
242...		3025.400	3.7	3951.283	7.0
7775.198	8.4	3025.470	1.9	3952.277	8.4
7784.145	13.6	3025.518	7.4	3952.354	15.6
8760.446	—1.0	3026.456	8.4	4118.434	10.8
9146.455	15.6	3032.368	11.4	4118.458	[8.4
9147.463	7.4	3033.330	10.5	4123.364	8.4
9162.348	11.6	3179.370	10.8	4127.441	1.7
9165.412	—1.0	3210.345	10.5	4128.371	11.6
9167.363	15.6	3212.311	8.4	4146.427	10.1
9172.383	13.6	3214.220	0.0	4333.257	—1.0
9229.303	14.6	3358.570	10.5	4681.215	6.5
9491.409	12.6	3360.560	8.4	5348.445	[8.4
9496.481	11.6	3361.462	11.6	5366.494	0.0
9519.344	14.6	3361.569	15.6	5394.421	3.7
9521.440	11.6	3369.461	13.6	5724.390	0.0:
9526.336	10.8	3369.518	2.4	6134.377	6.5:
9547.377	16.6				

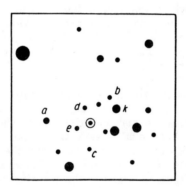

FIGURE 122. Comparison stars for EL Cephei.

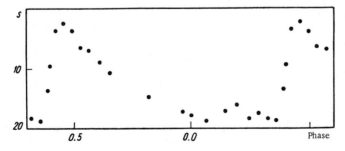

FIGURE 123. Light curve of EL Cephei.

233

V 455 Ophiuchi

Investigated by Jacchia /284/, who obtained

$$\text{Max hel JD} = 2428373.504 + 0.4529 \cdot E. \qquad (200)$$

Observed by no one since 1936, so that the period can hardly be improved. Tsesevich estimated the magnitude from several Odessa photographs, and then observed the star visually. To fit the observations with a single formula, Tsesevich estimated the magnitude from Moscow photographs. This "curve fitting", however, proved a difficult undertaking. Therefore, at Tsesevich's request, A. Filin estimated the magnitude from Dushanbe photographs. The resulting combined series of observations covered some 50 years (though with gaps).

The scattered photographic observations were combined into average seasonal curves, the average maxima were determined, and the behavior of the star over this long period was established.

The period is subject to pronounced fluctuations. All the observations from 1936 to 1958 can be fitted with the equation

$$\text{Max hel JD} = 2433179.164 + 0.45391817 \cdot E; \quad P^{-1} = 2.20304025. \qquad (201)$$

The average light curve from photographic observations was plotted using this equation. Old Moscow observations also have the same period, with the remarkable difference that the maximum is shifted by half a period! In the archives Tsesevich discovered visual observations from 1943. They were reduced to give an average visual curve, from which the maximum was determined. Tsesevich furthermore constructed a combined average curve from visual observations carried out in 1958. These curves were derived using the elements

$$\text{Max hel JD} = 2433179.202 + 0.45391256 \cdot E; \quad P^{-1} = 2.20306748. \qquad (202)$$

Although these elements are tentative, no further improvement is attempted, since the individual observation series are too short.

The Blazhko effect distinctly emerged from the average curves. The observations available, however, are insufficient for detailed study of this phenomenon. Figure 124 shows the limiting curve shapes.

Table 289 lists the average maxima and the $O - A$ residues. The first three maxima correspond to individual photographs on which the star was brightest. The table also includes Tsesevich's maxima obtained from his Harvard observations.

The $O - A$ residues were calculated from (202).

The period is subject to large secular fluctuations (Figure 125). Comparison stars (Figure 126): photographic scale $a = -5^s.9$; $A = 0^s.0$; $b = +5^s.0$; $c = +16^s.6$; visual scale $b = 0^s.0$; $c = 5^s.6$; $d = 12^s.4$; $e = 14^s.7$; $f = 16^s.3$. Tsesevich's light curves are given in Tables 290–292, the observations of Tsesevich and Filin are listed in Tables 293–297.

TABLE 289. List of maxima

Source	Max hel JD	E	O — A
Tsesevich (Mosc.)	2414868.38	— 40340	+0 .^{d}011
	6371.28	— 37029	+ .006
	9271.36	— 30640	+ .039
Tsesevich (Harv.)	20564.486	— 27791	— .032
	1925.316	— 24793	— .032
	6574.642:	— 14550	— .132:
	7998.616	— 11413	— .082
Jacchia /284/	8373.504	— 10587	— .126
Tsesevich (Mosc.)	9404.420	— 8316	— .045
Tsesevich (Harv.)	9458.448	— 8197	— .033
Tsesevich (visual)	30976.288	— 4853	— .076
Tsesevich (Harv.)	1062.553	— 4663	.055
	3079.767	— 219	— .028
Filin	3179.156	0	— .046
	3424.283	+ 540	— .032
	5358.445	+ 4801	+ .009
Tsesevich (Odessa)	6053.363	+ 6332	— .013
Tsesevich (visual)	6398.345	+ 7092	— .005

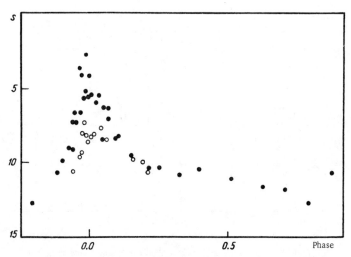

FIGURE 124. Extreme light curves of V 455 Ophiuchi.

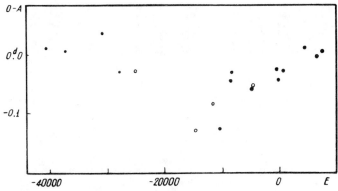

FIGURE 125. O—A residues for V 455 Ophiuchi (O Harvard observations).

235

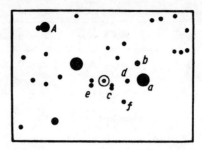

FIGURE 126. Comparison stars for V 455
Ophiuchi.

TABLE 290. Visual light curve from 1943 observations

φ	s	n	φ	s	n	φ	s	n
0^P.024	9.3	5	0^P.500	14.2	5	0^P.794	10.2	1
.121	11.1	4	.567	14.8	5	.829	5.1	3
.196	13.1	5	.653	15.2	5	.876	7.6	2
.321	15.4	4	.735	13.4	3	.922	9.0	2

TABLE 291. Visual light curve from 1958 observations

φ	s	n	φ	s	n	φ	s	n
0^P.006	5.5	5	0^P.257	10.3	5	0^P.932	9.1	5
.026	6.0	5	.327	10.9	5	.948	7.3	4
.048	8.5	5	.398	10.5	5	.966	6.6	4
.074	6.3	5	.514	11.1	5	.974	5.6	4
.106	8.3	5	.626	11.7	5	.987	5.3	5
.152	9.5	5	.707	11.8	5	.995	5.6	4
.193	10.1	5	.795	12.8	5			
.220	10.3	5	.884	10.7	5			

TABLE 292. Average photographic light curve, equation (201)

φ	s	n	φ	s	n	φ	s	n
0^P.024	2.9	4	0^P.281	13.7	4	0^P.640	15.2	4
.061	5.5	4	.334	14.2	4	.684	16.4	2
.110	7.0	5	.356	13.9	4	.748	17.8	4
.130	7.8	4	.428	15.2	4	.801	15.7	4
.185	12.0	4	.521	14.6	4	.890	13.9	3
.248	16.0	4	.555	14.9	5	.990	5.0	2

TABLE 293. Moscow observations (Tsesevich)

JD hel	s	JD hel	s	JD hel	s
241...		241...		241...	
4868.381	3.2	5524.406	14.5	9271.360	5.0
5159.387	7.0	6371.275	3.6	9653.299	17.6
5254.340	13.7	7847.258	15.6		

236

TABLE 293 (continued)

JD hel	s	JD hel	s	JD hel	s
242...		242...		242...	
6235.275	15.6	9404.438	3.0	9541.242	12.7
8045.288	10.8	9485.333	13.7	9548.197	14.5
9366.506	13.4	9486.334	14.8	9549.201	3.1
9396.464	15.4	9489.399	13.7	9556.176	14.5
9399.438	3.7	9495.360	14.0		
9403.440	14.3	9496.343	13.1		

TABLE 294. Photographic observations (Filin)

JD hel	s	JD hel	s	JD hel	s
243...		243...		243...	
3099.344	10.8	3454.389	10.0	4178.316	10.0
3113.192	14.3	3829.311	11.6	4180.350	16.6
3115.223	12.4	3849.201	8.9	4213.298	16.6:
3123.233	16.6	3852.209	15.0	4263.140	16.6:
3146.178	17.6	3861.259	16.6	4542.333	3.0
3155.153	12.2	3864.246	13.0	4562.289	9.0
3157.258	21.6	3880.186	16.6:	5358.204	12.2
3175.170	10.8	3883.209	4.0	5361.160	3.0
3179.156	5.0	3886.167	16.6:	5387.137	16.6:
3181.165	16.6	3914.154	16.6:	5391.142	5.0
3182.181	16.6:	4126.381	19.6:	5396.099	5.0
3381.401	18.6	4148.309	14.6	5400.099	15.0:
3415.310	16.6	4149.306	16.6	5689.231	16.6:
3424.337	7.0	4152.383	5.0	6049.233	16.6

TABLE 295. Odessa observations (Tsesevich)

JD hel	s	JD hel	s	JD hel	s
243...		243...		243...	
6050.406	15.5	6071.327	14.1	6080.314	14.3
6051.415	16.1	6074.372	14.3	6082.312	14.9
6053.380	4.4	6075.321	14.3	6488.235	0.3
6069.321	8.9	6076.307	14.9		
6070.326	10.3	6079.293	7.3		

TABLE 296. Visual observations, 1943 series (Tsesevich)

JD hel	s	JD hel	s	JD hel	s
243...		243...		243...	
0895.218	9.6:	0939.253	15.7	0993.165	11.7
.229	9.6:	0940.185	15.7	0994.133	10.8
.248	9.5:	.258	15.7	.150	13.2
.302	11.0:	.311	15.7	0996.193	14.7
.353	9.6:	.353	15.7	0999.225	14.7
.429	9.6:	0945.285	14.1	1000.176	13.4
0896.176	9.6:	0946.287	13.6	1001.120	14.1
.212	11.7:	0959.174	13.6	.211	13.4
0897.202	14.6:	0967.210	7.1	1002.114	13.2
.347	11.0:	.232	6.9	.195	9.5
0898.344	11.7:	.304	9.2	.209	8.6
0899.311	11.7:	.324	9.2	1019.117	11.7
.370	11.7:	0968.206	8.6	1020.191	14.7
.435	8.6:	.283	11.7	1021.091	14.7
0900.253	11.7:	0969.192	15.7	.138	14.7
.279	11.0:	.277	15.7	1024.082	11.7
0901.301	9.6:	.337	13.2	1048.090	9.5
0903.364	9.9:	0970.272	14.7	1057.084	4.6
0904.310	10.6:	0971.186	14.7	1058.080	7.4
0906.232	13.3:	.211	15.2	1061.078	15.7
0907.224	12.7:	.278	10.2	1062.074	3.6
0938.334	13.4	0976.306	8.2		

TABLE 297. Visual observations, 1958 series (Tsesevich)

JD hel	s	JD hel	s	JD hel	s
243...		243...		243...	
6339.438	10.7	6353.400	8.0	6398.354	2.8
.449	10.5	.403	7.3	.366	2.6
.461	10.7	.408	8.2	.380	3.3
.469	10.7	.410	8.5	.389	3.4
.492	10.1	.417	8.3	.398	7.3
.507	10.5	.423	8.1	.418	8.3
.517	10.3	.434	7.7	.445	9.2
.524	10.8	.441	8.6	.461	9.2
.534	11.0	.487	9.7	6399.334	11.0
6340.392	10.7	.503	9.9	6400.411	11.6
.435	10.1	.509	10.7	6404.386	11.2
.450	11.0	6373.376	4.2	6405.392	11.7
.472	10.1	.383	7.2	.431	11.4
6344.352	9.3	.406	9.0	.463	11.8
.361	10.1	.433	9.5	6406.359	11.7
.387	10.1	6375.495	12.4	.384	11.8
.418	10.1	6378.374	9.0	6407.386	9.9
.431	10.5	.381	8.8	.396	9.0
.443	10.5	6381.396	11.7	.401	7.2
.498	10.8	.431	11.3	.405	6.7
6345.328	10.5	.447	13.4	.412	3.7
6347.393	10.5	.475	12.0	.416	4.2
.439	10.8	.492	13.4	423	2.8
.466	10.1	6388.406	9.0	.428	4.3
.473	10.1	.448	10.6	.434	6.4
.486	8.4	6393.354	4.3	.442	5.6
.489	5.6	.369	7.3	.450	6.4
.496	3.7	.404	9.0	.457	7.1
.500	2.8	6395.322	11.5	.472	8.5
.505	2.5	.439	11.3	6408.388	7.4
.514	3.3	6396.398	13.4	.409	8.3
6351.406	11.0	.437	14.4	6422.367	9.3
.503	12.4	6397.355	11.7	6454.282	9.3
6353.386	10.7	6398.321	8.6	.297	10.4
.396	9.6	.328	8.6		
.399	9.3	.348	2.4		

RY Comae Berenices

Discovered by Guthnick and Prager on Berlin—Babelsberg plates
/247, 248/. Studied by Tsesevich /152/, Kukarkin /51/, Alaniya /2/,
and Dziewulski /225/. The star shows a secular inequality with a term
proportional to E^2. The period, however, varies in a much more complex
way. Tsesevich reduced all his observations and estimated the magnitude
from Simeise, Moscow, Odessa, and Harvard photographs. At Tsesevich's
request, L. V. Gvozdik estimated the magnitude from the latest Odessa
photographs. The maxima are listed in Table 298.

TABLE 298. List of maxima

Source	Max hel JD	E	O — A
Tsesevich (Harv.)	2415487.830:	— 20300	$+0^d.032$:
	6499.795:	— 18142	+ .006:
	7430.665	— 16157	+ .013
	8046.873	— 14843	+ .023
	8628.829	— 13602	+ .014
	9636.588	— 11453	+ .003
Kukarkin /51/	9658.176	— 11407	+ .021
Tsesevich (Harv.)	20596.519	— 9406	— .003
	1860.814	— 6710	+ .007
Guthnick and Prager /247, 248/	5007.450	0	.000
Tsesevich (Harv.)	5676.621	+ 1427	— .018

238

TABLE 298 (continued)

Source	Max hel JD	E	O — A
Tsesevich (Sim.)	2425742.30:	+ 1567	+0^d.008:
Kukarkin /51/	6013.327	+ 2145	— .017
Tsesevich /152/	6543.222	+ 3275	— .034
	6771.599	+ 3762	— .035
Tsesevich (Harv.)	6780.494	+ 3781	— .050
Tsesevich /152/	7484.380	+ 5282	— .056
Tsesevich (Harv.)	7484.850	+ 5283	— .054
	8608.926	+ 7680	— .047
Tsesevich (Mosc.)	8630.017	+ 7725	— .060
Tsesevich (Harv.)	9712.837	+ 10034	— .041
	30095.520	+ 10850	— .020
	1212.580	+ 13232	+ .005
	1748.622	+ 14375	+ .039
Tsesevich (Sim.)	2674.39:	+ 16350	+ .102:
Tsesevich (Mosc.)	3019.008	+ 17085	+ .044
	3384.310	+ 17863	+ .036
	4129.930	+ 19453	+ .028
Alaniya /2/	5187.443	+ 21708	+ .060
Tsesevich (Mosc.)	5942.842	+ 23319	— .015
Gvozdik	7020.489	+ 25617	.011
	7808.356:	+ 27297	+ .023:

The O — A residues were calculated from the equation

$$\text{Max hel JD} = 2425007.450 + 0.46894835 \cdot E. \tag{203}$$

The plot of O — A residues (Figure 127) shows that the period is subject to large irregular fluctuations, and not to a progressive secular change.

Comparison stars (Figure 128): visual scale (s_1) $k = 0^s.0$; $b = 11^s.6$; $A = 12^s.6$; $d = 19^s.2$; $e = 28^s.2$; Simeise photographic scale (s_2) $k = -8^s.0$; $a = 0^s.0$; $b = 5^s.7$; $d = 11^s.9$; $e = 22^s.9$; Odessa photographic scale (s_3) $k = = 0^s.0$; $a = 8^s.0$; $b = 13^s.3$; $d = 18^s.9$; $f = 24^s.3$; Moscow photographic scale (s_4) $u = 0^s.0$; $k = 10^s.8$; $d = 19^s.4$; $e = 28^s.4$.

Tsesevich's visual observations gave the average light curve (Table 299). The observations are listed in Tables 300 — 303.

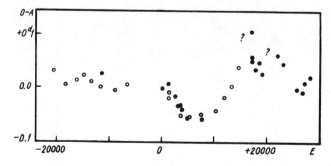

FIGURE 127. O—A residues for RY Comae Berenices (○ Harvard observations).

239

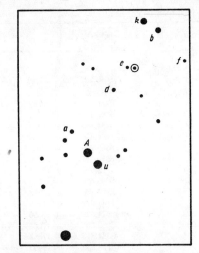

FIGURE 128. Comparison stars for RY Comae
Berenices.

TABLE 299. Average light variation curve

φ	s_1	n	φ	s_1	n	φ	s_1	n
$0^p.004$	10.5	5	$0^p.316$	22.0	5	$0^p.869$	25.5	5
.019	12.9	5	.383	26.4	5	.897	26.6	5
.031	13.2	5	.455	25.2	5	.909	25.8	5
.048	13.2	5	.558	27.1	5	.942	23.2	5
.066	14.7	6	.643	26.8	5	.960	16.8	7
.095	15.6	5	.694	26.9	5	.967	13.5	5
.126	15.9	5	.742	26.8	5	.973	14.5	5
.181	18.5	5	.811	26.4	5	.983	12.3	5
.227	20.9	5	.832	27.1	6	.993	11.2	6
.270	24.7	5	.855	24.8	5			

TABLE 300. Visual observations (Tsesevich)

JD hel	s_1	JD hel	s_1	JD hel	s_1
242...		242...		242...	
6538.224	24.2	6770.384	22.2	7424.467	16.2
6539.271	27.1	.427	25.2	.492	16.2
6540.219	26.4	6771.339	25.2	7425.428	24.6
.240	26.2	.386	27.2	.451	25.8
.265	26.0	.402	27.5	.474	25.8
6541.237	26.2	.425	27.3	7459.363	25.2
6542.226	26.2	.457	28.0	.381	26.4
.243	25.2	.485	27.8	.396	26.6
6543.212	10.6	.550	27.4	.432	26.9
6544.189	11.1	.598	10.2	.444	26.6
.227	13.5	6774.332	27.8	.493	24.9
6549.204	26.0	.347	27.6	.509	16.4
6552.191	14.6	.364	26.9	.513	14.4
6558.215	15.0	.376	25.7	.521	13.0
6559.198	14.0	6790.280	26.4	.530	10.6
.211	15.0	6809.353	26.8	.536	7.8
6560.191	17.5	6812.311	26.8	.540	10.6
6765.415	24.2	.328	26.0	.548	10.6
.436	23.2	.349	26.2	.558	13.2
.461	23.7	.368	24.2	7460.326	26.6
.486	14.4	.375	20.2	.378	27.3
.488	15.1	.382	15.8	.386	27.5
.495	11.1	.387	13.6	.397	27.4
.501	13.6	.396	11.6	.417	27.5
.510	22.2	7424.452	19.2	.421	27.5

TABLE 300 (continued)

JD hel	s_1	JD hel	s_1	JD hel	s_1
242...		242...		242...	
7460.442	22.8	7485.511	27.3	7509.421	26.8
.446	15.6	7490.339	27.3	7510.185	15.8
.451	13.1	.353	27.3	.197	16.3
.461	12.6	.409	27.3	.229	17.2
.472	10.1	.454	22.9	.283	23.3
.489	13.1	.460	14.8	.298	25.6
.507	16.2	.464	12.9	.348	26.0
.535	17.5	.469	9.6	.365	26.8
7461.328	27.2	.475	9.8	.415	27.3
.357	24.8	.481	10.6	.474	27.7
7484.303	26.6	.485	13.0	.487	27.6
.326	26.2	.491	12.9	7599.205	20.2
.358	21.3	7492.523	26.2	.212	20.2
.371	14.4	7508.218	26.9	7601.263	22.8
.377	10.6	.234	27.3	7602.196	21.4
7485.331	14.4	.254	26.9	7862.327	15.7
.335	14.4	.280	19.6	.332	14.9
.342	15.0	.287	14.4	.339	14.4
.351	15.6	.293	10.6	.352	13.4
.356	15.8	.298	9.6	.370	14.5
.363	16.0	.314	13.3	.391	14.7
.372	16.1	.334	15.2	.435	17.4
.395	16.9	.353	16.7	.482	18.2
.403	18.3	7509.329	22.5	7869.396	15.0
.430	24.2	.363	25.5		
.453	25.2	.394	26.7		

TABLE 301. Simeise observations (Tsesevich)

JD hel	s_2	JD hel	s_2	JD hel	s_2
241...		242...		242...	
8405.343	13.9	6061.370	16.5	6497.378	10.2
242...			14.1	6501.318	17.6
0985.519	16.9	6070.378	16.5		16.7
	17.4		16.3	6512.325	2.3
5716.295	16.9	6093.340	13.0		2.3
	16.9		14.1	243...	
5728.273	3.3	6127.373	9.4	2674.392	3.8
	4.7		7.5		7.8
5730.298	11.9	6421.369	3.6	2996.307	16.9
	13.9		5.7		16.5
5738.284	17.4	6424.361	14.3	3002.287	13.3
	16.9		15.9		11.9
5742.291	2.6	6451.382	6.8	4127.334	15.2
	—1.0	6497.378	8.0		17.4

TABLE 302. Odessa observations (Gvozdik)

JD hel	s_3	JD hel	s_3	JD hel	s_3
243...		243...		243...	
6288.452	3.5	6699.372	15.5	7373.557	18.9
6304.398	4.6	6702.396	21.1	7729.551	6.0
6313.447	16.7	6722.339	15.2	7758.468	21.1
6344.357	15.2	7015.492	18.9	7759.448	21.1
6345.349	17.8	7016.455	21.1	7761.491	5.0
6607.572	18.9	7017.461	20.7	7781.412	18.9
6608.596	21.6	7020.502	3.4	7790.373	21.6
6613.589	16.5	7044.379	11.5	7808.341	3.5
6660.488	17.0	7046.444	18.9	8085.563	10.1
6663.484	21.1	7052.407	18.9	8106.484	18.9
6667.417	11.2	7077.362	16.7		
6668.442	16.1	7079.336	18.9		

241

TABLE 303. Moscow observations (Tsesevich)

JD hel	s₄	JD hel	s₄	JD hel	s₄
242...		243...		243...	
7916.391	18.4	3361.417	15.0	5598.336	19.4
7918.392	22.8	.444	16.1	.362	19.4:
7925.396	23.4	3362.413	17.0	.389	23.4
8629.397	22.4	.441	20.2	.418	21.4:
8630.401	22.4	3382.324	21.2	.444	21.4
.445	23.0	.361	21.5	.483	23.4
.488	11.9	.391	16.0	5601.325	19.4:
8631.420	11.9	3389.334	20.5	.351	21.6
8652.343	20.0	.369	22.1	5907.374	19.4:
.387	23.3	.453	9.3	.399	16.2
8653.313	24.4	3408.338	17.9	.424	17.5
.401	23.2	.381	17.1	.455	20.5
8654.303	20.4	3410.452	19.4	.491	22.0
.390	12.9	3711.427	26.1	5918.436	11.9
243...		.458	} 24.8	.461	10.8
3005.413	8.1			.486	9.8
.441	10.8	.474		5924.438	21.8
.467	11.7	3718.459	23.9	.462	21.4
.550	17.6	.509	20.5	.485	20.5
3006.399	9.5	3733.390	19.4	.509	16.0
.426	12.5	3761.397	12.7	533	11.9
.454	14.9	4116.319	9.3	.558	10.8
3010.359	22.4	4118.348	19.4	5928.467	16.5
.386	23.0	.391	21.0	5929.443	20.5
.425	24.2	.416	23.0	.468	20.5
.455	22.9	4121.397	22.4	.493	18.4
.511	20.4	.420	20.5	5930.359	18.4
3011.402	21.1	4123.347	9.1	.388	19.4
.429	23.0	4124.361	11.8	.418	19.4
.456	22.4	4126.328	19.4	.443	17.8
3025.407	22.8	4127.365	21.4	.469	18.4
.435	22.5	4128.347	22.4	.494	18.4
3026.423	24.2	4130.381	10.8	5933.379	21.6
3029.310	9.2	4146.357	9.3	.413	17.3
3032.414	21.2	4454.387	10.3	.442	9.8
.435	18.5	.451	12.0	5954.328	19.4
3033.418	21.0	4455.465	14.6	.358	20.4
.443	20.3	4477.339	18.6	439	16.0
3357.435	21.2	.361	16.3	5956.330	21.4
.460	22.1	4480.375	21.4	.358	21.4
.489	21.2	4484.328	19.4	.383	19.4
.514	21.2	4507.336	19.4	.408	11.8
3358.400	21.9	5522.440	22.4	5957.343	13.0
.426	23.0	.468	24.4	.375	19.4
.460	22.4	.494	21.4	.408	19.4:
.489	13.3	.521	24.4	5976.356	22.4:
.518	7.8	.588	23.4	.385	23.0
3360.429	10.3	5550.385	11.9	.414	23.4
.456	10.8	5564.361	17.6		
.488	11.9	5598.310	19.4		

BD Herculis

Discovered by Albitzky /179/ on Simeise photographs. The most detail
study of its properties was carried out by Martynov /75/. Parenago /88/
studied the star from old Moscow photographs. Recently studied by
Lavrov /54/, Lavrova /55/, and Fridel' /140/. Lavrova, in particular,
found that in recent years the maximum markedly precedes the ephemeris
Tsesevich jointly with Karamysh examined the entire Odessa collection.
The observations of Karamysh were reduced to Tsesevich's power scale.
In Table 305 Karamysh's measurements are marked with K. Tsesevich
also revised the observations of Parenago and Albitzky and constructed
three seasonal light curves, from which more reliable maxima were

determined. All the maxima are listed in Table 304. The O−B residues were calculated from Martynov's equation

$$\text{Max hel JD} = 2430495.404 + 0.4739126 \cdot E. \tag{204}$$

TABLE 304. List of maxima

Source	Max hel JD	E	O—B	O—A
Parenago /88/	2415237.53	− 32196	+0^d.216	+ 0^d.070
	9271.37	− 23684	+ .112	+ .002
Albitzky /179/	24786.269	− 12047	+ .090	+ .030
Mokhnach /76/	6543.443	− 8339	− .004	− .048
Tsesevich /149/	6973.297	− 7432	+ .011	− .029
Martynov /75/	6983.244	− 7411	+ .006	− .034
	7345.308	− 6647	+ .001	− .036
	7562.378	− 6189	+ .019	− .016
Tsesevich /149/	7664.255	− 5974	+ .005	− .029
Martynov /75/	9368.438	− 2378	− .002	− .020
	9497.355	− 2106	+ .011	− .006
	9732.416	− 1610	+ .011	− .004
	9870.322	− 1319	+ .009	− .005
	30113.437	− 806	+ .007	− .005
	0196.370	− 631	+ .005	− .006
	0495.402	0	− .002	− .010
	0587.345	+ 194	+ .002	− .005
	0606.298	+ 234	− .002	− .009
	0671.227	+ 371	+ .001	− .005
	0840.412	+ 728	.000	− .005
	0850.365	+ 749	.000	− .004
	0932.351	+ 922	.000	− .004
	1017.187	+ 1101	+ .005	+ .002
	1342.282	+ 1787	− .004	− .004
	1360.295	+ 1825	+ .001	.000
Fridel' /140/	3064.477	+ 5421	− .007	+ .008
	3193.384	+ 5693	− .004	+ .012
Lavrov /54/	3209.972	+ 5728	− .003	+ .013
Fridel' /140/	3466.829	+ 6270	− .007	+ .012
	3829.844	+ 7036	− .009	+ .013
	3917.522	+ 7221	− .005	+ .018
	4233.624	+ 7888	− .003	+ .023
	4591.426	+ 8643	− .005	+ 025
Alaniya /1/	4593.304	+ 8647	− .022	+ .007
Fridel' /140/	4978.612	+ 9460	− .005	+ .028
	5038.799	+ 9587	− .005	+ .028
	5618.864	+10811	− .009	+ .029
	6097.041	+11820	− .010	+ .033
Lavrova /55/	6492.250	+12654	− .044	+ .002
	6662.378	+13013	− .051	− .003
Tsesevich (Odessa)	6809.282	+13323	− .060	− .010
	7142.431	+14026	− .071	− .019
	7497.394	+14775	− .069	− .013
	7880.317	+15583	− .067	− .008
	8260.393	+16385	− .069	− .006

The O−B residues are shown in Figure 129. Analysis of the curve shows that the period of BD Herculis undergoes cyclic fluctuations some 20,000 E long. From all the O−B residues the following improved equation was derived by the least squares method:

$$\text{Max hel JD} = 2430495.412 + 0.4739083 \cdot E, \tag{205}$$

from which the O−A residues were calculated. Comparison stars (Figure 130): photographic scale $k = 0^s.0$; $a = 6^s.5$; $b = 15^s.0$; $c = 20^s.1$. The observations are listed in Table 305.

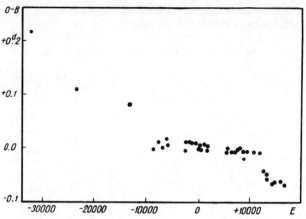

FIGURE 129. O—B residues for BD Herculis.

TABLE 305. Photographic observations (Tsesevich)

JD hel	s	JD hel	s	JD hel	s
243...		243...		243...	
6728.496	13.9	7140.432	17.0	7488.436	3.9
6730.500	17.8	7142.415	3.0	.459	7.1K
6734.504	3.8	.459	1.4K	7494.390	16.0
6749.474	18.4	7144.397	3.5	.413	19.3K
6750.450	17.9	.426	6.4K	.437	18.4K
6755.517	16.7	7145.431	12.2	7496.377	18.6
6756.467	14.1	7159.347	14.6	.400	10.8K
6757.473	16.3	7161.370	1.3	7497.406	0.0
6759.503	1.8	7162.352	—0.5	.433	3.0K
6760.476	3.0	7165.315	11.0	7498.398	5.0
6761.473	5.4	7166.342	13.1	7512.331	16.7
6777.428	17.0	7167.343	17.8	7519.351	13.0
6779.433	0.5	7169.360	16.7	.376	13.8K
6780.385	2.6	.389	15.7	7521.357	17.6
6786.436	17.2	7170.337	15.0K	.384	16.8K
6789.380	3.5	.364	5.8K	7523.333	16.5
6790.481	12.0	7172.3	4.0	.362	18.9K
6791.463	13.1	.355	6.4K	7524.338	14.1
6792.455	13.9	7174.352	13.3	7525.331	5.4
6804.305	14.6	7175.354	18.1	.363	1.4K
6805.334	14.4	7189.265	18.4	7526.328	2.4
6806.317	15.0	.289	17.0	.357	4.9K
6807.312	15.0	.314	10.0K	7545.292	2.4K
6809.316	1.3	7192.295	9.5	7547.286	14.0K
6813.377	17.8	.320	12.0K	7812.500	16.0
6814.363	17.0	7193.275	10.8	7844.474	14.6K
6815.332	17.0	.298	13.3K	7847.441	15.0
6817.292	10.1	7195.261	15.9	.466	16.4K
6834.247	13.8	1796.261	17.0	7849.468	10.5
6838.277	5.5	7197.274	16.0	7853.453	12.5
6840.243	12.2	.300	18.3K	7854.459	15.9K
6869.202	17.0	7198.259	17.0	7855.460	18.4K
7046.566	17.8K	.283	15.2K	7869.384	5.7
7052.537	14.0K	7199.300	—0.5	7871.368	5.6
7077.521	—1.7K	.324	2.4K	.399	7.9K
7079.499	6.4K	7218.222	6.5	7872.363	1.9
7088.501	6.4K	7228.213	—2.5	7873.420	14.6K
7119.498	13.3K	.238	5.9K	7878.354	17.0
7135.488	14.5	7458.469	17.3	.378	12.2K
7136.435	13.4K	7463.482	13.9	7880.360	2.6K
.464	14.2	.507	15.2K	7881.424	10.1
7137.425	12.2	7464.499	17.0	7882.355	11.1
.454	15.2K	7466.478	18.4	.380	14.0K
7139.436	17.6	7470.450	5.2	7883.396	16.0
.466	18.1K	.475	6.4K	.420	14.5K

244

TABLE 305 (continued)

JD hel	s	JD hel	s	JD hel	
243...		243...		243...	
7900.311	6.5	7955.249	6.4 K	8254.358	9.7
.337	13.2K	8198.447	9.7	.388	13.3K
7901.344	14.5K	.473	12.5K	8255.365	14.0
7902.312	13.5	8199.478	14.5	8260.370	2.5
.337	14.5K	8203.478	14.5	.394	−1.8K
7903.320	16.0K	8207.468	13.4K	8263.320	5.7
7904.312	16.0	8208.464	13.8	.347	11.8K
.342	17.1K	8209.444	16.0	8282.304	12.2K
7906.311	17.0	8225.462	13.3	8288.284	18.1
.338	11.7K	8226.432	13.0	.310	13.0K
7808.303	1.6	8228.423	17.0	8293.265	15.2 K
.331	5.3K	8230.433	15.9	8294.267	16.0 K
7910.297	11.5	8235.395	12.7		
7955.224	2.6	8236.378	12.7K		

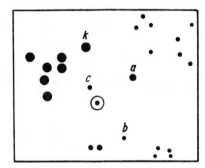

FIGURE 130. Comparison stars for BD Herculis.

AE Pegasi

The greatest number of visual observations were carried out by Lange. He compiled a summary table of maxima, which shows that the star has a variable period, but is not comprehensive enough to fix the correct count of epochs. Tsesevich estimated the magnitude from Harvard photographs and augmented Lange's table of maxima (Table 306). The O−A residues were calculated from

$$\text{Max hel JD} = 2426239.532 + 0.49673126 \cdot E. \qquad (206)$$

TABLE 306. List of maxima

Source	Max hel JD	E	O—A
Tsesevich (Harv.)	2415470.866	− 21679	−0^d.029
	6585.570	− 19435	+ .010
	7450.855	− 17693	− .011
	8719.546	− 15139	+ .029
Tsesevich (Mosc.)	9277.39:	− 14016	+ .043:
Tsesevich (Harv.)	20772.560	− 11006	+ .052
	5474.589	− 1540	+ .023
Lange	6161.544	− 157	− .001

245

TABLE 306 (continued)

Source	Max hel JD	E	O—A
Tsesevich (visual)	2426234.57:	— 10	$+0^d.005$:
Lange	6239.532	0	.000
Florya	6239.533	0	+ .001
Tsesevich (visual)	6239.532	0	.000
Florya	6248.472	+ 18	— .001
Tsesevich (visual)	6250.455	+ 22	— .005
Florya	6254.431	+ 30	— .003
	6261.381	+ 44	— .007
Tsesevich (Harv.)	6462.584	+ 449	+ .020
Okunev	6562.417	+ 650	+ .010
Lange	7326.380	+ 2188	.000
Tsesevich (Harv.)	7482.850	+ 2503	.000
	8059.525	+ 3664	— .030
	8646.666	+ 4846	— .026
	9164.761	+ 5889	— .021
	9530.827	+ 6626	— .046
	9877.545	+ 7324	— .047
	30446.819	+ 8470	— .027
	1197.876	+ 9982	— .027
	1920.551	+11437	— .096
	2632.809	+12871	— .151
	3535.821	+14689	— .196
Tsesevich (visual)	6128.534	+19909	— .421
Strelkova	6457.851	+20572	— .436
Lange	7204.441	+22075	— .434
	7605.269	+22882	— .468
	7606.267	+22884	— .463
	7878.451	+23432	— .488
	7882.423	+23440	— .490
	7883.416	+23442	— .490
	7884.411	+23444	— .489
	8259.397	+24199	— .535
	8260.392	+24201	— .533
	8263.368	+24207	— .538
	8266.354	+24213	— .532

The plot of O—A residues (Figure 131) shows that the period underwent large irregular fluctuations. During the observations it changed several times. The period generally decreased; temporarily it increased, only to drop back to the previous diminished value.

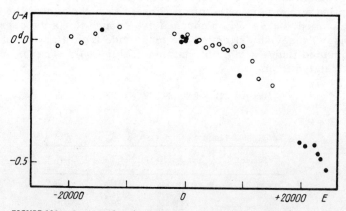

FIGURE 131. O—A residues for AE Pegasi (○ Harvard observations).

Discovered and investigated by Shapley and Hughes /350/. Tsesevich studied its behavior from Simeise photographs. A cursory examination of the enhancement times listed in Table 307 shows that the period decreased abruptly near JD 2425000. The table also lists the O−A residues calculated from

$$\text{Max hel } JD = 2425177.30 + 0.521265 \cdot E. \tag{207}$$

the O−B residues from

$$\text{Max hel } JD = 2425177.298 + 0.5212844 \cdot E; \ P^{-1} = 1.9183386, \tag{208}$$

and the O−C residues from

$$\text{Max hel } JD = 2425177.309 + 0.5212650 \cdot E; \ P^{-1} = 1.9184100. \tag{209}$$

TABLE 307. Times of enhancement

JD	E	O—A	O—B	O—C
2419677.25:	−10551	−0^d.18	+0^d.02	—
9713.23:	−10482	− .17	+ .04	—
20773.52:	− 8448	− .13	+ .03	—
0774.49:	− 8446	− .21	− .04	—
1495.42	− 7063	− .19	− .05	—
1517.33	− 7021	− .17	− .03	—
4763.43:	− 794	+ .01	+ .03	—
5177.30	0	.00	.00	−0^d.009
6247.491	+ 2053	+ .034	.00	+.025
8085.427	+ 5579	− .010	− .12	−.019
33206.358	+15403	+ .013	− .28	+.004

The maxima were determined with higher reliability from the seasonal light curves (Table 308).

TABLE 308. List of maxima

Max hel JD	E	O—B	O—D
2420773.477	− 8448	−0^d.010	−0^d.139
4763.414	− 794	+ .016	+ .042
6247.497	+ 2053	+ .003	+ .086
7749.220	+ 4934	− .095	+ .046
33206.370	+ 15403	− .272	+ .082

TABLE 309. Average light curve

φ	s	n	φ	s	n	φ	s	n
0^P.040	10.3	5	0^P.399	21.9	5	0^P.749	24.0	5
.062	8.8	5	.497	23.5	5	.811	22.8	4
.108	12.5	5	.542	22.7	5	.870	18.3	4
.147	14.4	5	.571	21.1	5	.921	11.7	4
.200	18.0	5	.604	22.1	5	.964	11.6	4
.255	18.0	5	.643	22.0	5			
.335	19.9	5	.696	23.6	5			

The data of Table 308 fully confirm the previous conclusion. Therefore, both equations (208) and (209) were used in constructing the average curve, each in its appropriate epoch. The average curve is given in Table 309.

Comparison stars (Figure 132): $k = 0^s.0$; $a = 10^s.0$; $b = 16^s.2$; $c = 18^s.7$; $d = 28^s.4$. Tsesevich's Simeise observations are listed in Table 310.

TABLE 310. Simeise observations (Tsesevich)

JD hel	s	JD hel	s	JD hel	s
241...		242...		242...	
9677.249	8.0	5500.533	8.9	8067.347	16.8
	10.0	5509.467	14.6		20.6
9713.226	8.0		15.8	8071.413	10.0
	8.9	5916.438	20.3		14.4
9722.348	23.5		18.7	8078.496:	24.0
	24.8	5939.262	23.5		24.5
242...			24.1	8095.406:	11.2
0017.514	21.7	6240.481	24.1		10.0
	22.7	6247.491	9.0	8098.505	11.7
0043.448	21.5		9.1		10.0
	22.2	6269.432	10.9	8099.359	24.5
0065.391	20.7		12.2		24.5
	24.1	6272.341	21.7	8122.219	20.7
0067.445	21.7		24.2		21.9
	20.7	6273.296	24.5	8433.527	24.5
0398.514	22.7		24.1		21.9
0773.517	10.0	6296.328	24.7	8435.480	21.9
	11.7		24.5		23.5
0774.488	10.0	6300.324	19.8	8438.526	20.9
	11.0	.324	18.7		21.9
0776.398	23.0	6596.513	21.6	9169.492	20.7:
	20.9	6597.518	20.7		22.7:
0805.454	21.6		20.6	9522.503	23.0
	21.3	6600.426	11.9		22.6
0837.194	18.2		14.8	9577.411	19.7:
	18.7	6602.475	9.0	243...	
0838.185	11.9		10.9	0259.503	24.5
	13.5	6603.497	11.9		24.1
1495.332	20.7		11.9	2804.526	10.0
.416	10.0	6627.530	12.9		7.0
1495.332	22.7		11.1	2808.506	24.9
.416	11.7	6630.379	20.6		24.9
1513.276	18.7		21.6	2824.475	21.6
	18.7	6648.287	16.1		21.6
1513.380	22.6		13.9	2828.427	10.9
	23.1	6652.231	24.5		14.8
1517.333	8.2		23.5	2832.398	23.1
	9.1	6654.319	23.5		22.6
2991.390	23.0		23.5	2851.306	23.8
	24.5	6655.338	24.1		24.5
3673.515	20.6		24.5	2854.319	24.0
	19.9	6656.303	17.0		23.0
3730.382	19.6	6657.264	14.8	3163.521	23.5
	17.3		14.4		23.0
4055.468	16.5	6975.473	24.1	3206.384	8.9
	14.8		20.9		7.8
4415.517	17.7	7392.284	18.7	3507.452	24.0
	20.7		18.7		22.7
4419.464	15.8	7398.188	16.8	3538.500	16.5
	13.9		17.6		20.6
4763.429	8.3	7725.429	20.5	3539.421	8.8
	9.0		20.6		10.0
4843.211	12.2:	7740.294	24.5	3894.518	18.7
4853.313	19.7		24.5		18.1
	20.2	7749.252	11.7	3897.487	16.1
4854.279	16.5		11.7		14.8
5148.509	20.7	7781.277	23.5	4663.359	18.7
	21.9		23.0		18.7
5500.533	6.4	7783.291	22.6		

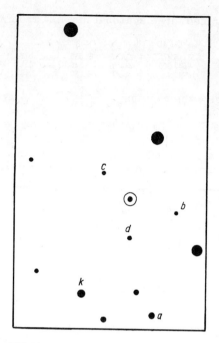

FIGURE 132. Comparison stars for SW Piscium.

·When the reduction of the Simeise observations had been completed, it turned out that one of the areas photographed by the seven-camera astrograph covered this variable star also. We thus had suddenly at our disposal 45 new observations (Table 311). The magnitude was determined on the same scale.

TABLE 311. New observations

JD hel	s	JD hel	s	JD hel	s
243...		243...		243...	
6824.541	19.7	7555.446	24.0	7912.492	24.0
6847.496	24.5	7560.444	11.1	7914.512	22.7
7165.510	20.7	7578.416	23.7	7915.534	18.7
7170.526	14.7	7582.392	15.2	7931.427	24.7
7172.523	7.5	7583.416	16.0	7942.430	9.5
7179.536	23.7	7605.359	21.6	7943.423	24.0
7193.474	13.2	7606.340	17.8	7946.479	22.7
7195.480	8.0	7636.363	25.5	7959.354	23.0
7196.524	7.5	7643.255	9.1	7961.352	15.8
7197.459	20.7	7645.252	24.0	7963.363	11.1
7197.537	14.0	7669.218	24.0	7968.394	23.7
7204.448	15.5	7907.567	9.0	7969.394	23.7
7204.499	17.8	7909.475	22.7	7973.424	20.5
7207.506	10.9	7910.502	23.7	7974.418	17.0
7522.544	14.7	7911.495	20.7	7986.303	10.0

249

Tsesevich determined three seasonal maxima from these observations:

Max hel JD	E	O—B	O—D
2437195.475	+ 23056	—0^d.556	—0^d.048
7643.268	+ 23915	— .546	— .021
7942.450	+ 24489	— .582	— .045

All the maxima give the equation

$$\text{Max hel JD} = 2425177.256 + 0.5212642 \cdot E. \qquad (210)$$

The O—D residues were calculated from this equation. The O—D curve (Figure 133) shows that the period fluctuates irregularly.

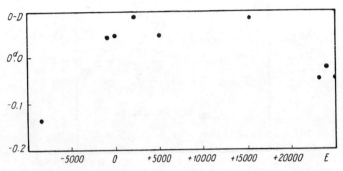

FIGURE 133. O—D residues for SW Piscium.

UU Hydrae

Discovered by Belyavskii [Beljawski] /189/ on Simeise photographs. First elements determined by Tsesevich, then improved by Gaposchkin /238/ and again revised by Payne-Gaposchkin /329/. Also observed by Gur'ev /27 and Lange.

To improve the star's properties, Tsesevich estimated its magnitude from Simeise photographs and reduced all his visual observations. The average seasonal curves yielded six maxima, which are included in the summary table (Table 312). The O—A residues were calculated from

$$\text{Max hel JD} = 2420564.442 + 0.5238585 \cdot E; P^{-1} = 1.908912426. \qquad (211)$$

These residues are plotted in Figure 134. We see that the period undergoe irregular fluctuations. The last six maxima are fitted with the equation

$$\text{Max hel JD} = 2426066.457 + 0.52386198 \cdot E. \qquad (212)$$

The star possibly has Blazhko effect.

TABLE 312. List of maxima

Source	Max hel JD	E	O—A	O—B
Tsesevich (Sim.)	2420564.442	0	0^d·000	—
	4559.356	+ 7626	— ·031	—
Tsesevich (visual)	5616.461	+ 9644	— .072	—
	6066.453	+10503	— .075	—0^d.004
	7490.311	+13221	— .064	— .003

TABLE 312 (continued)

Source	Max hel JD	E	O—A	O—B
Gur'ev /27/	2428194.915	+14566	$-0^d.050$	$+0^d.007$
Payne-Gaposchkin				
/329/	9729.835	+17496	— .035	+ .011
Tsesevich (Sim.)	32980.37:	+23701	— .042	— .017
Lange	7369.309	+32079	+ .010	+ .006

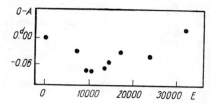

FIGURE 134. O—A residues for UU Hydrae.

Comparison stars (Figure 135): visual scale $r = 0^s.0$; $p = 7^s.5$; $q = 15^s.0$; $t = 18^s.2$; $s = 21^s.8$; photographic scale $p = 0^s.0$; $q = 7^s.9$; $t = 20^s.2$. Tsesevich's observations are listed in Tables 313, 314.

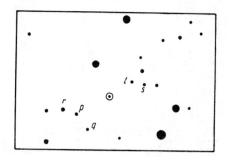

FIGURE 135. Comparison stars for UU Hydrae.

TABLE 313. Visual observations (Tsesevich)

JD hel	s	JD hel	s	JD hel	s
242...		242...		242...	
5616.418	12.5	5619.418	22.3	6011.468	6.6
.454	10.8	.422	21.3	.483	8.8
.506	11.7	.454	22.6	.505	9.6
.551	13.7	.474	22.6	6025.334	19.5
.679	19.1	.507	21.3	6039.317	11.8
.692	19.9	.541	26.8	.381	18.1
5617.429	22.3	5642.395	21.8	.405	18.9
.499	13.1	.401	21.3	.432	20.7

251

TABLE 313 (continued)

JD hel	s	JD hel	s	JD hel	s
242...		243...		243...	
6039.460	20.6	6325.625	21.4	7460 .480	11.9
.489	20.5	6336.457	19.5	.495	11.8
6053.272	22.8	.514	19.2	.506	13.1
6058.355	22.3	6774.277	13.3	.532	13.6
.383	22.3	.285	15.0	7484.367	19.3
.446	22.3	.316	17.5	7485.342	20.1
6062.439	16.7	7458.338	9.9	.369	20.3
6064.392	9.2	.342	10.5	.416	20.0
.423	13.9	.347	10.2	.426	20.4
6064.437	15.8	.362	11.1	.445	20.5
.456	16.4	.377	11.1	.482	20.4
6065.333	20.1	.393	12.9	7490.329	13.2
.357	21.0	7459.371	11.7	.374	13.3
.395	6.9	.382	11.5	.411	14.6
.402	7.5	.391	11.6	.487	19.0
.418	8.4	.407	11.4	7502.157	20.7
.433	8.4	.429	11.5	7507.150	13.8
.446	10.5	.441	11.9	.156	13.5
.472	10.8	.476	12.5	7508.170	12.5
6066.429	13.1	.520	14.3	.189	12.9
.437	8.6	.555	17.6	.238	13.7
.441	7.0	7460.201	22.8	.283	17.1
.466	7.5	.242	21.8	.346	19.1
6067.324	21.8	.261	18.9	.356	20.0
.379	21.0	.271	20.1	7509.321	13.8
.404	21.0	.294	20.8	7510.154	20.4
.422	21.0	.315	20.3	.170	19.5
.458	21.8	.325	21.0	.182	17.9
6068.348	18.9	.343	20.7	.186	13.7
.396	21.8	.352	20.4	.192	13.8
.429	21.0	.375	20.7	.201	12.7
.463	20.8	.386	20.3	.217	10.5
6071.297	17.3	.397	19.9	.225	9.6
.371	17.5	.404	18.9	.236	9.6
.401	21.0	.408	18.0	.244	9.5
6072.390	17.9	.414	16.4	.274	11.2
6073.384	13.7	.422	13.6	.295	12.2
6091.340	16.8	.429	11.2	.316	12.9
6310.525	18.5	.438	10.8	.337	13.0
.549	17.3	.447	10.9	.351	13.3
.582	10.0	.456	10.9	.370	13.8
6325.530	21.1	.468	11.6	.405	15.0

TABLE 314. Simeise observations (Tsesevich)

JD hel	s	JD hel	s	JD hel	s
241...		242...		242...	
9471.369	18.3	0922.377	16.4	4611.374	20.2
	18.1	0946.238	18.6	5380.289	11.3
9823.360	17.1		18.7		13.5
	16.1	1302.365	20.2	7841.494	18.6
9824.355	14.0		19.2		17.4
	14.0	1306.379	17.5	7859.423	18.6
242...			17.9		20.2
0161.442	22.2	1653.382	21.7	8194.520	14.8
	21.2		20.2	243...	
0564.443	5.5	4199.309	22.2	0048.335	9.4
	5.3	4205.347	15.7		7.9
0571.301	5.9		15.5	2980.354	7.9
	8.9	4210.348	20.2		9.9
0571.386	17.1		17.1	3660.495	15.8
	16.4	4557.470	17.4		14.7
0894.447	7.9		19.6	4396.423	11.3
	8.9	4559.372	5.7		11.7
0920.361	20.2		5.6		
0922.377	16.1	4611.374	18.0		

252

AT Virginis

Discovered by Hoffmeister /262/. First elements published by Lause /302/. Subsequent observations carried out by Tsesevich, Solov'ev /108/, Alaniya /1/, and Migach. Payne-Gaposchkin /329/ reduced the Harvard photographs.

Tsesevich collected all the published maxima and the results of his Simeise and Odessa photographic observations (Table 315).

TABLE 315. List of maxima

Source	Max hel JD	E	O—C	E_1	O—D
Tsesevich (Sim.)	2420580.515	—11888	$-0^d.079$	—	—
	4978.450	— 3524	+ .031	—	—
Lause /302/	6831.367	0	.000	—	—
Tsesevich (visual)	7485.470	+ 1225	— .008	—	—
Solov'ev /108/	7923.465	+ 2058	— .009	—	—
Tsesevich (Sim.)	8992.443	+ 4110	+ .016	—	—
Payne-Gaposchkin /329/	9360.500	+ 4791	.000	—	—
Tsesevich (Sim.)	33737.289	+13134	— .005	—6356	$-0^d.014$
Alaniya /1/	4483.426	+14553	+ .016	—4937	+ .017
Tsesevich (Odessa)	7079.278	+19490	— .034	0	+ .009
Migach	7427.360	+20152	— .035	+ 662	+ .013
	7458.345	+20211	— .073	+ 721	— .024

The O—C residues (Figure 136) were calculated from

$$\text{Max hel JD} = 2426831.367 + 0.5258053 \cdot E. \tag{213}$$

Figure 136 shows that the period is subject to irregular fluctuations. The latest maxima are apparently described by the equation (O—D)

$$\text{Max hel JD} = 2437079.269 + 0.5257971 \cdot E_1. \tag{214}$$

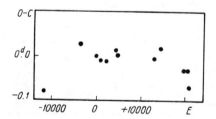

FIGURE 136. O—C residues for AT Virginis.

Comparison stars (Figure 137): visual scale $r = 0^s.0$; $n = 3^s.4$; $k = 3^s.8$; $q = 14^s.6$; $b = 22^s.3$; $b = 29^s.1$; Odessa photographic scale $n = 0^s.0$; $m = 5^s.1$; $c = 13^s.0$; $d = 17^s.0$; Simeise photographic scale $n = 0^s.0$; $m = 5^s.2$; $c = 13^s.6$; $d = 16^s.7$. Averale light curves are given in Tables 316, 317, Tsesevich's observations in Tables 318 — 320.

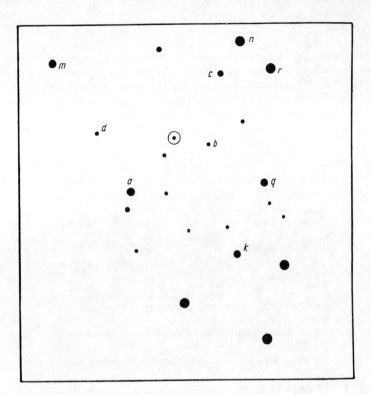

FIGURE 137. Comparison stars for AT Virginis.

TABLE 316. Average visual light curve

φ	s	n	φ	s	n	φ	s	n
$0^P.014$	10.9	4	$0^P.354$	22.5	5	$0^P.792$	26.4	5
.061	13.5	5	.381	23.5	5	.860	26.5	5
.131	16.4	3	.452	25.2	5	.923	25.0	3
.199	20.0	3	.507	25.7	5	.954	18.1	4
.254	21.3	4	.561	25.8	5	.969	13.5	3
.320	22.9	5	.662	25.7	5	.988	11.8	4

TABLE 317. Odessa photographic light curve

φ	s	n	φ	s	n	φ	s	n
$0^P.070$	5.3	4	$0^P.345$	11.1	5	$0^P.735$	11.8	5
.122	6.3	5	.426	12.4	5	.823	10.1	3
.166	8.8	5	.485	12.9	5	.910	1.0	5
.214	9.6	5	.541	12.8	5	.972	2.7	2
.277	10.3	4	.630	12.5	5			

TABLE 318. Visual observations (Tsesevich)

JD hel	s	JD hel	s	JD hel	s
242...		242...		242...	
7459.412	24.6	7485.405	26.4	7508.258	24.0
.426	26.8	.426	26.6	.301	24.8
.458	26.7	.438	22.3	.340	26.2
.472	25.7	.442	19.0	7509.327	26.2
.488	25.2	.449	16.3	.361	25.2
.498	25.7	.451	12.9	.388	26.2
.533	26.2	.454	11.7	7510.214	16.6
.554	26.2	.461	10.9	.224	15.8
7460.345	23.7	.467	10.1	.239	19.7
.389	24.8	.472	10.1	.275	18.0
.407	25.2	.483	10.6	.287	18.4
.499	26.2	.491	12.1	.305	19.7
.533	26.3	.510	11.2	.359	22.3
7461.352	13.4	7490.333	19.7	.376	23.9
.360	16.2	.342	20.6	.380	23.8
7484.318	26.0	.364	20.9	.449	24.0
.328	26.4	.369	20.9	.470	25.3
.336	26.4	.386	20.9	7842.463	20.6
.356	26.6	.396	21.5	.468	16.5
.370	26.2	.412	23.4	.475	15.8
7485.318	26.6	.455	24.6	.480	13.1
.333	26.4	.477	25.7	.482	13.1
.355	26.4	7492.480	20.6	.491	11.1
.364	26.8	.500	21.1	.502	11.7
.376	26.4	.504	23.1	.513	12.0
.396	26.8	7508.216	25.2	7843.337	23.8

TABLE 319. Odessa observations (Tsesevich)

JD hel	s	JD hel	s	JD hel	s
243...		243...		243...	
6286.456	4.6	6997.601	12.5	7406.443	8.0
6288.453	2.6	7015.493	13.0	7425.379	8.4
6304.396	9.8	7016.456	14.3	7426.392	8.7
6313.448	11.6	7017.462	13.8	7729.551	11.1
6340.365	13.0	7019.475	10.4	7734.518	6.0
6344.358	8.7	7020.503	11.2	7758.468	11.9
6345.350	5.1	7028.496	11.2	7759.449	13.0
6347.348	2.2	7044.380	13.0	7761.467	10.4
6607.571	11.2	7046.445	13.0	7779.387	13.0
6608.595	13.0	7052.408	8.5	7780.401	10.4
6612.635	11.9	7077.363	9.5	7781.413	11.3
6613.588	9.0	7078.360	9.8	7790.374	13.0
6660.489	11.6	7079.337	3.8	7808.342	14.1
6661.467	10.7	7100.369	3.4	8085.563	11.0
6663.485	6.5	7101.353	0.0	8090.485	6.9
6667.418	12.0	7105.346	11.0	.510	8.3
6668.443	13.0	7111.341	−2.0	.533	8.0
6702.398	9.1	7373.506	11.4	8106.484	13.8
6714.347	2.3	7377.538	8.6	8138.422	11.4
6715.357	10.5	7378.484	2.0	8143.359	10.6
6716.376	11.8	7398.458	3.4	.383	10.6
6722.340	5.4	7400.458	10.7	8165.325	11.6
6971.591	7.7	7405.427	9.4		

TABLE 320. Simeise observations (Tsesevich)

JD hel	s	JD hel	s	JD hel	s
241...		242...		242...	
9498.285	20.4	0215.41:	15.8	0247.420	16.7
	18.4		15.8		16.7
9513.463	18.5	0239.41:	7.3	0571.489	15.6
	15.3		10.4		16.7
9885.334	8.6	0243.354	18.8	0580.540	5.2
	9.4		16.7	0620.331	14.8

255

TABLE 320 (continued)

JD hel	s	JD hel	s	JD hel	s
242...		242...		242...	
0620.331	16.7	3522.374	15.2	9018.377	14.4
0929.568	17.7	4621.287	15.1	9339.460	3.9
	16.7		16.7		2.1
0956.320	19.7	4978.461	3.9	9348.504	9.6
	14.5		4.2		9.8
0957.373	16.7	5683.458	14.8	9371.348	20.4
	15.6		14.4		20.4
0958.347	20.4	5716.439	18.8	9366.414	13.6
	21.4		18.8		13.2
0959.472	20.0	6061.457	12.1	9721.446	13.6
	15.2		19.1		14.7
0960.326	14.8	6420.483	19.5		
	13.8		19.5	243...	
0975.356	7.1	6421.561	16.7	0107.449	18.8
	8.8		17.9		16.7
0977.322	15.4	6423.459	8.7	2999.416	19.5
	18.8	7165.455	15.1		19.7
0978.376	15.2		14.8	3031.322	11.8
	18.5	7543.368	10.4		11.9
0985.304	15.0		4.6	3034.324	5.2
	15.4	7886.374	15.1		3.8
0988.369	15.6	7898.478	13.8	3358.478	16.7
	17.9		14.6		14.4
1007.345	16.7	8246.514	13.6	3737.426	12.9
	15.7		12.9		12.5
1010.328	16.7	8981.450	2.4	4125.351	3.0
	16.7		5.2		5.2
1318.487	16.7	8992.488	5.2	4455.482	11.5
	15.6		5.2		11.5
3522.374	15.3	9018.377	12.9		

SZ Hydrae

Discovered on Harvard photographs by D. Wood and tentatively designated HV 3573. First investigated by Lause /300/, who obtained

$$\text{Max hel JD} = 2425683.441 + 0.537412 \cdot E. \tag{215}$$

Going over his visual observations of 1929, 1930, and 1932, Tsesevich discovered in 1933 substantial deviations of the maxima from this equation /150/ and proposed an improved expression

$$\text{Max hel JD} = 2425683.433 + 0.537424 \cdot E - 1.09 \cdot 10^{-8} \cdot E^2. \tag{216}$$

In 1935 Lange carried out a series of visual observations and established for three maxima an even more pronounced deviation from the linear elements. Tsesevich /154/ fitted all the observed maxima between 1929 and 1935 with the elements

$$\text{Max hel JD} = 2425683.430 + 0.5374319 \cdot E - 1.36 \cdot 10^{-8} \cdot E^2, \tag{217}$$

and in 1948 derived a constant-period equation based on the observations of 1929—1944 /147/

$$\text{Max hel JD} = 2425683.448 + 0.5373884 \cdot E \tag{218}$$

with an indication of a highly probable decrease of the period.

256

Solov'ev revised in 1958 the visual observations of Gur'ev and pointed to a rapid decrease of the period after 1933. He re-examined the Dushanbe plates for 1941 — 1956 and obtained 6 near-maxima. Solov'ev, however, failed in his attempts to assign definite elements to these 6 maxima, which suggested probable irregular fluctuations of the period /27/.

Lange resumed visual observations of SZ Hydrae in 1959 and, having determined the first maximum, reduced his 24 unpublished visual observations of 1937. From the 1937 observations he found the large decrement in the star's period after 1935, established the correct count of epochs for all the observed maxima between 1929 and 1959, and calculated the constant period elements fitting the observations from 1935 to 1939:

$$\text{Max hel JD} = 2428159.065 + 0.5372356 \cdot E. \qquad (219)$$

The period of SZ Hydrae thus decreased by $0^d.000176$ since the time of Lause's observations (1929 — 1932). The deviations of the O — D residues from Tsesevich's formula (218) reached almost three days. *

Table 321 gives Lange's list of the maxima of SZ Hydrae observed from 1929 to 1963. Figure 138 plots the O — F residues calculated from

$$\text{Max hel JD} = 2415065.782 + 0.53734133 \cdot E'. \qquad (220)$$

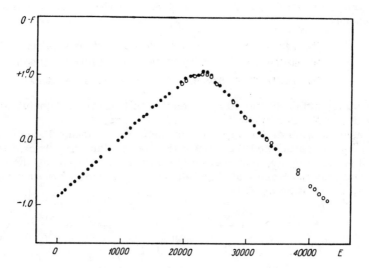

FIGURE 138. O—F residues for SZ Hydrae (O Harvard observations).

---

* At Tsesevich's request, L. Robinson studied the star on Harvard photographs. He confirmed Lange's conclusions and traced its behavior from the beginning of the 20th century.

TABLE 321. List of maxima

Source	Year	Max hel JD	E	O—D	Weight
Lause /300/	1929	2425707.616	+ 45	—0^d.014	2
	1930	6037.591	+ 659	+ .004	3
	1930	6095.631	+ 767	+ .006	2
	1932	6788.329	+ 2056	+ .010	2
Tsesevich /150/	1933	7424.567	+ 3240	— .019	4
Lange	1935	7861.416	+ 4053	— .067	3
Gur'ev /27/	1936	8205.283	+ 4693	— .129	4
Lange	1937	8651.165	+ 5523	— .279	1
Solov'ev /27/	1941	30080.208	+ 8183	— .689	0.5
	1944	1170.244	+10212	— 1.014	0.5
	1949	2973.222	+13568	— 1.512	0.5
	1950	3356.296	+14300	— 1.596	0.5
	1956	5542.265	+18350	— 2.260	0.5
	1956	5576.174	+18413	— 2.207	0.5
Lange	1959	6646.852	+20406	— 2.543	3
	1960	6994.991	+21046	— 2.632	2
	1961	7374.289	+21760	— 2.731	4
	1962	7749.826	+22458	— 2.828	4
	1962—1963	8025.974	+22974	— 2.898	2

WZ Hydrae

This remarkable star was little observed. Discovered by Hoffmeister, its period first determined by Lange /63/. Briefly observed by Gur'ev /31/ and recently again by Lange.

Payne-Gaposchkin /330/ derived three equations from the Harvard photographs:

$$\text{Max hel JD} = 2416169.681 + 0.53772640 \cdot E_1 \quad (2414000—2419000); \qquad (221)$$

$$\text{Max hel JD} = 2419811.736 + 0.53775858 \cdot E_2 \quad (2419000—2421000); \qquad (222)$$

$$\text{Max hel JD} = 2424618.630 + 0.53774101 \cdot E_3 \quad (2421000—2432000). \qquad (223)$$

The assignment of the star to Population II leaves no doubt, since according to A. Joy, its velocity is 315 km/sec.

Tsesevich studied the variation of the period. Using Payne-Gaposchkin's equations, he calculated the "extreme" maxima and the "break" seasons (in Table 322 they are enclosed in parentheses). To these, the actually observed maxima were added and the O—C residues were calculated from (223), the result being Table 322.

TABLE 322. List of maxima

Source	Max hel JD	E	O—C
Payne-Gaposchkin	(2414000.493)	—19746	(+0^d.097)
	6169.681	—15712	+.048
	(9000.273)	—10448	(—.039)
	9811.736	— 8939	—.027
	(21000.182)	— 6729	(+.011)
	4618.630	0	.000
Lange	7513.288	+ 5383	—.002
	7829.477	+ 5971	—.005
Gur'ev	8209.674	+ 6678	+.010
Payne-Gaposchkin	(32000.201)	+13727	(.000)
Lange	6613.441	+22306	—.040
	7780.310:	+24476	—.069:
	8144.346	+25153	—.084

258

The figures in the table reveal irregular cyclic variation of the period between wide limits. Lange proposes the equation

$$\text{Max hel JD} = 2436613.441 + 0.537726 \cdot E. \tag{224}$$

BK Eridani

One of the best examples of the determination of a false period is the case of BK Eridani. Tsesevich observed this star visually in 1960 and 1962. The 1962 observations showed that Solov'ev's period /106/ of $0^d.3540$ was wrong, and a tentative equation was proposed:

$$\text{Max hel JD} = 2437911.462 + 0.54816 \cdot E; \quad P^{-1} = 1.824285. \tag{225}$$

Reduction of the 1962 observations to a single light curve using this equation gave a reliable average curve (Table 323, Figure 139). The 1960 observations are consistent with this equation. Comparison stars (Figure 140): $a = 0^s.0$; $b = 2^s.3$; $c = 5^s.8$; $d = 14^s.6$; $e = 19^s.8$.

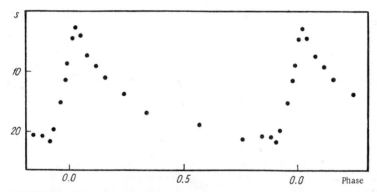

FIGURE 139. Light curve of BK Eridani.

TABLE 323. Average light curve, 1962 observations

φ	s	n	φ	s	n	φ	s	n
$0^P.010$	4.3	5	$0^P.248$	13.9	5	$0^P.905$	21.8	5
.022	2.8	3	.335	16.7	3	.924	19.7	5
.044	4.1	4	.570	18.4	5	.954	15.3	5
.086	7.5	5	.754	20.8	4	.979	11.6	5
.117	9.2	4	.840	20.2	4	.994	8.6	4
.158	11.4	5	.879	20.3	4	Max hel JD = 2437911.475.		

The 1960 observations are much less reliable, since a high-power ocular with a small field of vision was used. The comparison stars were situated far away, and the telescope had to be moved, estimating the magnitude from "memory". All the visual observations are listed in Table 324.

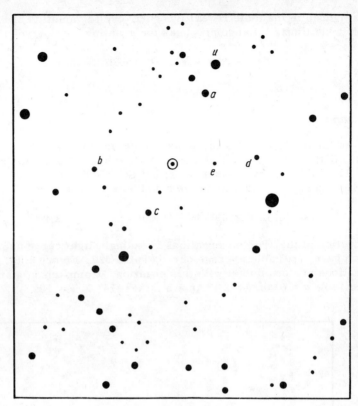

FIGURE 140. Comparison stars of BK Eridani.

TABLE 324. Visual observations (Tsesevich)

JD hel	s	JD hel	s	JD hel	s
243...		243...		243...	
7197.440	18.2	7911.419	21.8	7912.515	17.7
.465	17.7	.447	12.0	.520	17.2
.493	17.3	.453	10.4	.525	17.2
.524	17.5	.460	10.0	.530	17.2
7198.429	14.2	.470	2.7	534	15.9
.434	14.2	.478	2.9	537	13.5
.448	14.8	.484	4.6	.540	12.6
.466	16.3	.489	3.7	.543	12.1
.476	16.8	.498	3.7	.548	10.4
.488	17.8	.503	6.9	.553	8.4
.500	17.8	.511	8.9	.558	4.1
.540	17.8	.516	8.4	.562	2.2
.551	19.8	.523	11.0	.566	3.3
.565	19.4	.528	10.8	.571	2.9
.575	19.8	7912.408	21.8	.579	5.2
7204.410	13.4	.419	20.8	7913.409	19.8
.444	16.4	.429	19.8	.433	17.7
.461	15.8	.437	20.8	.445	18.6
.479	16.4	.454	21.8	7943.311	9.7
7904.511	16.7	.462	21.3	.318	10.2
7906.462	19.2	.466	21.3	.332	10.8
.482	17.7	.474	21.3	.339	11.4
7908.459	18.1	.479	18.8	.353	12.0
.480	17.7	.485	19.8	.373	13.0
7910.433	4.9	.491	21.3	.387	13.6
.443	10.8	.499	21.3	.395	13.4
.463	12.0	.505	21.3	.409	15.6
7911.413	21.3	.509	22.3	.430	16.3

TABLE 324 (continued)

JD hel	s	JD hel	s	JD hel	s
243...		243...		243...	
7943.456	17.2	7946.492	22.8	7946.547	7.0
7946.469	17.7	.508	24.3	.550	4.2
.479	19.8	.536	13.1	.552	3.9
.488	21.3	.538	11.7	.570	2.9

The Astrophysics Institute of the Tadzhik Academy of Sciences made available the unpublished photographic observations of Solov'ev, which are listed in Table 325.

TABLE 325. Observations of A. V. Solov'ev

JD hel	m	JD hel	m	JD hel	m
242...		242...		242...	
9134.400	[13.5	9145.348	[13.2	9170.371	13.17
.422	12.96	.393	12.96	.394	13.17
9135.417	12.96	.437	12.27	.416	13.17
.439	12.96	9160.309	12.34	.438	[13.20
9136.387	12.96	9163.319	12.96	.460	12.96
.410	12.96	.365	13.17	9171.317	12.55
.432	[13.2	.419	13.17	.339	12.58
.455	[13.2	.467	12.96	.367	12.75
9137.400	12.96	9166.324	11.93	.385	12.96
.422	12.96	.349	12.63	.407	12.96
.445	12.96	.372	12.58	.429	12.96
9138.366	11.93	.394	12.62	.451	13.17
.391	12.75	.418	12.96	9172.322	12.27
.417	12.75	9167.303	12.96	.344	11.93
.439	12.75	.326	12.96	.366	11.93
9139.374	12.96	.370	12.65	9220.280	13.17
.396	12.75	.392	11.93	.311	13.27
.419	12.27	.414	11.93	9231.296	12.96
9141.430	12.96	.460	12.96	9281.192	13.17
9142.421	13.17	9168.305	12.96	.253	12.96
.444	13.17	.351	[13.00	9552.299	12.96
.460	12.96	.396	[13.20	9558.348	12.75
9144.356	12.13	.419	12.96	.371	12.96
.378	11.93	.443	12.96	9559.416	12.96
.400	11.93	9170.291	12.45	.442:	12.75
.423	12.10	.313	12.75		
.445	12.75	.348	12.96		

Solov'ev's observations are in good agreement with the above equations, and they were used to construct the average light curve (Table 326).

TABLE 326. Average light curve from Solov'ev's observations

φ	m	n	φ	m	n	φ	m	n
0P.021	12.96	3	0P.250	12.14	2	0P.557	13.01	4
.051	12.96	4	.271	12.49	3	.614	13.10	3
.097	12.86	2	.310	12.76	4	.656	13.12	4
.133	12.40	3	.347	12.77	5	.699	13.10	3
.164	12.11	3	.386	12.76	4	.762	13.01	4
.204	11.93	3	.448	12.82	5	.831	13.01	4
.226	11.93	3	.501	12.89	3	.899	13.03	3

The age of the maximum $0^P.206$ and the time of the maximum Max hel JD = 2429167.409 were determined from this curve. Tsesevich estimated the magnitude of the star from Simeise observations (Table 327). They

261

are few, but quite reliable. Comparison stars (see Figure 140): $a = -5^s.0$; $c = 0^s.0$; $b = 3^s.9$; $d = 14^s.6$. The old and the new observations show a common period of $0^d.548148$. Therefore reduction of the scattered photographic observations to one period uses the equation

$$\text{Max hel JD} = 2429167.409 + 0.548148 \cdot E; \quad P^{-1} = 1.82432482. \tag{226}$$

This equation requires further improvement, however. It does not fit the oldest observations. Taking JD hel = 2420805.378 as the maximum, we can reduce the spread using the equation

$$\text{Max hel JD} = 2437911.475 + 0.5481494 \cdot E; \quad P^{-1} = 1.824320158. \tag{227}$$

This was used as the starting point by Tsesevich, although the O−C residues for the last two maxima are fairly large. The scatter is not eliminated, however, which is a result of variable period.

TABLE 327. Simeise observations (Tsesevich)

JD hel	s	JD hel	s	JD hel	s
241...		242...		242...	
8985.357	3.9	0805.378	−5.0	4435.449	8.4
	2.6	1158.556	7.5		9.7
9717.335	5.2		6.3	4792.518	4.9
	5.1	1188.369	9.7		3.9
9717.414	8.4		9.2	4824.428	3.9
	7.5	1194.325	9.2	4848.398	2.9
242...		1519.481	9.7		3.9
0462.441	10.5	3017.251	2.7	8466.488	8.8
	10.6		3.9		8.7
0475.296	5.0	3736.364	−3.8	9958.310	8.4
	3.9		−3.3		8.4

The magnitude was also estimated from Odessa photographs (Table 328, comparison stars $u = 0^s.0$; $a = 10^s.7$; $c = 14^s.7$; $d = 23^s.5$; $e = 29^s.0$) and Harvard photographs, which gave the seasonal maxima. The maxima are summarized in Table 329.

TABLE 328. Odessa observations (Tsesevich)

JD hel	s	JD hel	s	JD hel	s
243...		243...		243...	
7555.518	24.5	7605.433	29.0	7911.567	17.3
7560.512	[23.5	7606.437	22.5	7912.560	7.2
.535	21.8	7636.346	9.5	7943.507	22.4
7561.542	24.5	7637.358	[14.7	7946.551	6.8
7582.470	26.8	7643.361	24.5		
7583.520	27.2	7645.331	18.5		

TABLE 329. List of maxima

Source	Max hel JD	E	O—C
Tsesevich (Sim.)	2420805.378	− 31207	$+0^d.001$
	3736.364	− 25860	+ .032
Tsesevich (Harv.)	6520.907	− 20780	− .023
	7764.682	− 18511	.000
	9036.944	− 16190	+ .008

262

TABLE 329 (continued)

Source	Max hel JD	E	O—C
Solov'ev	2429167.409	— 15952	+ .013
Tsesevich (Harv.)	30098.701	— 14253	— .001
	1006.456	— 12597	+ .019
	2298.430	— 10240	+ .005
	3728.587	— 7631	+ .040
Tsesevich (visual)	7911.475	0	.000

The O — C residues indicate that the period undergoes small irregular fluctuations.

DZ Pegasi

Discovered by Hoffmeister in 1934 /266/. Subsequently studied by Solov'ev /105/, Van Schewick /359/, and Alaniya /2/. Tsesevich called attention to period variation and at his request Solov'ev estimated the magnitude from Dushanbe photographs and also communicated his unpublished visual observations. Bachmann estimated the magnitude from the Sonneberg photographs, and his averaged data gave the average maxima. Tsesevich estimated the magnitude from Moscow, Simeise, and Odessa photographs.

The Moscow and Simeise observations gave the average seasonal light curves. Lange observed the star visually. The maxima are listed in Table 330. The O — A residues were calculated from

$$\text{Max hel JD} = 2429127.192 + 0.6073534 \cdot E. \tag{228}$$

Their systematic variation shows that the period follows roughly the same trend as the period of TU Persei. It can be formally fitted with three equations: for JD 2419005 — JD 2428793 (O — B)

$$\text{Max hel JD} = 2429127.211 + 0.6073407 \cdot E; \tag{229}$$

for JD 2429166 — JD 2433500

$$\text{Max hel JD} = 2429127.189 + 0.6073761 \cdot E; \tag{230}$$

for JD 2433500 and later (O — D)

$$\text{Max hel JD} = 2433891.308 + 0.6073441 \cdot E_1. \tag{231}$$

The O — B residues are shown in Figure 141. Note that the period figuring in the first and the third equation is almost the same, i.e., the star recovered its previous state.

TABLE 330. List of maxima

Source	Max hel JD	E	O—A	O—B	O—D
Tsesevich (Mosc.)	2419005.264	— 16666	$+0^d.224$	$-0^d.007$	—
Tsesevich (Sim.)	20773.249	— 13755	+ .203	+ ·009	—
	5147.312	— 6553	+ .107	+ .005	—
Van Schewick /359/	7688.443	— 2369	+ .070	+ .022	—
	7697.482	— 2354	.000	— .049	—
Bachmann	8235.031	— 1469	+ .041	+ .003	—
	8793.190	— 550	+ .042	+ .016	—
Solov'ev	9116.683	+ 65	+ .014	— .005	—
Bachmann	30436.690	+ 2156	+ .044	+ .052	—
	2658.417	+ 5814	+ .072	+ .127	—
Solov'ev	3891.329	+ 7844	+ .057	+ .138	$+0^d.021$
Bachmann	4130.607	+ 8238	+ .038	+ .123	+ .005
Alaniya /2/	5394.464	+ 10319	— .012	+ .104	— .021
Bachmann	5670.839	+ 10774	+ .021	+ .139	+ .013
Solov'ev	5783.155	+ 10959	— .022	+ .097	— .030
Tsesevich (visual)	6438.520	+ 12038	+ .008	+ .142	+ .011
Bachmann	6511.363	+ 12158	— .032	+ .104	— .027
Lange	7908.290	+ 14458	— .018	+ .147	+ .008
	7925.288	+ 14486	— .025	+ .140	.000
	7931.366	+ 14496	— .021	+ .144	+ .005
	7939.260	+ 14509	— .022	+ .143	+ .004
	7956.265	+ 14537	— .023	+ .142	+ .003
	7959.299	+ 14542	— .026	+ .140	.000
	7962.339	+ 14547	— .023	+ .143	+ .003

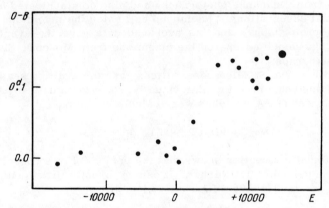

FIGURE 141. O—B residues for DZ Pegasi.

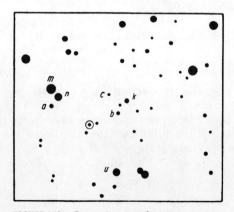

FIGURE 142. Comparison stars for DZ Pegasi.

264

Comparison stars (Figure 142):

	s_1	s_2	s_3
u	-10.0	—	0.0
m	—	0.0	—
k	0.0	6.1	—
n	—	10.1	14.3
a	5.8	—	18.6
b	12.3	—	—
c	16.9	21.2	25.3

The average light curve (Table 331) was constructed from Solov'ev visual observations using the equation

$$\text{Max hel JD} = 2429127.193 + 0.607344 \cdot E; \quad P^{-1} = 1.6465134 \tag{232}$$

Solov'ev's observations are listed in Table 332, Tsesevich's in Tables 333—335.

TABLE 331. Average visual light curve

φ	s_3	n	φ	s_3	n	φ	s_3	n
$0^P.009$	11.4	10	$0^P.361$	21.1	10	$0^P.881$	25.0	10
.039	11.5	10	.413	24.2	10	.906	24.8	3
.062	13.3	10	.511	23.7	10	.942	20.6	10
.097	14.8	10	.572	24.8	10	.961	16.8	4
.134	15.5	10	.647	25.0	10	.970	14.3	5
.174	15.7	10	.712	24.8	10	.983	12.2	9
.244	18.2	10	.772	25.0	10	.995	12.2	7
.296	20.2	10	.819	25.4	10			

TABLE 332. Visual observations (Solov'ev)

JD hel	s_3	JD hel	s_3	JD hel	s_3
242...		242...		242...	
9115.312	25.2	9133.270	12.3	9137.284	25.2
.325	25.2	.290	11.3	9138.168	14.3
.333	23.2	9134.182	21.9	.193	15.3
.341	23.2	.200	25.2	.214	17.3
.358	25.2	.255	25.2	.232	17.3
9116.202	25.2:	.270	23.4	.328	23.0
.274	12.3	.295	25.2	9139.206	25.2
.280	12.3	9135.192	14.3	.265	25.2
.294	14.3	.227	19.7	.272	25.2
.302	14.3	.254	20.7	.281	25.2
9117.317	23.4	.279	21.6	.295	22.2
.330	23.4	.318	21.6	.304	19.7
9118.305	21.6	9136.174	25.2	.308	16.5
9119.292	12.3	.205	27.2	.317	17.3
.300	14.3	.220	25.2	.326	12.2
9129.251	21.5	.240	23.4	.331	12.2
.313	25.2	.251	24.2	.340	11.2
9130.170	25.2	.259	22.7	9140.210	23.4
.194	21.6	.268	20.7	.286	27.2
.213	14.3	.274	19.7	.306	27.2
.229	12.3	.279	17.3	9141.168	14.3
.274	12.3	.282	16.1	.182	12.2
.300	14.3	.286	14.3	.191	12.2
.336	15.2	.290	12.3	.204	14.3
9131.179	25.2	.295	9.5	.220	15.2
.312	25.2	.300	10.1	.251	16.1
.333	25.2	.304	12.4	.318	17.9
9132.217	19.7	.334	12.4	9142.183	25.2
.285	20.8	.338	12.3	.206	25.2
.313	23.0	.359	14.3	.243	25.2
9133.186	25.2	9137.175	27.2	.273	25.2
.239	16.5	.204	25.2	9143.220	25.2
.249	14.3	.231	25.2	9144.273	13.3
.259	12.3	.276	25.2	.289	16.1

265

TABLE 332 (continued)

JD hel	s_3	JD hel	s_3	JD hel	s_3
242...		242...		242...	
9147.236	11.8	9188.148	19.7	9200.201	16.1
.245	12.5	.239	25.2	9201.192	25.2
9161.254	14.3	9189.133	14.3	.248	25.2
.266	14.3	.140	14.3	9213.154	23.0
9163.127	16.1	.152	12.2	9218.086	25.2
.138	17.3	.161	12.2	.096	25.2
.189	19.3	.181	12.2	.114	25.2
.230	19.7	.216	14.3	9220.100	14.3
9164.270	10.2	9190.156	25.2	.113	12.2
.279	9.5	.181	25.2	.125	10.2
9167.270	10.2	9191.118	20.8	.135	10.2
9168.183	22.2	.162	20.8	.141	10.2
9169.151	16.1	.195	22.3	.150	10.9
.183	16.1	.262	25.2	9221.180	25.2
.214	17.3	9195.097	25.2	.191	25.2
.268	19.7	.109	25.2	9224.079	23.0
.290	21.6	.122	25.2	.093	23.0
.334	21.6	.152	25.2	.107	23.0
9170.120	25.2	.185	20.8	9229.073	25.2
.154	23.4	.195	14.3	9231.087	17.0
.163	25.2	.203	13.1	9232.108	25.2
.198	25.2	.210	12.2	.137	25.2
.215	25.2	.277	13.1	9234.109	12.2
.251	25.2	9196.195	25.2	.233	16.5
9171.236	23.4	9197.103	18.7	.252	19.7
.316	25.2	.110	16.5	9255.199	25.2
9172.215	14.3	.115	14.3	.222	25.2
.291	22.2	.119	16.5	9256.078	18.3
.341	19.7	.130	14.3	.088	16.5
9173.092	23.4	.135	14.3	.098	16.3
.167	25.2	.192	14.3	.142	18.7
9187.092	25.2	.209	18.7	9257.145	14.3
.196	25.2	9198.227	16.5:	.153	14.3
.249	25.2	9199.102	23.0	.162	12.2
9188.117	20.8	.205	23.0	.197	14.3
.135	20.8	.240	25.2		

TABLE 333. Moscow observations (Tsesevich)

JD hel	s_1	JD hel	s_1	JD hel	s_1
241...		241...		242...	
8566.388	14.3::	9005.203	2.5	0021.445	0.7
8572.291	—1.8	9007.240	7.7	0063.276	—3.6
8599.214	7.7	9282.454	13.8	0743.393	3.2
8601.280	15.4	9309.377	15.8	1167.297	16.9:
8919.392	16.9::	9652.344	12.3::	1378.301	14.6
8924.387	16.1	9654.353	15.6		
8955.456	—0.9	9660.471	10.7		

TABLE 334. Simeise observations (Tsesevich)

JD hel	s_2	JD hel	s_2	JD hel	s_2
242...		242...		242...	
0773.245	6.5	5182.450	18.2	9145.496	7.9
	5.4		19.2		7.9
0779.382	8.7	5508.447	19.4	243...	
	10.1		18.7	2800.492	7.1
1464.313	21.2	6949.505	16.2		6.7
1466.533	17.5	7693.448	14.4	3151.530	7.3
5147.367	7.6		14.1		8.5
	11.9	8400.524	19.8	4274.310	18.4
5153.307	12.7		19.2		18.4
	16.2	8777.468	8.6		
5177.266	16.2		5.6		

266

TABLE 335. Odessa observations (Tsesevich)

JD hel	s_1	JD hel	s_1	JD hel	s_1
243...		243...		243...	
6053.511	1.5	6138.275	11.4	6463.411	—3.6
6056.523	—2.2	6139.271	1.2	6465.396	8.4
6057.522	14.9	6140.304	—3.0	6466.388	14.9
6072.439	3.9	6163.257	15.1	6468.432	8.6
6074.504	11.4	6164.254	9.5	6481.313	13.8
6075.458	7.1	6184.200	3.2	6482.299	—1.0
6076.449	14.9	6187.181	2.9	6483.320	11.0
6078.444	—1.0	6194.206	14.9	6484.344	9.0
6079.428	14.1	6397.496	11.4	6485.315	—2.7
6081.447	0.0	6411.531	11.2	6487.338	1.9
6082.433	14.1	6426.466	3.5	6488.355	—1.8
6083.432	9.3	6428.477	9.5	6489.345	9.5
6084.440	—4.0	6429.519	3.5	6490.365	14.9
6101.375	1.9	6430.473	13.4	6495.367	13.4
6102.348	13.5	6436.414	14.9	6518.288	11.0
6107.421	14.9	6438.490	—2.5	6538.206	—2.5
6110.401	9.0	6441.510	3.9	6541.194	—3.3
6111.350	7.1	6454.451	9.5	6542.205	11.2
6112.397	—3.6	6455.456	6.9	6543.210	7.4
6128.388	3.9	6461.420	15.6	6546.219	5.8
6129.369	—1.2	6462.417	15.4		

TU Persei

Discovered by Tserasskaya [Ceraski] /218/. Period determined by
Blazhko /196/ who also obtained an improved formula, from which the
O — A residues were calculated:

$$\text{Max hel JD} = 2418572.297 + 0.6070767 \cdot E; \, P^{-1} = 1.64723831. \qquad (233)$$

Tsesevich observed the star in 1961. The ephemeris corrections
reached $4^h.5$. This led to a revision of Blazhko's unpublished visual observations. The magnitude was then estimated from all the Moscow photographs (although on some of them the star appears at the very edge of the
plate). From these observations the average seasonal light curves were
determined. The maxima determined from the old observations by Lange
and Kanishcheva /71/ were also used. The list (Table 336) also includes
the maximum determined by Katina from Odessa photographs.

The O — A residues treated by the least squares method gave

$$\text{Max hel JD} = 2418572.324 + 0.6070706 \cdot E, \qquad (234)$$

from which the O — B residues were calculated.

The plot of O — B residues is shown in Figure 143. Comparison stars
(Figure 144): photographic scale $a = 0^s.0$; $g = 4^s.4$; $d = 12^s.2$; visual scale
(Blazhko) $c = 0^s.0$; $e = 3^s.8$; $d = 7^s.6$; $g = 12^s.2$; $h = 16^s.0$; $f = 17^s.0$;
visual scale (Tsesevich) $e = 0^s.0$; $h = 11^s.3$; $k = 16^s.8$.

The average light curve was constructed from Blazhko's visual
observations (Table 337).

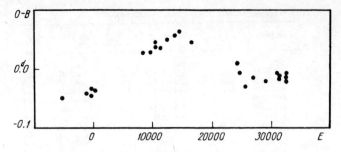

FIGURE 143. O—B residues for TU Persei.

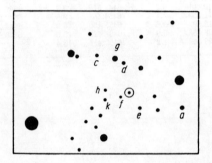

FIGURE 144. Comparison stars for TU Persei.

TABLE 336. List of maxima

Source	Max hel JD	E	O — A	O — B
Tsesevich (Mosc.)	2415261.307	— 5454	$+0^d.006$	$—0^d.054$
	7824.372	— 1232	— .007	— .041
Blazhko	8564.386	— 13	— .019	— .046
	8572.288	0	— .009	— .036
	8598.390	+ 43	— .011	— .038
	23675.385	+ 8406	+ .001	+ .026
	4468.221	+ 9712	— .005	+ .027
Lange /71/	4794.835	+ 10250	+ .002	+ .037
Blazhko	5160.300	+ 10852	+ .007	+ .046
	5713.330	+ 11763	— .010	+ .035
	6218.427	+ 12595	— .001	+ .049
Lange /71/	6976.059:	+ 13843	— .001:	+ .057:
Blazhko	7119.335	+ 14079	+ .005	+ .064
Tsesevich (Mosc.)	8573.248	+ 16474	— .031	+ .043
	33181.486	+ 24065	— .112	+ .008
	3588.206	+ 24735	— .133	— .009
	4250.496	+ 25826	— .164	— .033
	5069.449	+ 27175	$—0^d.157$	$—0^d.019$
Katina	6414.108	+ 29390	— .173	— .021
Tsesevich (visual)	7524.452	+ 31219	— .172	— .009
Lange	7960.318	+ 31937	— .188	— .020
	7991.282	+ 31988	— .184	— .016
	8286.314	+ 32474	— .192	— .021
	8289.362	+ 32479	— .179	— .008
	8295.427	+ 32489	— .185	— .014
	8298.459	+ 32494	— .188	— .017

TABLE 337. Visual light curve (Blazhko's observations)

φ	s	n	φ	s	n	φ	s	n
0^P.009	4.3	10	0^P.519	12.1	10	0^P.950	7.0	10
.030	4.5	10	.653	13.0	10	.960	6.2	10
.058	6.0	10	.810	13.6	10	.968	6.4	9
.118	8.0	10	.884	12.5	10	.977	4.9	9
.269	10.2	10	.914	10.5	10	.990	4.6	10
.365	11.3	10	.938	8.6	10			

TABLE 338. Visual observations (Blazhko)

JD hel	s	JD hel	s	JD hel	s
241...		241...		242...	
8513.394	3.8	8573.423	13.4	4468.211	5.7
8516.410	6.4	429	12.2	.214	4.8
8522.448	6.8:	.439	12.2	.219	2.8
8534.399	14.1	.448	11.2	.224	2.8
8537.401	10.9	.453	10.4	.227	2.4
.406	10.9	.457	10.4	.233	3.8
.416	12.2	8586.245	3.8	.250	4.9
8546.388	10.2	.269	1.9	.271	6.5
8548.396	12.2	8598.330	13.6	.296	7.6
8560.439	13.2	.363	9.1	.311	8.7
.458	13.4	.365	9.4	.411	9.1
8564.318	11.3	.371	8.5	5160.273	9.9
.371	7.6	.381	7.6	.284	7.6
.374	5.7	.384	3.8	.319	6.5
.378	5.7	.393	3.8	.366	9.9
.383	3.8	8663.329	5.3	5713.279	10.7
.394	5.1	.333	4.9	.287	10.7
.426	6.3	.342	3.8	.292	9.9
8566.269	7.6	.368	4.3	.295	9.9
.298	8.8	8876.375	12.2	.302	8.6
.350	9.9	.398	10.7	.306	8.6
.392	10.4	.404	9.7	.310	6.8
8567.233	12.2	.414	5.7	.315	5.3
.242	12.2	.425	3.8	.320	6.1
.262	13.4	242...		.324	5.4
.286	14.1	3407.269	12.2	.327	5.6
.312	14.6	.303	12.2	.341	5.3
.323	14.1	.362	12.2	.375	6.2
.331	13.2	3675.361	7.6	6218.359	13.5
.354	12.2	.365	5.3	.362	13.5
.380	10.7	.376	5.1	.382	12.2
.392	10.4	.382	4.8	.397	8.8
.414	6.3	.389	4.8	.399	8.8
8568.239	10.5	.398	4.8	.408	7.1
.250	10.1	.419	5.7	.419	4.8
.264	11.2	.438	6.3	.424	4.8
.295	12.2	.448	6.6	.442	3.8
.337	13.8	.463	8.7	.443	4.8
.362	13.2	.477	8.7	.453	5.7
8572.213	12.2	.488	8.7	7119.308	5.3
.226	13.4	.501	9.9	.350	1.9
.239	10.4	4180.181	12.2	.356	2.8
.244	10.4	.192	11.3	7150.269	5.7
.257	9.4	.197	11.3	.295	5.2
.266	5.7	.215	11.3	.326	5.7
.294	3.8	4389.487	11.0	7867.254	5.7
.301	3.8	.490	11.0	.257	4.8
.316	3.8	.494	11.2	.268	4.8
.342	8.5	.499	10.4	.276	5.1
.360	8.7	.503	10.4	.314	5.7
8573.294	14.1	.509	10.8	7870.248	6.1
.312	14.1	.514	10.8	.254	5.3
.323	13.6	4468.196	9.4	.273	5.7
.335	13.0	.198	9.9	.277	5.3
.353	13.9	.202	7.6	.293	5.1
.378	14.1	.205	7.6	.320	5.1
.417	14.6	.207	6.6		

TABLE 339. Moscow observations (Tsesevich)

JD hel	s	JD hel	s	JD hel	s
241...		243...		243...	
3594.212	2.5	3179.498	13.2	3570.563	5.5
4715.253	11.4	.525	11.4	.596	0.0
5261.421	7.9	3180.422	13.2	3588.215	0.0
6382.294	15.2	.450	13.2	.244	2.9
6460.174	8.9	.478	11.2	3645.238	2.5
6462.271	11.3	.503	13.2	3647.238	11.1
6491.310	3.3	.527	13.2	.265	7.0
6705.443	13.7	.552	11.2	3672.161	11.1
7068.392	16.2	3181.472	2.2	.186	13.2
.456	14.2	.533	2.8	3673.163	2.8
7266.296	11.3	3183.448	6.0	.190	1.9
7798.398	9.6	.473	9.6	3681.167	—1.0
7800.355	11.2	.498	13.2	3710.307	2.2
.444	12.2	.524	10.8	3900.447	17.2
7801.348	2.8	3189.471	5.5	.496	11.3
7824.386	1.8	.499	10.4	.524	6.4:
8292.284	13.2	.526	8.3	.550	10.9
8897.359	14.2	3209.335	14.2	3901.441	0.0
.432	12.2	.429	1.0	.472	2.7
9254.405	14.2	.455	2.9	.497	5.4
9629.457	12.2	.479	4.4	.527	4.4
242...		.534	10.8	3949.475	2.7
0720.419	14.2	3210.400	15.2	3951.371	2.9
7813.336	9.6	.425	13.2	.396	7.2
7869.314	8.9	.458	15.2	.421	6.4
8081.445	13.7	.477	14.7	.453	9.0
8539.283	2.4	.502	15.7	3953.462	11.2
8573.245	1.0	3211.378	10.4	3979.323	14.2
9165.478	12.2	3212.397	2.9	.385	9.6
9172.529	5.4	.479	3.1	4037.194	14.2
9192.506	0.0	.504	4.4	4250.484	0.0
9227.317	12.2	.528	6.5	4329.328	11.7
9283.208	9.8	.551	8.0	.397	3.2
9284.230	3.3	3214.317	4.4	4330.369	9.9
.276	4.4	.349	10.4	4331.348	6.1
9285.345	0.0	.382	10.5	4332.409	4.4
9286.264	10.9	.432	11.2	4421.196	2.7
9287.203	2.1	.487	11.8	.223	7.0
.258	3.5	3299.168	10.5	4426.218	11.2
9288.266	13.7	.223	5.7	.241	14.2
9306.314	12.2	3301.158	7.0	4450.271	2.2
9310.272	2.9	.314	9.2	4454.305	13.7
9318.233	3.9	.385	13.2:	4455.256	7.3
9335.219	2.9	3329.253	14.2	4680.273	2.9
9521.494	0.0	.304	15.2	.304	—0.5
9526.413	1.8	.367	14.2	4681.435	12.7
9527.500	10.8	3351.238	8.9	.465	8.5
9547.485	15.7	.287	14.2	.500	3.3
9556.383	13.7	3357.253	13.2	.528	1.1
9558.486	7.3	3362.268	17.2	4683.328	2.0
9559.487	13.2	3539.319	14.2	4768.163	15.2
9588.447	7.9	.388	14.2	.218	14.2
243...		.428	15.2	4776.231	0.0
3003.294	8.5	.471	10.9	4795.208	6.1
3004.266	2.0	.509	11.1	.233	10.0
.296	2.9	3541.428	5.5	.292	15.2
.325	4.4	.462	2.9	5041.492	5.4
.346	6.0	.496	2.8	.528	2.5
3005.280	10.0	.533	3.5	5069.306	11.2
.322	10.0	3542.385	11.1	.366	10.2
3006.355	12.7	3543.507	14.2	5069.403	3.3
3007.279	2.0	3569.365	0.0	5075.144	11.5
.304	2.4	.397	—1.0	.168	14.2
.327	6.1	.426	3.3	5076.484	11.1
.355	12.2	.456	3.2	5363.542	11.2
.388	11.7	.436	5.4	5365.541	12.2
3154.478	10.5	.531	8.3	5394.565	11.3
3156.471	14.2	3570.326	12.2	.585	15.2
3178.445	1.8	.365	10.4	5401.580	3.2
.544	4.0	.398	13.2	5402.497	10.8
3179.342	12.2	.429	15.2	5431.397	9.9
.397	15.2	.461	16.2	.422	9.3
.445	12.2	.493	9.2	.447	9.6
.471	11.2	.523	12.2		

TABLE 340. Visual observations (Tsesevich)

JD hel	s	JD hel	s	JD hel	s
243...		243...		243...	
7521.355	8.2	7523.491	14.0	7524.472	6.6
.366	9.7	7524.322	14.6	.485	6.5
.392	7.9	.343	14.6	.495	6.9
.402	6.2	.353	16.8	.501	6.5
.407	6.8	.371	14.3	.519	7.5
.429	4.4	.384	14.6	.531	9.2
438	6.2	.394	13.5	7525.380	14.0
7522.448	14.7	.399	8.5	7526.365	9.7
7523.344	9.8	.409	6.5	.381	7.9
7523.360	9.8	.413	6.3	.392	9.5
.380	9.7	.419	6.5	.415	10.2
.389	9.8	.432	5.9	.520	13.1
.419	10.7	.441	6.1	7528.355	13.5
.455	12.9	.448	6.3	7530.334	14.6
.468	10.4	.453	4.7	7531.368	14.6
.472	13.1	.460	4.6		

V 365 Herculis

Studied in detail by Tsesevich /161/. Blazhko effect detected and studied. Hoffmeister independently reached analogous, though less detailed conclusions. His maxima remarkably coincide with Tsesevich's maxima. All the maxima are listed in Table 341. It shows that the period undergoes considerable irregular fluctuations. The O−C residues were calculated from the equation

$$\text{Max hel JD} = 2429037.438 + 0.613138 \cdot E. \qquad (235)$$

The O−C residues are plotted in Figure 145.

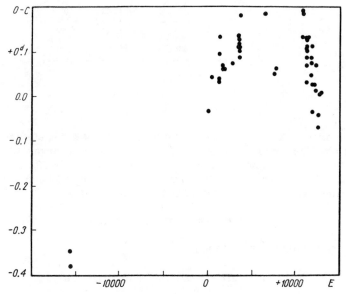

FIGURE 145. O−C residues for V 365 Herculis.

TABLE 341. List of maxima

Source	Max hel JD	E	O — C
Tsesevich	2419514.45	— 15531	—0^d.342
	9541.39	— 15487	— .380
Hoffmeister	29037.408	0	— .030
	9322.591	+ 465	+ .044
	9747.487	+ 1158	+ .035
	9752.396	+ 1166	+ .039
	9787.441	+ 1223	+ .135
	9790.465	+ 1228	+ .094
	30024.659	+ 1610	+ .069
	0072.475	+ 1688	+ .060
	0145.439	+ 1807	+ .061
	0793.542	+ 2864	+ .077
	1207 463	+ 3539	+ .130
	1223.413	+ 3565	+ .138
Tsesevich	1293.28	+ 3679	+ .107
	1314.20	+ 3713	+ .181
	1317.17	+ 3718	+ .085
	1320.26	+ 3723	+ .109
	1325.16	+ 3731	+ .104
	1328.23	+ 3736	+ .108
	1352.15	+ 3775	+ .116
Hoffmeister	3090.467	+ 6610	+ .187
	3736.580	+ 7664	+ .052
	3828.560	+ 7814	+ .062
	5694.468	+ 10857	+ .191
	5699.365	+ 10865	+ .183
	5700.545	+ 10867	+ .136
	5933.532	+ 11247	+ .131
	5946.410	+ 11268	+ .133
	5952.522	+ 11278	+ .114
	5960.453	+ 11291	+ .074
	5979.493	+ 11322	+ .107
	5984.430	+ 11330	+ .138
	5987.458	+ 11335	+ .101
	5990.515	+ 11340	+ .092
Tsesevich	6047.48	+ 11433	+ .036
	6339.35	+ 11909	+ .052
	6344.32	+ 11917	+ .116
	6347.36	+ 11922	+ .091
	6353.48	+ 11932	+ .079
	6372.37	+ 11963	— .038
	6388.38	+ 11989	+ .031
	6396.35	+ 12002	+ .030
	6399.40	+ 12007	+ .014
	6703.43	+ 12503	— .072
	6722.47	+ 12534	— .040
	6730.41	+ 12547	—0^d.070
	6838.40	+ 12723	+ .007
	6840.24	+ 12726	+ .008

SS Leonis

Discovered by Cannon /215/. Belyavskii [Beljawski] studied it from Simeise planetary photographs /189/. Observed by Florya /128/, Solov'ev /108/, Alaniya /1/, and Tsesevich. Payne-Gaposchkin /329/ examined the Harvard plates and established that the star's period changed at a certain instant; for JD 2414000 — JD 2422000 she gives

$$\text{Max hel JD} = 2415485.723 + 0.62633242 \cdot E, \qquad (236)$$

and for JD 2422000 — JD 2432000

$$\text{Max hel JD} = 2422342.814 + 0.62634552 \cdot E'. \qquad (237)$$

272

Table 342 lists all the available maxima.

TABLE 342. List of maxima

Source	Max helJD	E	O − C	O − D	O − A
Payne-Gaposchkin					
/329/	2415485.723	− 19927	$+0^d.053$	$+0^d.093$	$0^d.000$
Tsesevich (Sim.)	9482.350	− 13546	− .002	+ .026	.000
Belyavskii /189/	20191.364	− 12414	− .006	+ .020	+.006
Payne-Gaposchkin					
/329/	2342.814	− 8979	− .037	− .018	+.004
Tsesevich (Sim.)	4212.476	− 5994	− .003	+ .010	—
Tsesevich (visual)	4618.342	− 5346	− .006	+ .006	—
	5334.237	− 4203	− .019	− .009	—
	6093.376	− 2991	− .005	+ .003	—
Florya /128/	6830.576	− 1814	− .008	− .003	—
Tsesevich (visual)	7461.323	− 807	+ .013	+ .017	—
Solov'ev /108/	7966.768	0	+ .001	+ .003	—
	8291.214	+ 518	+ .002	+ .003	—
Tsesevich (Sim.)	33001.313	+ 8038	+ .017	+ .004	—
Alaniya /1/	4011.573	+ 9651	− .011	− .027	—
Tsesevich (Odessa)	7044.362	+ 14493	+ .035	+ .009	—

The O −C residues were calculated from the equation

$$\text{Max hel JD} = 2427966.767 + 0.626341 \cdot E. \tag{238}$$

Their variation shows that the period is greater than the above figure. For $-10000 < E < +20000$, a new formula was derived

$$\text{Max hel JD} = 2427966.765 + 0.62634289 \cdot E. \tag{239}$$

The period most probably changes. A definite conclusion, however, cannot be reached without examining the Harvard photographs, which remain unpublished. Table 342 shows that the old maxima fit the first equation of Payne-Gaposchkin (O −A residues). The period thus increased by $0^d.00001047$.

Comparison stars (Figure 146): visual scale $m = 0^s.0$; $t = 7^s.3$; $s = 11^s.2$; $n = 17^s.0$; $q = 21^s.9$; $p = 26^s.6$; $r = 31^s.2$; Odessa photographic scale $A = 0^s.0$; $t = 6^s.2$; $q = 16^s.5$; $p = 25^s.5$; Simeise photographic scale $a = 0^s.9$; $s = 7^s.2$; $n = 12^s.6$; $d = 20^s.2$.

The visual observations were reduced to give two average light curves (Tables 343, 344). The ages φ were calculated from

$$\text{Max hel JD} = 2420191.364 + 0.6263412 \cdot E; \quad P^{-1} = 1.596574.$$

The observations are listed in Tables 345 −347.

TABLE 343. Average light curve, 1928

φ	s	n	φ	s	n	φ	s	n
$0^P.011$	12.6	3	$0^P.336$	24.0	4	$0^P.897$	25.6	3
.030	13.7	3	.484	27.1	2	.920	18.7	3
.060	14.7	3	.638	29.3	3	.937	16.1	3
.086	15.0	3	.716	30.2	3	.956	12.0	2
.121	16.8	3	.795	29.4	5	.974	12.8	3
.180	21.4	3	.852	29.0	5	.994	11.9	4

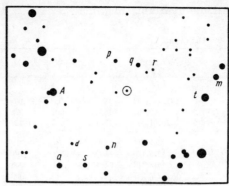

FIGURE 146.　Comparison stars for SS Leonis.

TABLE 344.　Average light curve, 1933—1934

φ	s	n	φ	s	n	φ	s	n
0^p.014	12.4	4	0^p.217	20.1	3	0^p.900	24.4	5
.028	12.8	4	.296	22.5	4	.930	22.0	5
.054	13.2	4	.397	22.6	3	.952	17.1	3
.075	12.8	5	.509	24.2	3	.966	15.0	4
.099	14.8	5	.560	24.2	2	.988	13.8	4
.122	16.4	4	.701	25.4	3	.996	12.8	3
.144	17.9	4	.803	25.2	5			
.179	19.4	4	.853	25.5	5			

TABLE 345.　Visual observations

JD hel	s	JD hel	s	JD hel	s
242...		242...		242...	
4615.493	23.3	5330.477	14.9	5347.454	13.8
.506	23.8	.490	11.2	.463	13.6
4617.317	25.2	.496	10.9	.474	15.5
.329	21.1	.507	12.4	.484	17.0
.336	22.4	.521	15.0	6093.358	11.0
.355	21.1	.535	14.6	.364	9.3
.367	21.5	5331.298	22.8	.370	6.2
.382	21.5	.328	24.7	.377	4.8
.402	23.3	.347	24.7	.385	6.2
.408	24.7	.398	27.1	.395	10.2
4618.329	10.2	.448	27.1	.406	11.8
.339	6.6	.500	28.4	.418	13.5
.347	6.6	.523	29.4	.430	13.9
.351	8.0	5332.317	18.9	7424.394	9.4
.360	9.8	.324	18.2	.406	10.3
.372	11.7	.332	15.0	.413	13.8
.382	13.1	.343	13.1	.417	14.6
.403	14.6	.355	11.7	.424	15.0
.423	16.1	.362	11.7	.432	15.0
.473	18.3	.368	12.6	.438	18.0
5326.336	23.7	.380	13.8	.447	18.1
5327.298	24.9	.388	13.6	.457	18.9
.336	10.9	.404	14.8	.470	19.8
.346	11.7	.426	17.5	.486	20.2
.361	12.1	.450	18.0	.513	22.5
.371	13.1	.476	18.9	.523	24.1
.401	14.8	.512	29.1	.557	22.5
.456	16.3	5345.355	29.8	7425.428	25.5
5330.282	30.0	.376	29.7	.440	26.1
.306	30.9	.388	29.4	.469	25.6
.317	29.8	.404	29.4	.497	26.1
.374	29.8	.411	28.9	.505	25.9
.411	30.3	.419	28.9	.525	26.2
.430	23.5	.433	29.0	7459.359	25.4
.446	18.9	.459	28.4	.380	24.6
.452	17.5	5347.304	28.4	.386	23.5
.456	15.7	.321	28.4	.394	19.1

TABLE 345 (continued)

JD hel	s	JD hel	s	JD hel	s
242...		242...		242...	
7459.400	14.3	7461.344	14.1	7508.272	14.7
.404	13.3	.350	14.8	.275	14.1
.409	14.0	.366	14.7	.280	14.1
.417	14.2	7485.372	23.2	.292	14.5
.425	9.7	.414	23.8	.303	14.6
.436	9.7	.436	24.2	.346	17.0
.445	9.9	7490.472	24.7	.358	20.8
.462	11.2	7492.521	25.2	7509.325	24.6
.471	12.6	7508.174	24.6	.422	24.7
.486	13.7	.187	25.0	7510.156	14.5
.502	15.0	.197	25.0	.169	13.1
.521	17.0	.214	24.0	.176	13.5
.540	19.3	.217	24.7	.195	14.5
.564	19.9	.223	23.5	.210	17.0
7460.296	22.8	.227	24.2	.246	18.6
.401	23.8	.233	21.4	.271	19.7
.524	24.9	.237	20.6	.297	20.2
7461.301	12.2	.241	20.1	.363	20.8
.312	12.2	.250	18.0	.398	21.9
.320	13.2	.257	17.6	.479	24.5
.332	12.9	.260	15.1		

TABLE 346. Odessa observations

JD hel	s	JD hel	s	JD hel	s
243...		243...		243...	
6288.424	7.2	6993.484	15.2	7373.454	14.9
6304.371	15.6	6995.523	1.7	7377.457	15.0
6306.399	17.5	.551	6.2	.485	18.1
6335.339	2.7	6996.493	13.9	7378.436	14.2
6338.353	15.5	6997.570	12.6	7396.384	3.7
6345.321	8.1	7015.424	19.1	.409	3.5
6607.531	15.6	.453	16.5	7397.393	14.6
6608.547	10.9	7016.394	14.6	.417	16.0
6612.593	15.3	.426	13.8	7398.382	11.6
6613.548	8.1	7017.399	3.4	.407	10.9
6660.425	3.5	.428	2.8	7400.385	14.0
6661.401	12.7	7020.427	15.5	.409	14.4
6663.457	16.0	.470	16.5	7405.356	13.0
6667.388	8.3	7041.391	9.6	.380	14.8
6668.416	16.5	7044.331	6.2	7406.378	3.9
6959.536	14.6	7046.409	8.5	.405	2.7
.569	12.4	7052.372	15.2	7424.341	17.5
6960.520	8.8	7326.638	16.5	7425.322	11.8
6971.550	13.1	7373.425	13.6	7426.340	17.5

TABLE 347. Simeise observations

JD hel	s	JD hel	s	JD hel	s
241...		241...		241...	
9481.313	13.9	9513.272	11.2	9852.304	16.6
	10.4		9.9		9.4
9481.395	13.7	9527.304	17.7	9858.315	5.9
	14.8		14.1		8.7
9482.332	—1.0	9835.474	16.0	9869.370	11.1
	—1.0		15.6		10.4
9505.298	16.6	9841.456	10.3	242...	
	16.1		11.1	0191.380	1.8

275

TABLE 347 (continued)

JD hel	s	JD hel	s	JD hel	s
242...		242...		242...	
0191.380	2.9	4234.283	17.7	8226.390	13.7
0196.	—1.4		14.5	8249.327	3.2
	2.4	4251.247	10.5		2.7
0606.277	15.8		5.9	8251.418	16.0
	13.4	4255.295	14.9		16.0
0929.453	11.5	4964.399	11.1	8605.456	13.7
	13.7		12.6	9320.448	16.9
1665.507	13.9	6445.370	1.8	.550	5.4
	13.5		—1.0	9320.448	12.6
3521.336	15.6	6769.356	10.8	.550	5.0
	12.6		7.2	243...	
3846.457	16.0	7896.461	10.1	3001.334	2.1
	14.8		17.3		2.9
3847.492	5.8	7923.414	3.2	3035.296	11.2
	8.3	7928.360	16.9	4476.305	9.9
4212.502	1.2		16.4		
	2.1	8226.390	14.8		

UY Bootis

Numerous observation series published. Payne-Gaposchkin /329/
studied the Harvard photographic series and detected variation of the period.
Tsesevich carried out a new series of observations and found that the maximum deviated from the ephemeris almost by 0.5 of period. At Tsesevich's
request, T. G. Nikulina estimated the magnitude from Dushanbe photographs.
Her results established the variation of the period during the last decade.
The observations of Nikulina and Lange were reduced by Klepikova.
Tsesevich estimated the magnitude from Odessa photographs taken with
the seven-camera astrograph. From the average seasonal light curves
a number of maxima were determined, which together with several
individual maxima (marked with an asterisk) are listed in Table 348.

TABLE 348. List of maxima

Source	Max hel JD	E	O — D
Payne-Gaposchkin			
/329/	2411459.819:	— 26945	—0^d.655:
Tsesevich (Harv.)	5114.386	— 21330	— .442
	5704.656	— 20423	— .466
	6513.022	— 19181	— .418
	7170.395	— 18171	— .373
	7870.686	— 17095	— .365
	8385.476	— 16304	— .373
	8938.725	— 15454	— .321
	9507.667	— 14580	— .196
	9803.827	— 14125	— .159
	20218.391	— 13488	— .167
	0612.785	— 12882	— .170
	1104.801	— 12126	— .174
	1875.417	— 10942	— .129
	2394.867	— 10144	— .033
	4108.532	— 7511	+ .023
	4786.694	— 6469	+ .030
Prager /340/	5391.375	— 5540	+ .100
Payne-Gaposchkin /329/	5688.836	— 5083	+ .136
Tsesevich (Harv.)	6247.141	— 4225	+ .037
	6700.142	— 3529	+ .068
	7158.329	— 2825	+ .077
	7524.702	— 2262	+ .039
	7902.815	— 1681	+ .025

TABLE 348 (continued)

Source	Max hel JD	E	O—C
Lange /62/	242 8002.392	— 1528	+0^{d}.027
Tsesevich (Harv.)	8266.592	— 1122	— .006
	8632.353	— 560	— .006
Gur'ev /27/	8704.590	— 449	— .010
Tsesevich (Harv.)	8980.562	— 25	+ .014
Prager /340/	8996.776	0	— .042
Gur'ev /27/	9046.912	+ 77	— .019
Tsesevich (Harv.)	9362.567	+ 562	— .012
Gur'ev /27/	9417.866	+ 647	— .033
Tsesevich (Harv.)	9744.578	+ 1149	— .032
	30080.364	+ 1665	— .069
	0868.512	+ 2876	— .064
Tsesevich (visual)	1178.286	+ 3352	— .081
Tsesevich (Harv.)	1792.638	+ 4296	— .102
Herbig /253/	2306.759	+ 5086	— .129
Tsesevich (Harv.)	2603.538	+ 5542	— .124
	3148.242	+ 6379	— .157
Nikulina	3327.225	+ 6654	— .149
Tsesevich (Harv.)	3567.344	+ 7023	— .183
	4214.319	+ 8017	— .123
Nikulina	4300.270	+ 8149	— .080
	4896.470	+ 9065	— .031
	5115.804	+ 9402	— .023
Tsesevich (visual)	6371.290	+ 11331	+ .031
Nikulina	6535.960	+ 11584	+ .043
Tsesevich (visual)	6734.455	+ 11889	+ .038
	7405.470	+ 12920	+ .058
Lange	7780.337*	+ 13496	+ .053
	7808.325*	+ 13539	+ .055
	7817.429*	+ 13553	+ .048
	7819.376*:	+ 13556	+ .024:
	8203.398*	+ 14146	+ .081
	8550.316*:	+ 14679	+ .112:
	8561.374*	+ 14696	+ .106
	8563.335	+ 14699	+ .112
	8576.348*	+ 14719	+ .111
	8591.318*	+ 14742	+ .112

The first and the oldest maximum of Payne-Gaposchkin greatly deviates
from Tsesevich's formula

$$\text{Max hel JD} = 2428996.818 + 0.65081997 \cdot E. \qquad (240)$$

This led to a revision of the magnitudes using the numerous Harvard photographs and construction of new seasonal light curves. The resulting maxima
are also included in Table 348.

The O—C residues are plotted in Figure 147. The period changed
abruptly twice: near $E = -5000$ and $E = +6000$. After the second jump,
the period returned to the initial value. The O—C residues moreover
show characteristic irregular, though quite significant wavy variation.
The observations are listed in Tables 349—351. Comparison stars
(Figure 148):

	s_1	s_2	s_3
k	0.0	—	—
a	9.5	0.0	0.0
b	15.6	8.0	13.0
c	20.0	14.6	18.7
d	32.5	—	28.3
e	—	22.2	—

The asterisk in Table 350 marks observations on orthochromatic plates,
y observations on panchromatic plates without light filter.

277

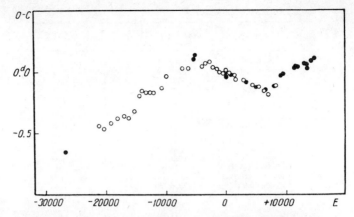

FIGURE 147. O—C residues for UY Bootis (o Harvard observations).

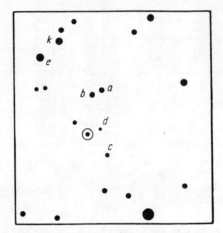

FIGURE 148. Comparison stars for UY Bootis.

TABLE 349. Odessa observations (Tsesevich)

JD hel	s_1	JD hel	s_1	JD hel	s_1
243...		243...		243...	
6286.485	21.8	6373.336	14.2	7028.524	22.9
6288.517	22.3	6376.338	23.1	7029.531	19.1
6292.536	7.6	6590.697	20.0	7044.406	14.7
6304.430	17.1	6607.602	10.7	7046.473	22.9
6313.476	12.6	6608.663	20.0	7052.437	22.2
6314.508	21.9	6613.626	18.9	7072.413	16.5
6335.441	13.3	6646.525	24.7	7073.403	23.0
6338.382	18.1	6660.513	21.1	7075.387	22.0
6339.411	13.3	6661.493	21.6	7373.558	11.9
6340.393	23.6	6662.502	21.6	7377.592	13.9
6344.383	19.4	6663.519	10.7	7378.524	26.2
6345.379	17.8	6667.486	11.5	7402.446	21.0
6347.383	18.9	6668.508	21.8	7405.472	6.5
6362.352	21.1	6971.634	21.5	7426.442	17.9
6364.344	19.3	7002.610	7.5	7427.382	21.0
6371.338	8.4	7019.504	9.5		
6372.336	22.3	7020.532	21.0		

TABLE 350. Photographic observations (Nikulina)

JD hel	s_2	JD hel	s_2	JD hel	s_2
243...		243...		243...	
3065.253	17.4	4544.258	8.0	5626.280	14.6
3053.254	2.7	4561.192	2.0	5627.255	17.6
3064.217	4.0	.215	3.4	5628.265	16.6
3354.350	22.2	4563.214	3.0	5633.261	3.0
3381.404	8.0 *	4565.219	4.6	5635.206	3.0
3382.275	18.4	4569.188	8.0	5637.222	5.7
3395.356	22.2 *	4574.186	11.0	5638.245	17.6
3408.240	17.1 *	4858.281	14.6y	5653.194	14.6
3409.289	4.0 *	.305	17.1y	5655.203	14.6
3416.265	20.0 *	4859.283	9.9y	5658.197	14.6
3439.220	8.0 *	4860.279	14.6y	5950.277	3.0
.245	7.0 *	4870.287	14.6y	5957.270	10.8
3449.237	19.7 *	4915.230	12.0y	5976.261	4.0
3763.286	4.0	4920.211	16.8y	5981.298	9.5
4102.331	4.0	4921.226	4.0y	6005.192	18.6
.356	11.0	4922.201	19.6y	6007.201	18.4
4125.275	5.0	4923.211	6.0y	6008.217	8.0
4126.199	18.4	4924.202 .	16.6y	6334.197	4.4
4148.289	18.4	4925.196	10.8y	6339.249	18.4
4152.270	21.1	5217.258	8.0	6364.290	14.6
.293	22.2	5220.346	14.6	6368.240	8.0
4157.260	8.0	5576.314	14.6	6376.236	16.6
.283	15.5	.336	18.0	6377.250	12.0
4178.202	22.2	5577.249	3.0	6398.192	18.4
.225	22.2	.274	4.0	6672.408	18.8
4181.206	6.0	5598.365	14.6	6674.426	19.7
.232	4.0	5599.313	5.0	6688.290	8.0
4185.194	18.4	5602.320	14.6	6699.376	5.7
4188.219	9.9	.383	15.6	6719.317	19.6
4485.377	11.0	5604.301	17.4	6745.218	22.2
4537.224	4.0	5608.220	14.6	6748.200	4.0
.247	8.0	5609.192	3.4		
4538.227	14.6	5623.248	16.6		

TABLE 351. Visual observations (Tsesevich)

JD hel	s_3	JD hel	s_3	JD hel	s_3
243...		243...		243...	
6703.359	15.8	6720.393	18.7	6722.449	20.2
.372	19.8	.413	19.1	6732.336	22.2
.394	19.8	.422	19.7	.376	20.9
.417	20.4	.431	19.7	.384	21.6
.460	21.8	.446	20.1	6734.423	15.2
.474	22.2	.464	20.1	.434	9.7
6717.380	21.7	6722.329	20.1	.455	7.0
.407	21.2	.365	20.1	.464	9.8
.430	20.9	.396	20.6		
.439	20.6	.419	20.1		

Farquhar /233/ published radial velocity measurements for this star. Using the equation

$$\text{Max hel JD} = 2428996.804 + 0.65079733 \cdot E; \ P^{-1} = 1.52657668, \qquad (241)$$

Tsesevich calculated the ages corresponding to Farquhar's observations (Table 352).

These observations are plotted in Figure 149. The light maximum lies halfway along the decreasing branch of the radial velocity curve, and the radial velocity minimum lags $0^P.1$ behind the light maximum.

TABLE 352. Radial velocity curve from Farquhar's data /233/

JD hel	v, km/sec	φ		JD hel	v, km/sec	φ
243...				243...		
2312.788	+106	$0^P.264$		2319.852	+ 99	$0^P.118$
.826	+119	.322		.874	+112	.152
2319.684	+140	.860		.912	+107	.210
.718	+177	.912		.932	+102	.241
.746	+155	.955		2320.687	+122	.401
.768	+139	.989		.737	+160	.478
.783	+138	.012		.803	+167	.579
.800	+124	.038		.858	+212	.664
.830	+ 98	.084		.915	+175	.751

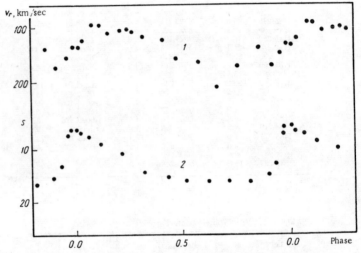

FIGURE 149. Variation of radial velocities (1) and light (2) for UY Bootis.

TV Leonis

One of the most complex objects among variable stars ever studied by the author. Discovered by Shain on the Simeise plates in 1929 /348/. Its period was first determined by Selivanov /96/ from his visual observations of 1934—1935. Selivanov's period ($0^d.402$), however, was wrong. Determination of the true period is obstructed by three factors: first, the observations are scattered in time; second, the star is observed in winter, when clear nights are scarce; three, the period is variable. Among the various observers of this star (Lange /60/, Iwanowska /281/ Dziewulski /225/ and Alaniya /1/), only Lange questioned the validity of Selivanov's figure.

Tsesevich observed the star visually for a few seasons and also estimated its brightness from Odessa and Simeise photographs. At his request, G. S. Filatov estimated the magnitude from the Dushanbe plates. Complete reduction of all observations, except those of Dziewulski and Alaniya (never published), is given in Tables 353—359.

Comparison stars (Figure 150):

	m_s	s_{T1}	s_{T2}	s_L	s_{ph}
a	10.25	0.0	—	—	0.0
t	—	—	0.0	—	6.5
s	11.55	—	11.0	—	12.3
b	—	25.0	19.7	—	21.3
k	—	11.6	—	0.0	—
a'	—	21.0	—	13.0	—
c	—	—	—	26.0	—
d	—	—	—	35.0	—

(m_s estimated stellar magnitude for Selivanov's observations, s_{T1} power scale for Tsesevich's observations of 1945, s_{T2} power scale for Tsesevich's observations of 1958—1959, s_L power scale for Lange's observations, s_{ph} power scale used in the reduction of photographic observations).

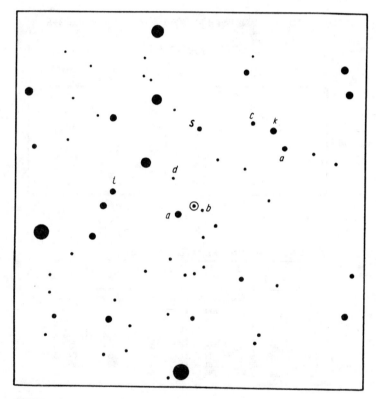

FIGURE 150. Comparison stars for TV Leonis.

After the search for the true period, which was found to be close to $0^d.673$, the seasonal light curves were constructed and the maxima determined (Table 353).

281

TABLE 353. List of maxima

Source	Max hel JD	E	O—C
Tsesevich (Sim.)	2419852.318	0	$+0.032^d$
	23521.336	+ 5453	+ .044
	6036.390	+ 9191	+ .016
Selivanov	7591.282	+ 11502	— .030
Lange	7858.388	+ 11899	— .043
Selivanov	7914.228	+ 11982	— .048
Tsesevich (Sim.)	8249.325	+ 12480	— .027
Tsesevich (visual)	31589.345	+ 17444	+ .007
Filatov	3033.235:	+ 19590	— .022:
Tsesevich (visual)	6702.279	+ 25043	+ .016
	7041.390	+ 25547	+ .015
Lange	7432.328	+ 26128	+ .031
	7734.421	+ 26777	+ .018
Pachinskii	8102.46	+ 27124	+ .013

The O—C residues were calculated from

$$\text{Max hel JD} = 2419852.286 + 0.6728418 \cdot E; \quad P^{-1} = 1.48623347. \tag{242}$$

The O—C curve in Figure 151 reveals irregular variation of the period.

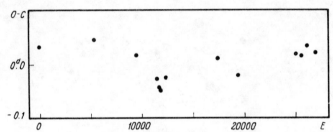

FIGURE 151. O—C residues for TV Leonis.

TABLE 354. Average visual light curve (Selivanov)

φ	m	n	φ	m	n	φ	m	n
0.032^p	11.09	4	0.428^p	11.71	5	0.827^p	11.56	5
.146	11.47	5	.614	11.68	5	.856	11.42	5
.183	11.59	5	.636	11.72	5	.877	11.01	5
.216	11.57	5	.660	11.74	5	.905	10.73	5
.243	11.54	5	.686	11.75	5	.929	10.59	5
.280	11.68	5	.716	11.77	5	.940	10.56	5
.319	11.65	5	.744	11.77	5	.954	10.69	5
.354	11.67	5	.761	11.72	5	.985	10.68	5
.382	11.64	5	.785	11.72	5			

TABLE 355. Average visual light curve from Lange's old observations (Figure 152)

φ	s	n	φ	s	n	φ	s	n
0.002^p	16.8	5	0.316^p	28.8	5	0.862^p	21.7	5
.031	18.4	5	.449	30.0	5	.869	20.0	5
.072	20.4	5	.544	30.0	5	.883	19.2	2
.099	21.5	5	.617	29.6	5	.903	13.0	3
.145	20.8	5	.680	30.8	5	.931	12.4	5
.197	23.1	5	.767	31.5	5	.953	14.3	5
.243	24.4	5	.832	30.8	2	.974	13.5	5

TABLE 356. Visual observations (Lange)

JD hel	s_L	JD hel	s_L	JD hel	s_L
242...		242...		242...	
7829.388	31.0	7858.364	13.0	7863.378	29.4
.405	26.0	.371	10.4	7864.187	28.6
.406	20.8	.384	9.1	.200	30.0
.407	19.5	.391	9.1	.227	29.4
.408	18.6	.404	13.0	.261	30.3
.409	19.5	.415	18.2	.293	30.5
.410	19.5	.450	19.5	.350	31.0
.415	19.5	7860.405	9.9	7865.151	13.0
.420	18.8	.417	13.0	.163	13.0
.435	13.0	.445	18.8	.210	19.5
.463	18.2	.455	19.5	.252	20.8
.472	18.2	.496	23.4	.290	24.7
.510	19.5	.500	22.1	.387	31.4
7842.242	16.7	7861.179	26.0	.481	31.0
.242	16.0	.187	20.2	7867.156	9.1
.260	10.8	.192	19.5	.187	15.4
.271	12.0	.206	19.5	.221	18.3
.309	16.7	.216	20.8	243...	
.335	18.6	.223	21.4	7425.311	35.0
.366	19.5	.239	21.7	.318	30.0
.405	20.2	.290	23.4	.346	32.0
.452	20.8	.327	26.0	.350	30.5
7843.235	31.0	.393	28.6	7427.291	29.0
.277	29.0	.438	30.5	.297	30.0
.294	26.0	.499	33.0	.307	30.0
.318	30.5	7862.174	32.0	.316	30.0
7848.489	26.0	.212	30.5	.329	30.5
.502	26.0	.221	30.5	.346	30.5
.541	31.0	.227	27.8	7429.365	26.0
7857.215	23.8	.233	30.5	.372	26.0
.235	23.4	.283	31.0	7432.302	19.0
.275	26.0	.324	32.0	.309	13.0
7858.199	31.4	.378	19.5	.315	9.8
.238	31.4	.401	13.0	.319	9.3
.262	32.3	.451	17.6	.328	8.0
.281	31.4	.475	19.5	.341	8.3
.315	30.5	.506	20.8	.348	10.2
.335	23.4	.524	22.1	.357	13.0
.342	22.1	7863.270	23.4	.372	18.2
.352	19.5	.304	26.0	.385	20.8

TABLE 357. Visual observations (Tsesevich)

JD hel	s_{T1}	JD hel	s_{T1}	JD hel	s_{T1}
243...		243...		243...	
1584.344	22.0	1586.344	2.1	1589.367	9.3
.386	19.5	.369	20.0	1591.338	8.9
1585.364	15.4	1586.381	20.1	.361	9.7
.376	10.4	1588.329	19.8	1592.358	19.5
.382	9.7	.360	19.3	1607.345	19.7
.392	10.5	.396	20.1	1615.333	16.8
1586.317	20.0	1589.352	9.5		

JD hel	s_{T2}	JD hel	s_{T2}	JD hel	s_{T2}
243...		243...		243...	
5960.347	10.0	6699.331	14.9	6702.319	6.0
6339.327	12.6	.336	15.6	.327	4.0
.350	12.6	.347	14.9	.337	5.5
.376	13.8	.371	16.1	.346	6.0
6340.332	4.8	.383	14.0	.353	5.9
.340	4.3	.389	15.0	.364	5.0
.354	4.4	6701.298	15.8	.376	6.4
.367	4.4	.307	15.4	.387	7.0
6344.315	3.3	.314	15.8	6717.332	8.8
.327	3.3	.340	15.8	.345	14.0
.340	3.8	.355	14.9		
.356	3.3	6702.314	3.7		

283

TABLE 358. Simeise observations

JD hel	s_{ph}	JD hel	s_{ph}	JD hel	s_{ph}
241...		242...		242...	
9482.332	5.3	4210.455	9.7	7149	5.3
	5.2		9.7	8226.390	7.3
9835.377	18.6	6036.406	5.9		7.1
9841.456	11.7		3.5	8249.327	3.6
	11.7	6422.466	12.3		3.2
9852.304	3.5		12.3	.426	5.9
.263	6.5	6445.369	12.2		5.8
.346	3.5		11.0	8605.456	10.6
242...		6769.467	12.3	9320.498	14.1
3521.336	3.2		14.8		13.3
	2.7	7149.333	7.0		

TABLE 359. Odessa observations

JD hel	s_{ph}	JD hel	s_{ph}	JD hel	s_{ph}
243...		243...		243...	
6959.536	10.1	7020.427	11.0	7398.382	11.8
6993.484	12.3	7041.390	2.0	7400.384	12.3
6995.523	13.3	7373.426	11.0	7405.356	5.8
6996.493	10.5	7377.457	10.3	7406.378	11.6
7016.393	12.3	7396.384	9.8		
7017.398	9.2	7397.393	4.7		

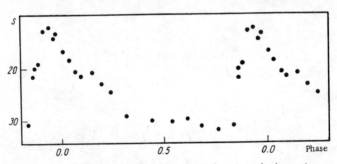

FIGURE 152. Average light curve of TV Leonis from Lange's observations.

AT Serpentis

Discovered by Hoffmeister /267/ and first studied by Solov'ev /101, 102, 352/. Initially Solov'ev proposed a period of $0^d.7407$. Later /109/ he came to the conclusion that the period was $0^d.427$. The short series of observations carried out in 1943 led Tsesevich to the equation /158/

$$\text{Max hel JD} = 2431232.448 + 0.7456 \cdot E. \qquad (243)$$

Gaposchkin /238/ and later Payne-Gaposchkin /329/ studied the star from Harvard photographs and discovered abrupt variation of its period. Since then only one maximum was observed (Alaniya /2/).

284

Tsesevich estimated the magnitude from Odessa, Simeise, and Harvard photographs and derived a summary table of maxima (Table 360). Gaposchkin's maxima are not included in the table, since the details of his study were not published and we could only try to recover the maxima from his elements. Tsesevich used the same Harvard sky photographs to construct seasonal curves. The O—C residues were calculated from

$$\text{Max hel JD} = 2422461.660 + 0.746581 \cdot E. \qquad (244)$$

TABLE 360. List of maxima

Source	Max hel JD	E	O—C
Tsesevich (Harv.)	2415353.783	−9521	+0.321 ^{d}
	6621.463	−7823	+ .306
	7524.789	−6613	+ .269
	8462.435	−5357	+ .209
	9630.050	−3793	+ .172
	20486.326	−2646	+ .119
	1213.481	−1672	+ .104
	2242.986	− 293	+ .074
	3602.458	+1528	+0.022
	4488.665	+2715	+ .038
	5178.457	+3639	− .011
	5918.308	+4630	− .022
	6687.264	+5660	− .044
	7403.246	+6619	− .034
	8337.260	+7870	+ .008
Solov'ev /101, 102, 352/	8343.220	+ 7878	− .005
Tsesevich (Harv.)	9243.650	+ 9072	+ .007
	9775.895	+ 9797	− .019
	30495.618	+10761	.000
Tsesevich /158/	1232.448	+11748	− .046
Tsesevich (Harv.)	1620.663	+12268	− .053
	2722.595	+13744	− .074
	3610.996	+14934	− .105
Alaniya /2/	5271.359	+17158	− .138
Tsesevich (Odessa)	7400.597	+20010	− .149

The period changed markedly. From $E = -9000$ to $E = +5000$ the period remained almost constant. Then it increased abruptly, but after $E = +9000$ it returned to its previous value (Figure 153). The recent observations fit the equation

$$\text{Max hel JD} = 2428343.221 + 0.7465682 \cdot E; \qquad P^{-1} = 1.339462356. \qquad (245)$$

The observations are listed in Tables 361, 362. Comparison stars (Figure 154): $m = 0^s.0$; $p = 7^s.2$; $q = 12^s0$.

TABLE 361. Simeise observations

JD hel	s	JD hel	s	JD hel	s
241...		242...		242...	
8878.306	8.0	3193.463	10.4	7595.384	8.8
	7.9	5066.341	13.0	7597.360	9.1
8880.329	12.0		13.0	3395.516	10.8
	15.0	6122.463	13.5		13.0
8901.283	11.0		13.0		
242...		7573.416	9.6		
3193.463	9.0	7595.384	13.0		

285

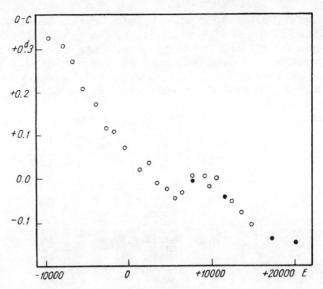

FIGURE 153. O—C residues for AT Serpentis (○ Harvard observations).

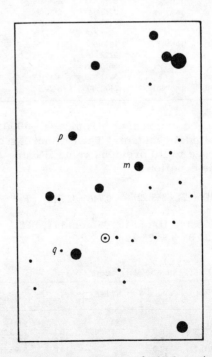

FIGURE 154. Comparison stars for AT Serpentis.

TABLE 362. Odessa observations

JD hel	s	JD hel	s	JD hel	s
243...		243...		243...	
6288.543	11.3	6730.384	11.0	7464.397	12.0
6292.566	12.1	6732.388	8.4	7733.622	8.2
6313.504	9.0	6773.326	14.5	7758.565	12.0
6336.505	12.0	7015.579	10.4	7761.562	13.5
6338.441	12.0	7016.564	13.0	7763.536	7.2
6345.408	11.2	7019.550	12.5	7764.529	11.5
6347.413	10.4	7020.562	7.2	7811.462	9.3
6364.374	3.8	7028.553	13.5	7814.420	11.3
6373.367	8.4	7029.568	3.9	7817.386	10.4
6376.374	7.2	7044.461	5.2	7839.355	12.0
6391.358	9.9	7046.503	13.0	7849.377	9.1
6395.351	11.3	7052.465	13.5	8090.621	11.6
6396.335	12.0	7071.456	8.3	8143.483	5.9
6400.326	9.8	7080.372	7.2	8144.492	11.3
6405.340	9.8	7373.605	10.8	8164.409	7.2
6647.605	10.8	7378.572	10.8	8165.411	12.0
6661.520	5.4	7400.556	4.6	8170.421	8.4
6690.518	10.1	7402.508	14.0	8172.426	13.0
6702.460	11.0	7404.526	9.8	8173.431	10.4
6703.469	8.8	7405.522	13.0	8194.364	9.6
6715.412	9.6	7406.529	3.8	8195.349	13.0
6722.427	13.0	7432.389	13.5	8198.360	12.0
6726.385	5.1	7457.383	5.9	8200.394	12.0
6728.409	14.0	7458.367	10.1	8203.368	10.1
6729.407	4.1	7462.399	15.0		

CONCLUSIONS

Unfortunately, it is very difficult to select a mathematical criterion for describing the extent of variation of an irregularly fluctuating period. We will therefore only present the summary table (Table 363) with some additional data from various sources. The additional objects are marked with an asterisk (*). The table lists the name of the star, its period P, the interval covered by observations ΔE, the spectral type at the minimum from calcium lines (Preston's Sp Ca II), the difference in hydrogen and calcium spectral types ΔS, the radial velocity v_r, and the galactic coordinates l and b.

TABLE 363. Stars with irregular fluctuations of the period

Star	P	ΔE	Sp Ca II	ΔS	v_r, km /sec	l	b
Stars with small fluctuations							
RS Boo*	d 0.377	—	F3	2	− 10	17°	+66°
SW Aqr	0.459	39700	A9	5	− 5	20	−33
RV UMa	0.468	43000	A6:	8:	−180	75	+62
BB Vir	0.471	35700	—	—	—	310	+63
DX Cep	0.526	44500	A9	7	—	87	+22
SVS 540 Cnc	0.543	34350	—	—	—		
TU UMa	0.558	29586	A9	6	+105	167	+73
TZ Aqr	0.571	30000	F1	5	+ 25	22	−46
RU CVn	0.573	30850	—	—	—	19	+73
ST Boo	0.622	30500	—	—	—	24	+54
Stars with large fluctuations							
RV CrB	0.322	61500	—	—	—	16	+43
AQ Lyr	0.357	64637	—	—	—	23	+14
V 759 Cyg	0.360	65200	—	—	—	51	+ 9
SS Tau	0.370	46132	—	—	—	148	−37
RR Gem	0.397	57600	F3	3	+ 94	155	+21
ST Vir	0.411	39600	—	—	—	315	+52
EL Cep	0.417	51800	—	—	—	75	+12
RW Dra*	0.433	—	F1	3	−125	54	+40

287

TABLE 363 (continued)

Star	P	ΔE	Sp Ca II	ΔS	v_r, km/sec	l	b
RR Leo*	0.452	—	A7	8	+ 65	176	+55
V 455 Oph	0.453	47400	—	—	—	8	+12
AR Her*	0.470	—	A7	6	−335	40	+47
BD Her	0.474	48600	F4	2	—	15	+ 6
AE Peg	0.497	38200	A7	7	—	49	−35
BF Peg*	0.496	—	—	—	—	58	−31
SW Psc	0.521	35000	—	—	—	88	−57
WZ Hya	0.524	45000	—	—	—	223	+35
AT Vir	0.526	32000	—	—	—	275	+57
SZ Hya	0.537	45000	—	—	—	208	+27
UU Hya	0.537	32000	—	—	—	199	+39
BK Eri	0.548	31200	—	—	—	143	−50
RR Lyr*	0.567	—	F0	6	− 70	42	+11
DZ Peg	0.607	31200	—	—	—	62	−42
TU Per	0.607	38000	—	—	—	111	− 4
V 365 Her	0.613	28300	—	—	—	10	+31
SS Leo	0.626	35000	A8	8	+145	235	+58
UY Boo	0.650	40400	—	(10)	+140	324	+67
TV Leo	0.673	26800	A5	10:	− 55	232	+50
AT Ser	0.747	29200	A6	9	− 70	346	+41

The frequency of the periods is as follows:

δP	n	δP	n
0.25—0.30	0	.50— .55	8
.30— .35	1	.55— .60	4
.35— .40	5	.60— .65	5
.40— .45	3	.65— .70	2
.45— .50	9	.70— .75	1

Relatively few stars were investigated, and one can hardly draw general statistical conclusions. Figure 155 shows the above distribution in graphic form. The group of non-stable stars has periods of about $0^d.36$. If it is omitted, we obtain a reliable distribution curve peaked near $0^d.45$, which does not differ from the distribution curve of stars with "jumping" period.

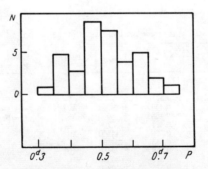

FIGURE 155. Frequency of periods in stars with irregular variation of the period.

The spectral data on these stars are very scanty. They nevertheless must not be ignored. The period can be correlated with the spectra type, dividing all the stars into four groups.

δP	$\overline{P}$	n	$\overline{Sp}$	$\overline{\Delta S}$	$\overline{\lvert v \rvert}$, km/sec	v Max
d d	d					
0.3—0.4	0.387	2	F3.0	2.5	57	94
0.4—0.5	0.465	7	A8.8	5.6	242	335
0.5—0.6	0.556	4	A9.8	6.0	67	105
0.6	0.674	4	A6.3	9.2	102	145

There is clearly a definite correlation between the period and the spectral type. Stars with irregular fluctuations of the period are remarkable in that their ΔS is very high, increasing with the increase in period and almost reaching one whole type. In this they fundamentally differ from constant-period stars.

Chapter 5

"EXTREME" AND "STRANGE" STARS

"EXTREME" STARS

In previous chapters we considered stars with periods from 0.25 to $0^d.75$. These are the "central" stars of the entire ensemble of RR Lyrae variables.

Of no less interest are stars with periods shorter than $0^d.25$ and longer than 1^d. These stars, especially the short-period category, are distinguished by certain peculiar features, which indicate that the RR Lyrae variables are not a homogeneous group, but actually consist of different subgroups. On the other hand, stars with periods over 1^d approach the cepheid field, which as we know also breaks into two groups $C\delta$ and CW.

The light variation period of one of the fastest fluctuating stars — CY Aquarii — is $0^d.0610$, i.e., 88^m. Complex superposition patterns, a sort of interference, were also observed in the ultrashort-period stars SX Phoenicis and AI Velorum. This constitutes a peculiar and hard-to-investigate Blazhko effect.

The magnitude of SX Phoenicis ranges from 6.7 to $7^m.5$ with a period of $0^d.05496420$, according to F. Walraven. The shape of the light curve and the "height" of the maxima are both variable with a period of $0^d.192836$.

AI Velorum changes its magnitude from 6.40 to $7^m.13$ with a period of $0^d.11157396$. On this fluctuation Walraven found superimposed another interfering fluctuation with a period of $0^d.08620767$. The beating period is $0^d.379188$, but the superposition is not linear.

Both stars are situated in the Southern Hemisphere. A similar star, however, is also found in the North Sky: this is VZ Cancri. Its magnitude varies from 7.19 to $7^m.94$ with a period of $0^d.17836376$. The shape of the light curve changes with a period of $0^d.716292$. It would thus seem that all stars of "ultrashort" periods are subject to extremely complex oscillations. CY Aquarii, however, shows nothing of the sort. This startling exception led us to a careful re-examination of the entire class of ultrashort-period stars.

STARS WITH LIGHT CURVES OF CONSTANT SHAPE

We will first consider those objects which on the basis of long and reliable observations can be regarded as a "prototype" of constancy.

Observed visually by Tsesevich in 1957. Light variation range very small, and comparison stars situated most inconveniently, especially for observations through a 12-in. reflector. Yet individual light curves gave more or less reliable minima, from which the period was determined to five places. The magnitude was further estimated from Odessa photographs of 1952—1956. The photographic observations, though few in number, nevertheless helped to improve the elements. Finally, the 1939 series of Moscow plates gave the light variation curve. All these combined, gave highly reliable elements.

The photographic and visual observations yielded the approximate minima (Table 364). The O—A residues were calculated by the least squares method:

$$\text{Min hel JD} = 2436006.3719 + 0.104090998 \cdot E;\ P^{-1} = 9.6069796. \qquad (246)$$

Reduction of all the visual observations to one period using this equation gave the average visual light curve (Table 365). From this curve the average maximum was determined, Max hel JD = 2436050.3400. Reduction to one period of the estimates obtained from Odessa photographs showed that although the average curve could not be determined for lack of observations, the resulting smooth curve was sufficient to estimate the average maximum, Max hel JD = 2434926.3665.

TABLE 364. List of minima

Source	Min hel JD	E	O — A
Tsesevich (photogr.)	2434128.47	—18041	+0.004
	4178.33	—17562	+ .004
	5224.45	— 7512	+ .010
	5227.45	— 7483	— .009
	5243.40	— 7330	+ .015
	5248.45	— 7281	— .035
Tsesevich (visual)	6006.372	0	.000
	6008.446	+ 20	— .008
	6009.397	+ 29	+ .006
	6013.340	+ 67	— .006
	6013.454	+ 68	+ .004
	6020.420	+135	— .004
	6027.389	+202	— .009
	6040.316	+326	+ .010
	6040.415	+327	+ .005
	6050.408	+423	+ .006
	6069.348	+605	+ .001
	6072.362	+634	— .004
	6105.266	+950	+ .008

TABLE 365. Average light curve from visual observations

φ	s_2	n	φ	s_2	n	φ	s_2	n
0.032	8.8	10	0.335	6.1	10	0.743	8.5	10
.076	8.3	10	.389	5.9	10	.814	8.2	10
.141	8.4	10	.461	6.0	10	.878	8.9	10
.184	8.2	10	.536	6.3	10	.936	9.2	10
.226	7.5	10	.610	6.8	10	.986	9.2	9
.282	6.7	10	.676	7.1	10			

Estimates from Moscow photographs were reduced to one period using (246). The average light curve is given in Table 366.

TABLE 366. Average photographic light curve from Moscow plates

φ	s_1	n	φ	s_1	n	φ	s_1	n
0.019	5.7	4	0.443	9.8	5	0.835	8.8	5
.121	6.9	5	.549	11.1	5	.888	7.5	3
.162	7.2	3	.619	10.0	4	.982	6.0	3
.239	8.3	2	.703	11.0	4			
.337	8.4	5	.779	9.5	3			

The corresponding average maximum

$$\text{Max hel JD} = 2429403 \cdot 3603.$$

List of all the maxima:

Max hel JD	E	O − A	O − B
2429403.3603	− 63858	+ 0.0633^d	0.0000^d
34926.3665:	− 10798	+ .0011	− .0096
6050.3400	0	.0000	.0000

The old maximum deviates from the ephemeris by more than half a period (Figure 156). The two extreme maxima give a new equation

$$\text{Max hel JD} = 2436050.3400 + 0.104090008 \cdot E. \qquad (247)$$

Comparison stars (Figure 157):

	s_1	s_2
a	8.3	0.0
b	—	10.6
c	11.7	—
k	0.0	—

When the present study was completed and published in /162/, Ahnert /173/ compiled all the maxima available to him. Analysis of this compilation led Ahnert to propose an improved equation of the period

$$\text{Max hel JD} = 2435282.766 + 0.10409160 \cdot E. \qquad (248)$$

The new observations of Migach confirm the validity of (248). Table 367 is a complete list of maxima.

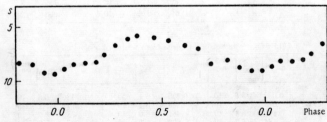

FIGURE 156. Light curve of YZ Bootis.

292

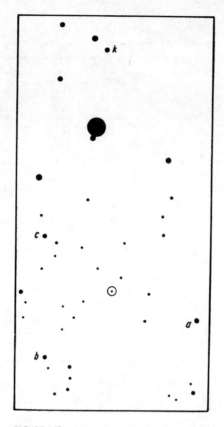

FIGURE 157. Comparison stars for YZ Bootis.

The period, however, is apparently variable. Migach's maximum deviates by $0^p.1$ from Ahnert's formula, which clearly cannot be attributed to observation errors.

Tsesevich's observations are listed in Tables 368—370.

TABLE 367. List of maxima

Source	Max hel JD	E	O—C
Tsesevich	2429403.3603	— 56483	+ 0.0001 (d)
	34926.3665:	— 3424	+ .0101
Eggen	5282.766	0	.000
	5282.870	+ 1	.000
Broglia, Mazani	5688.409	+ 3897	— .002
	5689.4497	+ 3907	— .0022
	5695.3862	+ 3964	+ .0011
	5695.4900	+ 3965	+ .0008
	5698.4032	+ 3993	— .0006
	5699.341	+ 4002	.000
Tsesevich	6050.3400	+ 7374	+ .0035
Spinrad	6428.817	+ 11010	+ .002
	6429.752	+ 11019	+ .001
Strelkova	6452.338	+ 11236	— .0012

TABLE 367 (continued)

Source	Max hel JD	E	O—C
			d
Antal, Tremco	2436603.48	+ 12688	+0.000
	6603.585	+ 12689	+ .001
	6606.50	+ 12717	+ .001
	6607.44	+ 12726	+ .004
	6607.54	+ 12727	.000
	6613.47	+ 12784	— .003
	6613.575	+ 12785	— .002
Ahnert	6673.53	+ 13361	— .004
	6674.58	+ 13371	+ .005
	6689.56:	+ 13515	— .004:
	6700.39	+ 13619	.000
	6701.43:	+ 13629	.000:
	6725.475	+ 13860	— .001
Broglia	7077.5112	+ 17242	— .0022
	7098.4349	+ 17443	— .0009
	7120.3975	+ 17654	— .0016
	7120.5010	+ 17655	— .0021
	7137.3654	+ 17817	— .0006
	7168.385	+ 18115	.000
Migach	7846.449	+ 24629	+ .011
Sakharov	8266.3482	+ 28663	+ .0047

TABLE 368. Moscow photographic observations, UT series (Tsesevich)

JD hel	s_1	JD hel	s_1	JD hel	s_1
242...		242...		242...	
9334.329	9.1	9373.237	11.3	9423.301	10.5
9335.308	8.3	.395	5.8	9424.255	11.3
.342	12.3	.453	12.3	9425.305	9.8
.366	10.0	9375.295	7.3	9427.302	5.8
9336.376	12.5	.375	6.3	9438.295	8.3
9337.366	6.3	.448	6.3	9429.312	8.3
9338.294	8.3	9376.209	6.3	9455.211	8.3
.385	9.1	.298	5.3	9456.227	7.3
.456	9.1	.359	7.3	9457.218	10.3
9339.298	9.3	.427	8.3	9460.230	8.3
.352	5.8	9377.219	6.8	9461.219	9.3
.412	13.0	9395.348	5.8	9462.215	9.0
.467	9.0	9396.357	9.8	9463.227	9.3
9340.419	9.1	9397.410	9.4	9464.198	11.3
9371.347	9.3	9403.393	9.1	9465.193	5.8
.402	5.3	9404.207	5.3	9466.197	10.6
.466	10.8	9421.240	9.5	9468.206	6.3

TABLE 369. Odessa photographic observations (Tsesevich)

JD hel	s_1	JD hel	s_1	JD hel	s_1
243...		243...		243...	
4117.376	5.2	5217.431	6.2	5246.398	8.3
4126.435	4.4	5224.413	8.3	5248.450	11.7
4128.473	9.8	5224.490	9.1	5252.356	7.8
4178.334	10.0	5225.448	6.2	5253.426	4.6
4209.333	6.2	5227.452	8.3	5255.406	5.3
4460.587	6.4	5240.403	4.6		
4926.331	7.8	5243.399	9.1		

TABLE 370. Visual observations (Tsesevich)

JD hel	s₂	JD hel	s₂	JD hel	s₂

JD hel	s_2	JD hel	s_2	JD hel	s_2
243...		243...		243...	
6006.3319	6.5	6013.4865	7.9	6050.2816	8.7
.3402	6.2	.4942	6.8	.2872	8.3
.3701	9.6	6018.3634	5.8	.3038	7.4
.3718	8.5	.3787	5.3	.3399	5.9
.3839	8.7	.4620	8.0	.3510	5.3
.3930	7.4	.4683	6.8	.3687	5.8
.4027	6.4	.4745	6.1	.3816	7.8
.4103	7.1	6019.3509	7.1	.3920	8.7
.4194	6.4	.3870	8.1	.4059	8.9
.4298	6.1	6020.3376	11.6	.4215	8.5
.4339	6.8	.3432	5.8	.4329	7.1
.4408	6.2	.3779	5.3	6069.3021	6.2
6007.3735	7.4	.3866	7.4	.3111	8.1
.3770	6.8	.4098	8.9	.3177	8.5
.4187	6.2	.4150	11.6	.3268	8.1
.4367	7.4	.4216	8.9	.3316	9.1
.4426	7.4	.4369	8.9	.3424	8.1
.4485	6.2	.4473	8.2	.3494	8.0
.4631	6.5	.4582	6.2	.3636	8.7
6008.3485	6.2	.4626	6.5	.3740	8.1
.3666	6.5	6027.3288	6.2	.3903	6.0
.3881	5.8	.3392	6.1	6071.2971	11.1
.4055	6.2	.3500	7.1	.3034	8.9
.4249	7.4	.3579	7.6	.3235	11.6
.4437	9.3	.3677	9.1	6072.2916	6.2
.4510	10.6	.3746	9.1	.3030	5.5
.4590	8.3	.3788	10.0	.3117	5.8
.4698	7.4	.3840	10.0	.3193	7.6
.4839	7.4	.3878	9.1	.3284	7.8
6009.3193	5.3	.3934	10.1	.3367	8.5
.3332	6.5	.3992	9.0	.3430	7.7
.3422	6.8	.4072	9.1	.3575	8.9
.3501	5.9	.4128	8.2	.3756	8.9
.3689	6.4	.4184	8.3	.3864	7.4
.3811	8.5	.4267	8.2	6075.2843	11.6
.3873	8.9	.4343	5.8	.2961	8.7
.3930	8.7	.4427	5.5	.3048	7.8
.3998	8.9	.4621	6.4	.3204	4.2
.4057	8.7	.4746	8.5	.3322	6.4
.4127	8.1	.4892	10.6	.3593	8.1
.4200	6.5	6040.3023	7.4	.3913	9.1
.4255	6.2	.3127	8.5	6076.3057	10.6
.4422	5.3	.3252	8.7	.3280	9.3
6013.3205	8.5	.3315	7.1	.3429	8.5
.3344	8.3	.3520	6.2	.3541	6.1
.3428	9.6	.3574	5.3	.3745	6.2
.3535	7.9	.3818	7.1	6095.2821	9.6
.3574	7.6	.3940	8.9	6102.2391	8.5
.3754	6.5	.4092	8.8	6105.2257	5.9
.3851	5.5	.4203	8.1	.2470	5.7
.4074	5.8	.4290	9.1	.2584	7.4
.4178	8.4	.4446	6.1	.2634	7.7
.4268	7.9	6041.3029	6.2	.2755	7.7
.4397	9.1	.3286	8.1	.2856	5.6
.4476	10.1	.3925	5.9	.2977	5.3
.4553	10.1	.4112	8.3		
.4695	8.8	.4327	7.0		

CK Aquarii

Discovered and studied by Shapley and Hughes /350/. Their elements

$$\text{Max hel JD} = 2425425.57 + 0.12406 \cdot E. \qquad (249)$$

The Simeise photographs are inadequate for a star of such a short period, but Tsesevich nevertheless proceeded to estimate the magnitude from these plates. The seasonal light curve requires a more exact equation of the

295

period. Visual observations (neither particularly accurate nor reliable, since the star is somewhat too faint for a 12-in. reflector) gave two average maxima, which were linked up with the latest photographic observations by the equation

$$\text{Max hel JD} = 2437547.329 + 0.1240624 \cdot E. \qquad (250)$$

This equation was in fact used to construct the seasonal light curves from Simeise observations. The length of each season was 1000 days. Table 371 lists all the maxima obtained by Tsesevich. They are fitted with the equation

$$\text{Max hel JD} = 2437547.319 + 0.1240624507 \cdot E; \ P^{-1} = 8.060456604. \qquad (251)$$

TABLE 371. List of maxima

Source	Max hel JD	E	O — A
Tsesevich (photogr.)	2421074.305	— 132780	−0.002^d
	3968.440	— 109452	+.004
	5501.360	— 97096	+.009
	6240.381	— 91139	—.010
	33478.315	— 32798	—.004
	4239.438	— 26663	—.004
Tsesevich (visual)	7197.335	— 2821	—.004
	7547.329	0	+.010

Comparison stars (Figure 158): photographic scale $k = 0^s.0$; $a = 11^s.1$; $b = 21^s.0$; visual scale $c = -9^s.6$; $d = 0^s.0$; $e = 9^s.3$. The light curves are given in Tables 372, 373; Tsesevich's observations in Tables 374, 375.

TABLE 372. Average photographic light curve (Figure 159)

φ	s	n	φ	s	n	φ	s	n
0.018^p	8.7	5	0.222^p	8.5	6	0.681^p	9.8	4
.032	7.3	4	.251	9.2	5	.721	10.0	5
.059	9.3	4	.320	10.9	5	.745	9.7	5
.070	7.8	5	.408	10.5	5	.795	8.9	5
.095	8.2	5	.450	10.0	5	.833	8.4	5
.130	8.9	5	.555	14.0	5	.854	8.4	5
.166	8.0	4	.610	11.5	6	.928	8.4	5
.206	9.1	5	.666	11.0	4	.967	7.0	5

TABLE 373. Average visual light curve

φ	s	n	φ	s	n	φ	s	n
0.040^p	— 2.4	4	0.504^p	+ 3.7	5.	0.797^p	+ 0.3	5
.140	— 2.2	5	.598	+ 5.4	5	.846	+ 0.1	4
.222	— 3.1	5	.657	+ 2.8	5	.959	— 0.3	5
.337	— 2.2	5	.710	+ 4.0	5			
.418	+ 2.8	5	.774	+ 2.7	5			

TABLE 374. Simeise photographic observations (Tsesevich)

JD hel	s	JD hel	s	JD hel	s
242...		242...		242...	
0372.389	6.1	5479.353	12.9	9105.457	9.5
0715.422	9.8	5482.365	7.4		9.5
0740.304	11.1		8.1	9114.454	11.1
	9.1	5495.293	6.5	243...	
0741.337	9.1		9.7	0207.473	9.5
	9.5	5496.304	8.9	2775.378	5.8
1070.441	9.1:		13.3		6.1
1074.483	15.4	5501.359	6.0	3129.404	12.9
	10.1		8.3		12.1
1084.456	6.1	5505.336	6.4	3132.453	6.8
	9.1		9.1		8.9
1137.291	10.1	5800.469	8.1	3475.474	5.0
1427.392	7.4:		11.1	3478.471	6.0
1431.324	8.6	5826.460	14.4		7.4
3255.356	9.1	5827.491	9.1	3486.438	9.7
	8.3		8.0		12.0
3968.445	7.1	5828.450	15.0	3869.380	6.0
3994.492	8.9	5830.398	11.1		6.0
4350.519	9.1	5834.464	11.1	3886.325	9.3
4358.467	9.5	6180.480	8.1		8.7
	9.1		7.4	3892.310	7.4
4362.511	9.1	6187.396	6.1		7.4
	8.1	6191.456	13.3	4222.454	5.1
4412.318	9.5		12.2		7.1:
4416.267	12.1	6239.318	10.5	4238.329	5.0
	11.1		10.6		5.6
4730.401	9.3	6240.268	7.4	4239.344	9.1
5445.482	12.1		7.8		8.7
	12.5	8379.483	13.1	4601.380	9.5
5449.486	8.1		9.9		8.5
5475.353	10.5	8747.454	9.5	4605.375	8.5
			12.1		9.5

TABLE 375. Visual observations (Tsesevich)

JD hel	s	JD hel	s	JD hel	s
243...		243...		243...	
7196.306	+3.0	7531.412	+5.0	7549.352	+7.2
.318	−5.0	.432	+4.0	.366	+6.5
.324	−6.0	.453	+2.5	.390	+6.2
.344	−7.0	.466	−1.7	.400	+6.5
.361	−8.0	7543.345	−2.6	.414	+7.3
.371	−6.0	7545.307	−1.7	.422	+2.1
.376	−5.0	.357	−1.4	.426	−1.7
.380	−5.0	.380	+5.2	.429	−2.1
.388	−5.0	.384	+6.2	7550.268	+2.1
.394	−5.0	.394	+5.9	7557.320	+6.0
.402	+2.0	.412	+7.2	7561.384	+4.0
.409	+1.0	.423	+3.0	.408	+6.0
7197.295	+2.0	.432	+2.5	.422	+6.0
.303	+1.0	7547.264	+7.2	7573.303	+6.0
.312	0.0	.278	+5.9	7578.230	+4.0
7198.270	+1.0	.282	+5.6	.245	+4.0
.284	+2.5	.296	+1.6	.290	+2.0
.310	−4.5	.300	+3.7	7583.238	+6.5
.341	−8.0	.311	+1.3	.246	+6.8
.376	−7.0	.319	+2.1	.254	+6.5
7528.313	−2.9	.328	−1.7	.282	−1.4
7530.346	−1.9	.334	−1.5	.290	−2.5
7531.380	+4.0	.339	−2.1	.299	−1.5
.386	+4.0	.352	−1.1	.306	−1.5
.395	+6.0	.362	+0.7	.325	+2.0
.405	+4.0	.377	+3.7	.340	+3.7

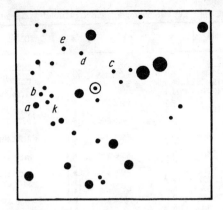

FIGURE 158. Comparison stars for CK Aquarii.

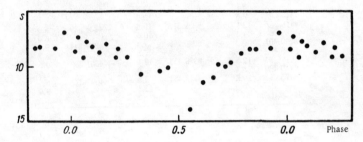

FIGURE 159. Light curve of CK Aquarii.

CW Serpentis

Discovered by Hoffmeister in 1935. Observed by Solov'ev /111/ and Tsesevich /159/. Solov'ev, however, never published his observations, so that only Tsesevich's observations are available, with estimates based on Simeise, Harvard, and Odessa photographs.

Tsesevich obtained the equation

$$\text{Max hel JD} = 2431212.280 + 0.18914 \cdot E, \tag{252}$$

which was subsequently improved to

$$\text{Max hel JD} = 2431212.280 + 0.1891495 \cdot E; \ P^{-1} = 5.28682339. \tag{253}$$

Reduction of observations to seasonal curves gave the maxima listed in Table 376.

TABLE 376. List of maxima

Source	Max hel JD	E	O — C
Tsesevich (Harv.)	2416084.789	— 79976	+0.009
	8295.765	— 68287	+ .005
	20155.873	— 58453	+ .007
	1828.902	— 49608	.000
Tsesevich (Sim.)	3193.469:	— 42394	+ .035:
Tsesevich (Harv.)	3785.872	— 39262	+ .019
	5060.902	— 32521	— .015
	5925.889	— 27948	— .013
	6876.947	— 22920	— .004
	8132.891	— 16280	— .019
	9418.929	— 9481	— .015
	30902.802	— 1636	— .010
Tsesevich (visual)	1212.280	0	.000
Tsesevich (Harv.)	2516.842	+ 6897	— .009
Tsesevich (Odessa)	6773.309	+ 29400	+ .004
	7733.239	+ 34475	— .004
	8143.317	+ 36643	— .005

These maxima led to the final equation

$$\text{Max hel JD} = 2431212.280 + 0.1891505 \cdot E. \tag{254}$$

The period is possibly somewhat variable.

Average light curves were constructed using (253) from visual and photographic observations (Table 377, Figure 160, 1 and 2 respectively).

TABLE 377. Average photographic light curve

φ	s	n	φ	s	n	φ	s	n
0.027	+ 1.1	5	0.318	+ 0.3	5	0.732	+4.1	5
.065	— 0.1	5	.404	+1.0	5	.862	+5.7	5
.137	— 0.9	5	.561	+3.0	5	.956	+3.3	5
235	— 1.0	5	.623	+3.6	5			

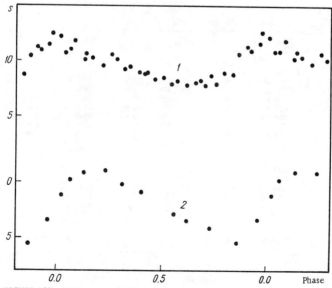

FIGURE 160. Light curve of CW Serpentis.

299

Tsesevich's observations are listed in Tables 378, 379. Comparison stars (Figure 161): $m = 0\overset{s}{.}0$; $p = 3\overset{s}{.}3$; $r = 9\overset{s}{.}9$; $q = 14\overset{s}{.}0$.

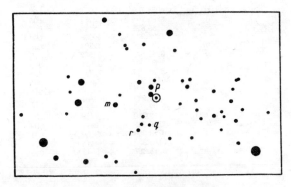

FIGURE 161. Comparison stars for C W Serpentis.

TABLE 378. Simeise observations (Tsesevich)

JD hel	s	JD	s	JD hel	s
241...		242...		242...	
8878.306	2.2	4297.363	1.9	7607.356	2.6
	4.3	5066.341	8.1		2.7
8880.329	6.3		8.6	243...	
	2.2	6122.463	2.2	3393.481	7.6
8884.336	7.3		2.1		2.7
242...		7573.416	1.6	3395.516	1.6
3193.463	—1.0	7595.384	1.1		1.3
	—1.0		6.9	3747.514	2.2
4297.363	1.6	7597.360	8.6		2.5

TABLE 379. Odessa observations (Tsesevich)

JD hel	s	JD hel	s	JD hel	s
243...		243...		243...	
6702.460	1.6	7071.456	1.6	7811.462	2.5
6703.469	3.2	7080.372	0.0	7814.420	0.9
6715.412	0.4	7373.606	3.9	7817.386	0.8
6722.427	—2.2	7378.572	—0.8	7839.355	—1.0
6726.385	—1.5	7400.556	0.8	7849.377	0.8
6728.409	3.6	7402.509	6.0	8090.621	1.8
6729.407	—1.2	7404.526	—1.2	8143.483	—1.0
6730.384	—0.2	7405.522	3.4	8144.492	1.5
6732.387	4.7	7406.529	7.4	8164.409	9.9
6773.326	—2.5	7453.381	2.5	8165.411	3.3
7015.579	1.0	7457.383	1.8	8170.421	1.1
7016.565	0.1	7458.367	2.8	8172.426	1.3
7019.551	—0.5	7462.399	1.2	8173.431	1.3
7020.563	—0.4	7464.397	5.2	8194.364	5.5
7028.553	2.0	7733.622	—1.5	8195.349	—1.0
7029.568	1.6	7758.565	0.0	8200.394	5.5
7044.461	2.6	7761.562	7.3	8203.368	5.8
7046.503	3.2	7763.536	0.0		
7052.465	—1.2	7764.529	6.8		

BC Eridani

Discovered by Hoffmeister /268/. First studied by Gur'ev /27, 30/. Also observed by Ashbrook and Gossner /182/. Recently, Tsesevich carried out visual observations. The maxima are listed in Table 380.

TABLE 380. List of maxima

Source	Max hel JD	E	O — A	O — B
			d	d
Ashbrook and	2410661.557	— 68998	—0.036	—0.038
Gossner /182/	3448.855	— 58436	+ .008	+ .005
	4605.786	— 54052	+ .025	+ .022
	4942.756	— 52775	+ .001	— .001
	6044.782	— 48599	+ .004	+ .001
Gur'ev /27, 30/	28869.791	0	.000	— .005
Ashbrook and	31798.771	+ 11099	+ .014	+ .009
Gossner /182/	1802.701	+ 11114	— .014	— .020
Tsesevich (visual)	7198.597	+ 31561	+ .030	+ .023

The O — A residues were calculated from

$$\text{Max hel JD} = 2428869.791 + 0.26389458 \cdot E.$$

Application of the least squares method gave the improved equation

$$\text{Max hel JD} = 2428869.796 + 0.263894627 \cdot E. \tag{255}$$

Since the maxima were determined not very accurately owing to the "fuzzed" light curve, we have no grounds to maintain that the period is variable.

Tsesevich's visual observations are listed in Table 381. Comparison stars (Figure 162): $a = 0^s.0$; $b = 17^s.5$.

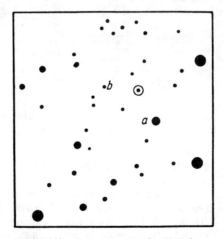

FIGURE 162. Comparison stars for BC Eridani.

TABLE 381. Visual observations (Tsesevich)

JD hel	s	JD hel	s	JD hel	s
243...		243...		243...	
6495.538	19.5	7197.483	14.3	7198.554	13.1
.549	13.0	.502	15.0	.567	11.7
.573	13.4	.518	13.6	.574	11.7
.590	14.0	7198.462	15.4	.579	11.0
.605	14.0	.488	15.0	.584	10.5
7196.530	13.1	.498	15.0	.592	8.5
.561	14.0	.517	14.6	7204.459	12.2
.582	15.0	.534	12.4	.468	12.5
.594	13.1	.546	13.1	.482	12.4

V 729 Ophiuchi

Discovered and studied by Hughes-Boyce /275/, who obtained the equation

$$\text{Max hel JD} = 2427924.55 + 0.319 \cdot E. \tag{256}$$

Tsesevich estimated the magnitude from Odessa, Simeise, and Moscow photographs. The star is too faint for Simeise photographs; the observations are therefore too few for the construction of seasonal light curves. The curves were constructed from Moscow and Odessa observations only. Table 382 lists not only the firm maxima but also the times of enhanced magnitude from Simeise photographs. The seasonal curves were constructed using the equation

$$\text{Max hel JD} = 2437109.367 + 0.318847 \cdot E. \tag{257}$$

TABLE 382. List of maxima

Source	Max hel JD	E	O — C
			d
Tsesevich (Sim.)	2423554.366	— 42512	—0.003
	3942.463:	— 41295	+ .052:
	5770.348	— 35562	— .036
	5771.349	— 35559	+ .008
Hughes-Boyce /275/	7924.55	— 28806	+ .008
Tsesevich (Sim.)	34540.370	— 8057	— .014
Tsesevich (Odessa)	6757.362	— 1104	+ .007
Tsesevich (Mosc.)	7109.369	0	+ .002
Tsesevich (Odessa)	7406.550	+ 932	+ .014
	7790.398	+ 2136	— .035

The last two maxima in Table 382 were determined after the completion of this study. The O — C residues were calculated using the improved equation

$$\text{Max hel JD} = 2437109.367 + 0.3188511 \cdot E; P^{-1} = 3.13626015. \tag{258}$$

The fitting of Moscow observations with this formula is shown in Figure 163. These are the most accurate observations, since the photographs were taken with the 16-in. astrograph and the stellar image is properly exposed. Comparison stars (Figure 164): $p = 0^s.0$; $q = 9^s.6$; $r = 18^s.2$; $s = 24^s.9$. Tsesevich's observations are listed in Tables 383 — 385.

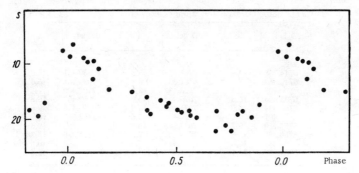

FIGURE 163.   Light curve of V 729 Ophiuchi.

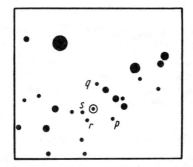

FIGURE 164.   Comparison stars for V 729
Ophiuchi.

TABLE 383.   Simeise observations (Tsesevich)

JD hel	s	JD hel	s	JD hel	s
242...		242...		243...	
3554.366	7.5	5770.348	8.5	3766.488	13.9:
3942.463	8.1	5771.349	7.7	3771.500	12.6
5382.442	15.2	5802.357	11.6:	3802.362	15.6
5385.465	14.5	243...		3803.364	18.2
5388.489	11.6	2686.452	13.9	3806.452	11.6:
5764.404	13.1	3034.505	14.5	4540.370	7.5:

TABLE 384.   Moscow observations (Tsesevich)

JD hel	s	JD hel	s	JD hel	s
243...		243...		243...	
7052.547	19.2	7100.394	19.7	7138.297	21.2
7074.462	18.2	7102.381	9.6	7139.305	17.1
7077.490	6.4	7103.344	12.8	7140.300	8.7
7078.482	10.7	7106.423	22.0	7143.312	17.2
7079.511	16.0	7109.360	7.5	7144.302	19.3
7080.472	19.2	7132.362	9.6	7145.299	18.2
7087.457	15.0	7133.341	14.3	7161.282	18.2
7088.468	17.7	7135.330	16.3	7163.276	8.7
7089.441	18.2	7136.327	18.2	7165.281	18.2
7099.355	19.7	7137.327	22.0		

TABLE 385. Odessa observations (Tsesevich)

JD hel	s	JD hel	s	JD hel	s
243...		243...		243...	
6744.358	[9.6	6761.379	15.6	7080.451	16.8
6749.363	7.9	6781.330	8.4:	7099.361	[15.6
.405	8.4	7073.437	13.6	7101.400	7.5
6755.424	15.6	7075.464	[15.6	7102.407	16.6
6756.344	16.6	7077.396	14.6:	.456	9.6
.401	9.6	.452	[15.6	7107.369	15.6
6757.362	6.7	7078.388	[9.6	.423	[9.6
6758.363	11.6	7079.367	7.5	7111.370	15.3
.389	10.6	.418	7.9	.396	16.8
6759.389	14.6	.467	[14.6	.424	15.6

V 727 Ophiuchi

Discovered by Hughes-Boyce /275/, who determined its period, $0^d.5007$. Tsesevich studied the star on Moscow photographs taken with the 16-in. astrograph, on Odessa photographs taken with the seven-camera astrograph, and on Simeise planetary photographs.

Faint star, with a period close to 1^d, which greatly complicates the observations. After a few trials, the following tentative elements were obtained:

$$\text{Max hel JD} = 2437087.464 + 0.333590 \cdot E; \quad P^{-1} = 2.9976918. \qquad (259)$$

This equation applied to Moscow and Odessa observations gave the average light curve (Figure 165, Table 386).

TABLE 386. Average light curve

φ	s	n	φ	s	n	φ	s	n
p 0.016	12.2	5	p 0.455	21.3	5	p 0.802	20.3	5
.069	10.4	5	.496	21.9	5	.840	18.6	5
.198	18.6	5	.540	20.7	5	.942	11.6	4
.267	20.4	5	.607	22.6	5	.982	10.6	4
.338	19.3	5	.658	19.9	5			
.398	20.9	5	.744	21.8	5			

The light curve is characteristic of RRc stars. Comparison stars (Figure 166): $k = 0^s.0$; $a = 11^s.7$; $b = 19^s.9$; $c = 24^s.6$. The Simeise photographs do not give any reliable maxima, since their distribution in time is poor. Two significant enhancements were nevertheless determined (Table 387).

TABLE 387. Times of enhancement

Source	Max hel JD	E	O — A	O — B
Tsesevich (Sim.)	2419920.46	— 51461	d −0.13	d + 0.05
	25027.34	— 36152	— .18	— .04
Hughes-Boyce /275/	7960.30	— 27360	— .14	-- .04
Tsesevich (Mosc.)	37087.464	0	.000	+ .018
Tsesevich (Odessa)	7454.413	+ 1100	.000	+ .014

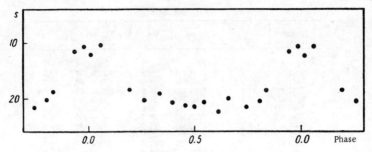

FIGURE 165. Light curve of V 727 Ophiuchi.

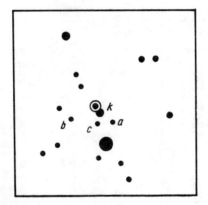

FIGURE 166. Comparison stars for V 727
Ophiuchi.

The O — B residues were calculated from the new equation

$$\text{Max hel JD} = 2437087.446 + 0.3335932 \cdot E. \qquad (260)$$

The observations are listed in Tables 388 — 390.

TABLE 388. Simeise observations (Tsesevich)

JD hel	s	JD hel	s	JD hel	s
241...		242...		242...	
9556.324	21.9	5382.442	19.1	9071.384	23.9
9920.450	11.7:	5385.465	18.3		24.6
242...			21.5	243...	
0253.497	20.9	5388.489	19.9	2686.452	18.3
0636.497	22.9	5764.404	22.9		19.1
	24.6		19.1	3034.505	19.1
1006.424	[17.7	5770.348	20.7		21.8
1369.461	19.1		16.6	3766.488	23.4
	20.9	5771.349	17.4		20.9
3554.366	[19.9	5799.345	21.9	3802.362	23.9
3937.477	[19.9		21.1	3803.364	16.2
	19.9	5802.357	17.4		21.8
3931.370	20.9	5820.338	21.9	3806.452	20.9
3942.463	[19.9		19.1	3771.500	21.9
4312.376	[19.9	5826.354	19.1	3806.452	21.9
	19.9		23.4		22.9
5027.341	12.7	6860.355	[19.9	4183.358	19.1
	13.1				20.9

305

TABLE 389. Moscow observations (Tsesevich)

JD hel	s	JD hel	s	JD hel	s
243...		243...		243...	
7052.547	7.2	7100.394	22.2	7138.297	19.9
7074.462	11.7	7103.344	21.8	7139.305	22.7
7077.490	11.2	7106.423	19.9	7140.300	22.7
7078.482	10.7	7109.360	21.8	7143.312	22.7
7079.511	14.0	7113.351	24.6	7144.302	23.4
7080.472	9.6	7132.362	22.7	7145.299	21.5
7087.457	9.7	7133.341	21.9	7161.282	22.2
7088.468	9.7	7135.330	22.0	7163.276	19.9
7089.441	11.7	7136.327	21.8	7165.281	22.5
7099.355	20.9	7137.327	21.8	7462.453	8.7

TABLE 390. Odessa observations (Tsesevich)

JD hel	s	JD hel	s	JD hel	s
243...		243...		243...	
6744.358	22.9	7079.367	19.1	7404.547	21.1
.405	21.9	.418	18.4	7406.550	22.3
6755.424	20.9	.467	10.7	7427.472	18.7
6756.344	20.9	7080.400	22.7	7454.392	8.7
6756.401	22.7	.451	10.2	7457.407	11.7
6757.362	20.9	7099.361	20.9	7458.338	14.3
6758.363	18.2	7101.400	21.9	7464.418	14.0
.389	19.1	7102.407	24.6	7764.549	24.6
6759.389	18.4	.456	10.5	7789.422	18.9
6761.379	22.9	7107.369	18.2	7790.398	17.3
6781.330	17.8	.423	17.8	7793.419	15.6
7073.437	15.4	7111.370	18.7	.474	14.6
7075.464	11.7	.396	23.0	7794.439	19.9
7077.396	19.9	.424	17.7	7810.402	20.7
.452	10.8	7373.624	19.1	7812.357	14.7:
7078.388	19.9	7402.533	20.9	7813.432	25.6

BI Virginis

Discovered and studied by Boyce /206/. Tsesevich investigated it using Simeise photographs. The star is unfortunately very faint and the observations highly unreliable. The equation

$$\text{Max hel JD} = 2426840.35 + 0.33565 \cdot E \qquad (261)$$

was used to construct the seasonal light curves. The approximate maxima determined from these curves:

Max hel JD	E	O—C d
2420608.581	− 18566	−0.008
6033.457	− 2404	+ .015
6840.35	0	− .006
8246.411	+ 4189	− .003
33737.387	+ 20548	− .004

The O—C residues and the average light curve were calculated from

$$\text{Max hel JD} = 2426840.356 + 0.3356548 \cdot E; \; P^{-1} = 2.9792513. \qquad (262)$$

The period apparently remained constant. Comparison stars (Figure 167) $a = 0^s.0$; $b = 7^s.8$; $c = 17^s.2$. The average curve (Figure 168) is given in Table 391, and the observations are listed in Table 392.

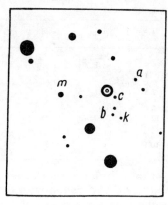

FIGURE 167. Comparison stars for
BI Virginis.

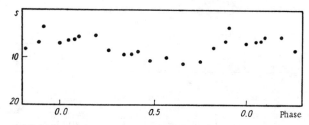

FIGURE 168. Light-variation curve of BI Virginis.

TABLE 391. Average light curve

φ	s	n	φ	s	n	φ	s	n
0.045ᵖ	6.6	5	0.377ᵖ	9.4	6	0.818ᵖ	8.0	4
.075	6.3	6	.412	8.9	6	.886	6.6	4
.095	5.9	4	.479	10.8	5	.905	3.8	4
.190	5.6	4	.568	10.0	6	.995	7.0	5
.260	8.2	6	.662	11.1	5			
.343	9.4	5	.748	10.8	5			

TABLE 392. Simeise observations

JD hel	s	JD hel	s	JD hel	s
241...		242...		242...	
9482.436	6.9	4231.516	7.8	7896.347	5.2
	6.9	6030.541	7.8		6.1
9498.285	5.6		9.7	7898.478	6.8
	7.3	6033.509	6.2		10.4
9511.414	8.8		6.7	8246.514	7.8
	6.4	6060.483	11.6	8246.514	5.2
9863.356	7.8	6397.562	12.9	8249.534	7.8
9865.424	7.8		9.0		9.3
	4.9	6420.483	7.8:	8257.512	8.8
242...			5.2	8278.340	4.4
0608.392	9.8	6423.459	9.5		6.5
	7.8	6424.458	4.3	8612.512	[10.8:
0929.568	9.7		4.9		12.8:
	9.5	7510.552	9.8	8624.339	3.9
0958.347	3.3		12.5		2.2
	4.5	7513.478	7.3	8631.399	10.1
3848.528	12.8		9.0		7.8
4231.516	13.0				

TABLE 392 (continued)

JD hel	s	JD hel	s	JD hel	s
242...		242...		243...	
8635.336	9.8	9348.564	7.8	3002.339	8.7
	10.8	9362.331	7.1		7.8
8637.398	7.8		11.6	3031.322	9.0
	7.8	9721.446	9.9		9.9
8981.450	6.8		9.8	3034.324	10.1
	9.5	243...			8.7
8992.488	12.0	2650.373	10.4	3358.478	6.7
	12.5		12.5		3.9
9339.460	6.7	2996.403	12.0	3737.426	4.3
	10.1		10.8		5.9
				4455.482	10.8

When the main study of this variable was concluded, Tsesevich examined new Moscow photographs (Table 393) taken with the 16-in. astrograph. These observations gave the maximum Max hel JD = 2437102.347, which deviates from (262) by $+0^d.017$ for $E = 30573$; this result lies within the margin of error. The above equation is currently valid. Comparison stars (Figure 167): $k = 0^s.0$; $m = 6^s.9$; $c = 17^s.5$.

TABLE 393. Moscow observations (Tsesevich)

JD hel	s	JD hel	s	JD hel	s
243...		243...		243...	
3052.341	7.9	7078.321	8.8	7102.316	4.6
3061.458	9.0	7079.295	8.8	7103.317	5.0
3388.379	6.4	7080.297	7.9	7106.315	8.6
4077.423	5.4	7087.301	5.5	7113.315	8.8
7051.338	5.0	7099.323	5.0	7128.301	6.9
7052.331	3.4	7100.318	6.9		

AU Virginis

Discovered by Hoffmeister /262/. Lause determined its elements in 1932 /302/. Later observed by Solov'ev /108/ and Tsesevich /150, 151, 153/. Payne-Gaposchkin /329/ derived from the Harvard photographs

$$\text{Max hel JD} = 2428188.945 + 0.34323 \cdot E. \tag{263}$$

Tsesevich estimated the magnitude from Simeise photographs. Reduction to one period has shown that the minima are more prominent than the maxima. Tsesevich nevertheless determined the maxima, like all other observers. The complete list of maxima is given in Table 394.

TABLE 394. List of maxima

Source	Max hel JD	E	O — A	O — B
Tsesevich (Sim.)	2420249.353	— 23132	$+0.004^d$	$+0.002^d$
	6087.366	— 6123	+ .018	+ .012
Lause /302/	6782.726	— 4097	— .006	— .013
Tsesevich (visual)	7485.349	— 2050	+ .025	+ .018
Solov'ev /108/	7935.620	— 738	— .021	— .029
Payne-Gaposchkin /329/	8188.945	0	.000	— .008
Solov'ev /108/	8312.534	+ 360	+ .026	+ .018
Tsesevich (Odessa)	36667.428	+ 24702	+ .016	+ .002

Payne-Gaposchkin's formula was improved by the least squares method:

$$\text{Max hel JD} = 2428188.953 + 0.34323023 \cdot E; \; P^{-1} = 2.913496285. \qquad (264)$$

The average light curves were constructed from this equation (Tables 395, 396; Figure 169, 1 — Simeise observations, 2 — visual observations). Comparison stars (Figure 170):

	s_1	s_2	$m_{\text{вн3}}$
a	0.0	0.0	10.44
b	10.6	—	—
c	13.6	—	11.35
d	21.0	—	—
e	—	9.7	11.99
f	—	—	11.13

The observations are listed in Tables 397 — 399.

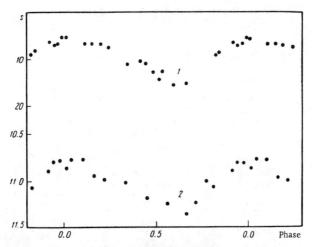

FIGURE 169. Light variation curves of AU Virginis.

TABLE 395. Photographic light curve from Simeise photographs

φ	s_1	n	φ	s_1	n	φ	s_1	n
$\overset{p}{0.006}$	5.6	3	$\overset{p}{0.444}$	10.8	6	$\overset{p}{0.840}$	8.2	5
.106	6.8	4	.479	12.8	5	.913	6.1	5
.149	6.6	6	.515	14.1	6	.934	7.2	7
.192	6.8	4	.529	12.3	6	.962	6.6	4
.244	7.6	4	.593	15.9	6	.986	5.3	4
.344	11.1	6	.653	15.6	2			
.410	10.2	4	.817	9.2	6			

TABLE 396. Visual light curve

φ	m	n	φ	m	n	φ	m	n
$\overset{p}{0.004}$	10.93	3	$\overset{p}{0.331}$	11.01	6	$\overset{p}{0.773}$	11.00	4
.048	10.88	4	.454	11.10	3	.818	11.04	4
.100	10.88	5	.558	11.13	3	.906	10.95	5
.162	10.98	4	.652	11.18	5	.943	10.90	3
.214	11.00	4	.716	11.12	5	.978	10.90	4

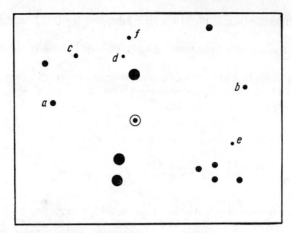

FIGURE 170. Comparison stars for AU Virginis.

TABLE 397. Simeise photographic observations (Tsesevich)

JD hel	s_1	JD hel	s_1	JD hel	s_1
241...		242...		242...	
9505.393	13.0	1007.345	8.3	7165.559	11.4
	14.6		8.3		12.3
9513.463	4.5	1010.407	13.6	7241.352	6.2
9885.334	7.4	1050.315	17.3		4.4
	8.3		17.7	7245.355	8.8
242...		1318.487	6.5	7543.368	7.1
0243.354	12.3		6.7		6.1
	13.0	1319.544	5.7	9018.377	12.6
0249.354	5.3	1338.408	6.8		15.7
	5.8		7.1	9348.504	10.2
0266.345	16.1	1342.472	8.9		10.6
	16.1		9.7	9366.414	12.1
0580.540	7.1	3521.527	15.1		11.6
	5.8		12.1	9371.348	5.2
0600.336	15.6	3522.374	7.8	9721.446	6.2
	15.6		8.7		6.2
0620.331	8.3	3529.416	12.1	243...	
	7.7		10.6	0107.449	12.4
0956.400	5.8	4610.429	5.3		12.6
	5.7	4978.461	7.7	2644.407	8.3
0957.373	11.9		7.4	2646.428	8.0
	7.4	5003.314	17.7		8.3
0958.506	6.2		13.6	2647.370	14.7
	8.0	6069.553	7.4		14.7
0959.472	5.8		7.7	3001.500	8.5
	6.5	6087.341	5.9	3035.297	4.0
0977.408	5.9		6.4	3422.324	12.1
	7.8	6087.438	7.4		12.1
0983.462	8.7		7.7	4119.436	11.6
	8.7	6799.383	9.6		14.7
0985.304	5.9	6803.382	7.3	4455.482	10.6
	7.7		6.5		13.6

TABLE 398   Odessa photographic observations (Tsesevich)

JD hel	s_2	JD hel	s_2	JD hel	s_2
243...		243...		243...	
6288.453	1.9	6722.340	3.2	7079.337	2.2
6304.396	5.2	6971.591	3.5	7100.369	6.5
6313.448	7.9	6997.601	2.3	7101.353	6.7
6344.358	6.5	7015.493	3.9	7105.346	4.2
6345.350	7.9	7016.456	3.6	7111.341	7.3
6347.348	7.0	7017.462	2.2	7373.506	4.3
6608.595	9.7	7019.475	10.7	7377.538	3.9
6612.635	5.8	7020.503	7.8	7378.484	7.9
6661.467	9.7	7028.496	2.3	7398.458	4.1
6667.418	2.3	7044.380	1.7	7400.458	7.2
6702.398	4.2	7046.445	5.2	7405.427	4.0
6714.347	10.7	7052.408	10.7	7406.443	3.2
6715.357	6.8	7077.363	6.9		
6716.376	8.0	7078.360	4.4		

TABLE 399.   Visual observations (Tsesevich)

JD hel	m	JD hel	m	JD hel	m
242...		242...		242...	
7459.472	11.29	7485.454	10.99	7509.366	10.98
.486	11.27	.473	11.02	.389	10.99
.531	11.22	.484	11.04	.429	11.04
.554	10.90	.510	11.04	7510.216	11.26
.564	10.95	.540	11.05	.240	11.32
7460.351	11.06	7490.336	11.04	.276	11.19
.380	11.07	.371	11.04	.288	11.07
.389	11.10	.388	11.04	7591.275	11.19
.403	11.10	.397	11.01	.305	11.02
.498	11.19	.404	10.97	7595.186	10.84
.532	11.16	.414	10.99	.269	10.90
7461.352	11.10	.420	11.02	7100.362	11.63
.360	11.02	.459	10.87	243...	
7484.329	10.83	.466	10.87	7100.372	11.64
.336	10.91	.477	10.90	.393	11.64
.357	10.93	.481	10.99	7101.325	11.03
.373	10.99	7492.471	10.95	.338	11.10
7485.320	10.85	.500	10.95	.352	11.17
.324	10.90	.507	10.99	.361	11.17
.333	10.82	.529	10.95	.375	11.25
.365	10.90	.540	10.97	.384	11.27
.375	10.80	7508.258	11.00	7105.328	11.49
.396	10.87	.300	10.97	.341	11.56:
.405	10.97	.340	10.84		
.427	10.97	.351	10.86		
.439	10.92	7509.328	10.95		

V 1176 Sagittarii

Discovered by Luyten /313/ in 1936. Erleksova /37/ determined 13 approximate maxima and derived the equation

$$\text{Max hel JD} = 2433856.241 + 0.35481 \cdot E. \tag{265}$$

Erleksova's maxima, however, are highly inaccurate. To improve her elements, Tsesevich made magnitude estimates from Simeise plates. The seasonal light curves gave the maxima. Erleksova's observations were completely revised; average seasonal light curves were constructed and more exact maxima found. The list of all maxima is given in Table 400.

311

TABLE 400. List of maxima

Source	Max hel JD	E	O — C
			d
Tsesevich	2421432.378:	0	—0.030:
	5445.353	+ 11310	— .010
	6882.375	+ 15360	+ .012
Erleksova	31269.651	+ 27725	+ .003
	3112.554	+ 32919	— .002
	3782.446	+ 34807	— .001

The O — C residues were calculated from

$$\text{Max hel JD} = 2421432.408 + 0.3548148 \cdot E;\ P^{-1} = 2.818371725. \qquad (266)$$

Tsesevich's observations are listed in Table 401. Comparison stars (Figure 171): $a = 0^s.0$; $b = 7^s.9$; $c = 12^s.9$; $d = 13^s.9$. Stellar magnitudes of a and c according to Erleksova are $m_a = 12.1$; $m_c = 12.9$.

TABLE 401. Simeise photographic observations (Tsesevich)

JD hel	s	JD hel	s	JD hel	s
241...		242...		242...	
9207.408	10.3	4324.481	2.4	6510.462	6.1
9208.465	6.9	4325.373	11.2	6512.460	9.9
	6.9		11.9		10.9
9209.409	4.9	4325.467	12.9	6535.383	0.0
	3.2		11.9		—1.0
242...		4352.378	11.2	6544.360	10.9
1039.417	12.9	4358.365	9.6	6854.452	10.3
1428.358	12.9	4360.366	2.6		6.9
	13.9		4.5	6882.356	0.0
1432.362	7.9	4379.418	12.2		—1.5
	1.8		12.4	6894.391	4.9
3225.426	7.9	4387.309	1.4		4.0
	10.9	5032.414	5.9	8722.416	5.3
3582.399	12.4	5057.433	11.9	9101.380	3.2
	13.9	5436.386	13.9	9816.356	3.4
3589.475	10.9	5439.414	5.9		1.1
	7.9		5.5	9818.389	12.5
3960.400	6.5	5445.378	0.6		12.9
3966.425	11.3		1.1	243...	
	7.9	5774.454	10.5	0177.473	12.2
3992.344	5.5		12.7		12.9
	4.7	5801.356	11.2	0199.379	11.2
3995.330	6.3		10.3		10.9
3998.336	12.4	5803.347	3.5	2731.362	11.5
	10.7		2.3	3832.425	10.5
4019.301	8.9	5830.308	2.3	3855.342	11.9
	6.8		0.0		10.9
4319.397	12.9	6155.418	6.9		
	12.1	6510.462	6.1		

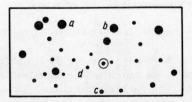

FIGURE 171. Comparison stars for V 1176 Sagittarii.

312

BS Virginis

Discovered and investigated by Boyce /206/. Tsesevich observed it from Odessa and Simeise photographs. The seasonal curves gave the following maxima:

Max hel JD	E	O — C
		d
2420608.407	0	+0.002
6420.461	+ 15413	+ .008
8637.343	+ 21292	— .007
33358.465	+ 33812	— .019
6347.294	+ 41738	+ .015

The O —C residues were calculated from a new equation

$$\text{Max hel JD} = 2420608.405 + 0.3770874 \cdot E; \ P^{-1} = 2.6519051. \qquad (267)$$

It was used to plot the average light curve (Figure 172, Table 402). Note that the ephemeris maximum 2426839.40, omitted from the improved equation, gives a deviation O —C = $+0^d.003$ for $E = 16524$. The most reliable maxima (the last two) give significant deviations of $-0^d.019$ and $+0^d.015$, which apparently indicate a recent abrupt change in the period. Tsesevich's observations are listed in Tables 403, 404. Comparison stars (Figure 173): $k = 0^s.0; \ a = 14^s.0; \ b = 22^s.3.$

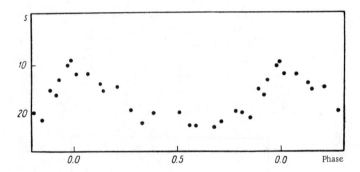

FIGURE 172. Light curve of BS Virginis.

TABLE 402. Average light curve

φ	s	n	φ	s	n	φ	s	n
p			p			p		
0.013	11.6	2	0.386	18.8	3	0.808	18.9	5
.073	11.7	4	.511	18.6	6	.850	20.0	4
.129	13.6	6	.561	21.2	5	.890	14.6	4
.146	14.6	5	.592	21.1	5	.917	15.6	6
.208	14.0	5	.685	21.4	6	.930	12.7	4
.280	18.2	5	.704	20.6	4	.976	10.0	5
.340	20.6	5	.787	18.4	6	.991	9.2	4

TABLE 403. Simeise observations (Tsesevich)

JD hel	s	JD hel	s	JD hel	s
241...		242...		242...	
9498.285	13.5	6397.562	21.4	8992.488	20.6
	15.7		19.5		20.5
9511.414	16.8	6420.483	8.6	9339.460	15.2
9513.463	20.6		9.0		12.7
9865.424	19.8	6423.459	9.0	9348.504	17.7
	19.0	7510.552	24.3		18.6
242...			19.8	9362.331	23.3
0215.41:	15.0	7513.478	19.8		22.3
	18.7		20.5	9721.446	16.1
0239.41:	12.0	7886.374	17.3		16.8
0608.392	12.4	7896.347	11.6	243...	
0929.568	26.3		11.2	2650.373	19.8
	22.3	7898.478	22.3		15.0
0958.347	11.2		21.5	2996.403	14.0
	12.0	8246.514	19.2	3031.322	20.5
0977.322	19.5		21.1		21.0
	20.8	8249.534	21.1	3034.324	20.4
0978.376	11.2		20.5		21.5
	14.0	8257.512	23.3	3358.478	10.5
0983.367	15.0		22.3		10.5
4231.516	14.0	8612.512	14.0	3734.402	11.2
	12.4		13.5		13.0
4621.287	18.2	8631.399	17.7	3737.426	12.0
	12.9		16.1		12.4
6030.541	11.7	8635.336	22.3:	3771.314	18.5
	12.8	8637.398	14.8		19.5
6033.509	17.1		14.0	4455.482	12.4
	18.6	8981.450	22.3		14.0
6084.396	24.3		19.9		
	22.3				

TABLE 404. Odessa observations (Tsesevich)

JD hel	s	JD hel	s	JD hel	s
243...		243...		243...	
6286.456	19.5	6347.347	11.5	6660.488	21.3
6288.452	4.7	6607.572	14.0	6661.466	12.6
6304.398	22.3	6608.596	8.2	6663.484	18.7
6344.357	14.0	6612.636	17.3	6667.417	7.6
6345.349	16.8	6613.589	11.8	6668.442	19.0

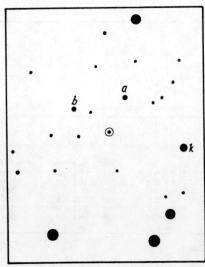

FIGURE 173. Comparison stars for BS Virginis.

314

Sixty-six photographs were obtained in 1959—1962 on the seven-camera astrograph. A favorable distribution of photographs in time gave a tentative period, $P = 0^d.38613$. Measurements of Simeise plates gave the improved value $P = 0^d.386218$. The light curves for lengthy seasons gave the following maxima:

Max hel JD	E	$O - C$
		d
2420633.436	— 34097	—0.004
5745.435	— 20861	+ .006
33802.332	0	— .004
7079.402	+ 8485	+ .001
7406.524	+ 9332	— .004
7793.531	+ 10334	+ .012
8143.437	+ 11240	+ .004

The O—C residues were calculated from the equation

$$\text{Max hel JD} = 2433802.336 + 0.38621862 \cdot E; \; P^{-1} = 2.58920712. \qquad (268)$$

The average light curves were constructed using this equation (Tables 405, 406; Figure 174, 1 — Simeise observations, 2 — Odessa observations). The star is clearly of subtype RRc. The period apparently remained constant between 1911 and 1960. The observations are listed in Tables 407, 408. Comparison stars (Figure 175): $a = -4^s.2$; $b = 0^s.0$; $c = 10^s.4$.

TABLE 405. Average light curve from Odessa observations

φ	s	n	φ	s	n	φ	s	n
p			p			p		
0.044	—0.6	1	0.542	+6.8	4	0.806	4.6	3
.167	+3.8	3	.578	+5.8	3	.841	4.0	3
.232	+2.8	1	.676	+7.9	4	.913	0.6	2
.364	+5.2	2	.706	+6.8	2	.957	0.0	2
.438	+7.4	3	.762	+4.9	3			

TABLE 406. Average light curve from Simeise observations

φ	s	n	φ	s	n	φ	s	n
p			p			p		
0.035	1.5	4	0.375	8.7	5	0.770	5.2	6
.071	1.6	4	.494	7.5	6	.822	5.2	4
.115	1.3	5	.554	7.6	6	.879	5.4	4
.132	3.1	6	.592	8.0	4	.919	2.4	4
.165	3.2	5	.657	7.2	6	.935	1.6	6
.204	5.6	6	.703	6.9	5	.974	0.2	4
.259	5.8	3	.737	5.9	6			

TABLE 407. Simeise observations

JD hel	s	JD hel	s	JD hel	s
241...		242...		242...	
8884.333	7.5	0253.457	4.1	0636.497	1.9
	6.9	.537	10.4		4.7
9556.324	1.0	0253.497	5.6	1008.429	1.0
	1.0	0630.492	7.8		3.5
9888.504	5.2		6.7	1369.461	6.9
	6.2	0633.457	0.9		5.2
9920.450	1.0		0.9	3193.376	3.8
	1.9	0634.464	6.4		3.8
			7.8		

TABLE 407 (continued)

JD hel	s	JD hel	s	JD hel	s
242...		242...		243...	
3554.366	6.6	5771.349	2.6	3031.494	0.9
	6.1		0.9		0.0
3931.370	2.8	5774.384	0.0	3034.505	0.9
	1.5		0.0		0.9
3937.477	6.6	5799.345	8.3	3412.426	6.6
	5.7		7.3		6.9
3942.463	5.7	5802.357	8.2	3445.356	6.0
	5.7	5820.338	1.0	3447.353	5.4
4296.366	2.5		2.1		3.5
	2.1	5826.354	7.8	3766.488	3.8
4312.376	7.3		7.8		3.5
	7.9	6486.488	5.8	3802.362	1.0
5027.341	8.5		6.6		0.9
	7.5	6860.355	7.8	3803.364	6.3
5382.442	0.0		6.9		6.9
5385.465	0.0	7716.219	1.9	3806.452	7.2
	0.9		2.1		8.5
5388.489	4.8	7931.490	4.7	3807.410	5.2
	3.5	7961.433	8.5		7.8
5744.497	7.8		8.1	3823.343	10.4
	8.5	9071.384	8.5	4131.496	5.6
5745.470	1.0		7.8		6.1
5764.404	2.6	243...		4183.358	6.1
	1.9	2686.452	4.3		7.8
5770.348	7.5		3.8		
	8.5				

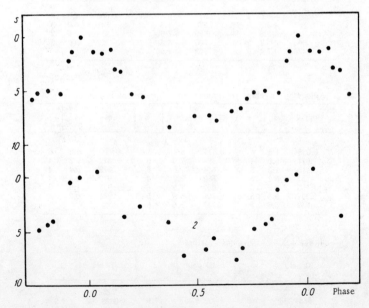

FIGURE 174. Light curves of HV 10539.

316

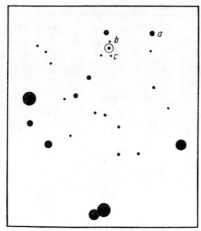

FIGURE 175. Comparison stars of HV 10539.

TABLE 408. Odessa observations

JD hel	s	JD hel	s	JD hel	s
243...		243...		243...	
6744.358	6.9	7080.451	7.3	7789.480	10.4
6749.363	6.6	7087.453	6.1	7790.403	7.2
.405	3.9	7099.361	7.3	.501	5.7
6755.424	6.1	7101.400	0.0	7793.425	8.9
6756.344	6.9	7102.407	6.6	.480	0.0
.401	5.7	.456	6.2	.526	3.2
6757.362	2.1	7107.370	8.9	7794.444	2.3
6758.363	5.5	7111.370	5.2	7810.407	7.6
.389	3.1	.396	2.7	.454	1.3
6759.361	4.2	.424	—0.8	7812.363	8.1
.389	6.6	7128.353	4.1	7813.389	5.7
6761.379	7.3	7373.624	3.5	.437	5.7
6781.330	2.8	7378.594	10.4	7847.393	7.6
7073.437	6.6	7402.536	8.1	8113.559	4.3
7075.464	5.7	7404.551	4.8	8114.507	—0.5
7077.396	4.7	7406.555	1.9	.597	0.0
.452	0.0	7427.477	2.1	8143.510	—1.4
7078.388	6.1	7454.398	—1.6	8144.515	0.0
7079.367	2.1	7457.412	7.8	8164.434	0.0
.418	— 0.6	7458.393	7.3	8165.435	—1.2
.467	3.1	7764.553	—1.7	8170.447	—1.7
7080.400	6.2	7789.427	5.2	8173.465	5.7

CONCLUSIONS

Table 409 is a summary of data on constant-period stars.

The ratio $\varepsilon = (\text{Max} - \text{Min})/P$ is the skewness of the light curve. SS Psc displays certain anomalies. The shape of the light curve is highly variable, but so far no regular features in this variation were detected. The spectral type F2 is markedly different from the spectra of the "normal" stars YZ Cap and U Com. The range of stars with almost symmetric light curves thus includes objects with periods ranging from 0.1 to $0^d.386$. These objects possibly form a homogeneous group.

317

TABLE 409. Constant-period "extreme" stars

Star	P	ΔE	ε	Sp	ΔS	v_r, km/sec
YZ Boo	0.104	85150	0.4	—	0	—
CK Aqr	0.124	132780	0.4	—	—	—
CW Ser	0.189	89000	0.4	—	—	—
BS Aqr*	0.198	—	—	F1	0	+ 50
DH Peg*	0.256	—	—	A7	0	− 55
BC Eri	0.264	100400	—	—	—	—
YZ Cap*	0.273	—	—	A4	0	− 75
SS Psc*	0.288	—	—	F2	2	+ 5
U Com*	0.293	—	—	A5	5	+ 15
V 729 Oph	0.319	44600	—	—	—	—
V 727 Oph	0.334	52500	—	—	—	—
BI Vir	0.336	39000	—	—	—	—
AU Vir	0.343	48000	0.4	—	7:	+ 105
V 1176 Sgr	0.355	34800	—	—	—	—
BS Vir	0.377	41738	—	—	—	—
HV 10539	0.386	45300	—	—	—	—

It would be wrong to suggest that all the stars with almost symmetric light curves have constant periods without any secular fluctuations. Several stars are known which are indistinguishable from the preceding as far as the shape of the light curve is concerned, but their periods are highly variable (Table 410).

TABLE 410. Variable-period "extreme" stars

Star	P	Sp	ΔS	v_r, km/sec
SX UMa	0.307	—	6:	− 135
RZ Cep	0.309	A4	5	0
TV Boo	0.313	A1	8	− 85
T Sex	0.325	—	—	+ 10
RU Psc	0.390	A6	7	− 115
YZ Tau	0.411	—	—	—

The periods of stars listed in Table 410 range between 0.307 and $0^d.41$, their calcium spectral types are A1−A6, and $\Delta S \sim 7$. Stars with skew light curves and light variation periods close to $0^d.36$ also have non-stable periods. Unfortunately, no spectral data are available for these stars at present.

STARS WITH VARIABLE SHAPE OF LIGHT CURVE

Besides SX Phoenicis, AI Velorum, and VZ Cancri (see Table 2), this group also includes SZ Lyncis, a star studied by Lange.

SZ Lyncis

According to Lange's report, SZ Lyncis was included in 1963 in the program of visual observations of short-period cepheids. The observations were carried out with 5-in. refractor at the Kryzhanovka Station of the

Odessa Astronomical Observatory from 1 March to 15 May. Thirteen definite maxima were determined (Table 411). After the heliocentric time of maximum, the table lists the maximum magnitude on Lange's power scale, the values of E and $O-C$ calculated from the equation

$$\text{Max hel JD} = 2437368.403 + 0.12053473 \cdot E, \qquad (269)$$

and the number of magnitude estimates n. Comparison stars (Figure 176): $a = 0^s.0$; $b = 16^s.0$; $c = 24^s.0$.

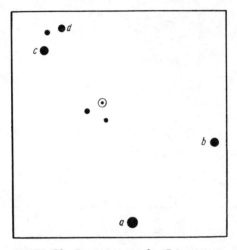

FIGURE 176. Comparison stars for SZ Lyncis.

The individual maxima gave the average maximum with the following rms error:

$$\text{Max hel JD} = 2438150.\,3013, \quad O-C = -0^d\,0105, \quad E = 6487;$$
$$\pm .0014 \qquad\qquad\qquad \pm .0014.$$

TABLE 411. List of maxima

Max hel JD	s	E	O — C	n
243...			d	
8090.2803	8.0	5989	— 0.0052	15
8144.281	8.0	6437	— .0041	12
.395	8.0	6438	— .0106	14
8145.354	8.0	6446	— .0159	25
8150.309:	8.0	6487	— .0028:	8
8152.354	8.0	6404	— .0069	21
8155.362	8.0	6529	— .0123	36
8159.3395	8.0	6562	— .0124	22
8160.3000	8.0	6570	— .0162	22
8161.3949	8.0	6579	— .0061	22
8162.3538	8.0	6587	— .0115	26
8163.3132	8.0	6595	— .0163	25
8164.2782	8.0	6603	— .0156	19

The considerable deviations of the individual maxima from the average calculated for constant-period elements apparently cannot be attributed to

319

observation errors. Photoelectric observations led K. Hefferts and B. Scheide to the conclusion that the light curve changed from cycle to cycle. Figure 177 shows Lange's visual observations of two maxima (10 and 11 May 1963) with phases calculated from equation (269). The relative phase shift of the entire light curve is clearly visible.

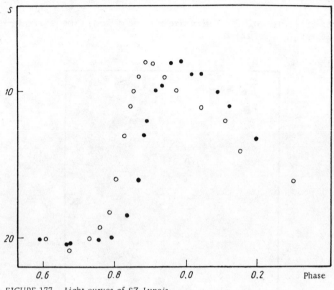

FIGURE 177. Light curves of SZ Lyncis:

○ 10 May, ● 11 May 1963.

Lange's 268 visual observations were reduced using (269) to two average light curves (Table 412).

TABLE 412. Average light curves

φ	s	n	φ	s	n	φ	s	n
First								
$\overset{p}{0.022}$	10.7	10	$\overset{p}{0.466}$	19.8	10	$\overset{p}{0.835}$	13.8	11
.059	11.7	10	.533	20.3	10	.857	10.1	11
.095	11.1	10	.607	20.5	10	.881	8.9	11
.122	13.8	10	662	20.8	10	.900	11.2	11
.154	13.5	10	.698	19.4	10	.920	10.3	11
.196	14.9	10	.740	18.0	10	.931	9.5	11
.247	16.0	10	.767	17.5	10	.961	8.9	11
.290	17.1	10	.794	14.6	10	.987	9.8	11
.354	19.6	10	.817	14.4	10			
Second								
$\overset{p}{0.023}$	10.75	31	$\overset{p}{0.570}$	20.43	20	$\overset{p}{0.880}$	10.07	33
.124	12.80	30	.680	20.13	20	.937	9.58	33
.244	15.99	30	.754	17.74	20			
.410	19.68	20	.815	14.24	31			

In the first curve, where the average points combine 10—11 observations each, the points are markedly spread around the smooth light curve. This

may be attributed to different deviations of the individual maxima from
the elements and to variation in the shape of the light curve. The second,
smoother curve consists of average points combining from 20 to 33 estimates
each. This curve gave the normal maximum

$$\text{Max hel JD} = 2438150.3031, \quad O-C = -0\overset{d}{.}0087, \quad E = 6487.$$

In 1956, Tsesevich carried out 97 visual observations regarding this
star as a W UMa-type eclipsing variable with a period of $0^d.274$. The
observations were very difficult, since the comparison stars did not
fit inside the reflector field of vision. Nevertheless, reduction of observations using (269) gave a fairly reliable average light curve, though with
substantial scatter of the individual estimates around the smooth curve
(Table 413).

TABLE 413. Average light curve

φ	s	n	φ	s	n	φ	s	n
0.054	7.16	6	0.443	13.35	6	0.803	9.15	6
.123	7.08	6	.492	12.95	6	.849	7.78	6
.186	8.05	6	.554	14.25	6	.913	5.82	6
.275	10.57	6	.609	12.43	6	.976	6.58	6
.335	10.02	6	.670	13.00	7			
.400	10.65	6	.733	11.72	6			

Comparison stars in Tsesevich's observation (Figure 176): $a = 0^s.0$;
$b = 11^s.4$; $c = 15^s.7$; $d = 24^s.7$. The average curve gave the normal
maximum

$$\text{Max hel JD} = 2435759.7408, \quad O-C = -0\overset{d}{.}0057, \quad E = -13576.$$

E and $O-C$ were calculated from (269).
    The constancy, or otherwise, of the average period of SZ Lyncis was
not investigated, and further observations are required to this end.
Table 414 lists Lange's observations, Table 415 Tsesevich's.

TABLE 414. Observations of G. A. Lange

JD hel	s	JD hel	s	JD hel	s
243...		243...		243...	
8090.2427	24.0	8144.2626	22.7	8144.3892	12.0
.2477	24.0	.2721	20.0	.3921	10.0
.2578	24.0	.2754	16.0	.3943	8.0
.2659	24.0	.2781	12.0	.3960	8.0
.2728	20.0	.2790	10.0	.4000	9.0
.2751	16.0	.2808	8.0	.4043	10.0
.2772	9.6	.2831	8.0	.4097	11.4
.2798	8.0	.2857	10.0	.4192	16.0
.2832	8.0	.2884	11.0	8145.2608	10.0
.2843	9.6	.2949	13.0	.2622	11.0
.2875	9.6	.3021	16.0	.2678	12.0
.2897	9.6	.3087	18.7	.2767	16.0
.3001	16.0	.3460	21.3	.2886	20.0
.3103	20.0	.3757	21.0	.2980	20.0
.3339	24.0	.3816	20.0	.3130	21.0
8106.2324	16.0	.3850	16.0	.3188	21.0
.2382	20.0	.3872	14.0	.3251	22.0

TABLE 414 (continued)

JD hel	s	JD hel	s	JD hel	s
243...		243...		243...	
8145.3302	22.0	815 .3798	10.7	8162.2641	13.3
.3344	20.0	.3812	12.0	.2732	16.0
.3374	19.0	.3878	12.4	.2807	17.6
.3408	16.0	.3962	16.0	.2854	19.2
.3429	15.0	.4034	17.6	.2939	19.2
.3448	13.0	.4104	19.2	.2998	19.6
.3472	12.0	8159.2636	16.0	.3052	20.0
.3497	10.0	.2664	18.0	.3106	20.0
.3529	8.0	.2732	20.0	.3161	20.0
.3550	8.0	.2837	20.8	.3219	20.8
.3573	9.0	.2942	20.8	.3255	19.2
.3612	9.0	.3108	20.0	.3276	20.0
.3624	10.0	.3147	20.0	.3333	16.0
.3668	12.0	.3204	20.8	.3375	12.8
.3722	14.0	.3258	20.0	.3404	11.2
.3786	16.0	.3291	19.2	.3434	9.6
8150.3051	11.4:	.3325	16.0	.3464	9.6
.3062	9.1:	.3353	12.0	.3502	8.5
.3080	8.0	.3376	10.0	.3520	8.0
.3105	8.0	.3395	8.0	.3547	8.0
.3136	9.1	.3409	8.0	.3584	8.0
.3235	10.7	.3434	9.0	.3622	9.6
.3245	11.4	.3467	10.0	.3674	11.2
.3443	18.0	.3518	12.0	.3759	12.8
8152.2600	10.0	.3587	12.0	.3812	14.4
.2641	11.0	.3663	13.0	.3892	16.0
.2700	12.0	.3735	14.0	8163.2632	20.0
.2830	16.0	.3840	16.0	.2673	20.0
.2997	20.0	8160.2542	19.6	.2737	20.0
.3141	20.0	.2564	20.0	.2788	20.8
.3208	20.8	.2594	20.0	.2830	20.8
.3409	21.3	.2641	20.0	.2870	20.0
.3491	16.0	.2678	20.0	.2916	19.2
.3516	10.0	.2764	20.8	.2931	18.4
.3535	8.0	.2814	20.0	.2948	16.0
.3550	8.0	.2844	19.2	.2990	12.4
.3578	8.0	.2876	18.4	.3008	11.6
.3598	9.0	.2898	16.0	.3041	10.7
.3653	10.0	.2930	13.0	.3060	9.6
.3704	12.0	.2950	11.0	.3091	9.6
.3770	13.0	.2955	10.0	.3131	8.0
.3863	14.0:	.2974	9.0	.3162	8.0
.3972	16.0	.2996	8.0	.3195	8.8
8155.2612	10.0	.3019	8.0	.3249	9.6
.2638	9.6	.3060	9.0	.3291	10.4
.2669	10.7	.3114	10.0	.3365	12.4
.2692	11.2	.3200	11.0	.3424	12.4
.2742	11.2	.3267	12.0	.3460	13.3
.2758	11.2	.3342	14.0	.3494	14.0
.2774	12.0	.3496	16.0	.3550	16.0
.2855	12.4	8161.3118	16.0	.3641	18.4
.2902	16.0	.3221	18.4	8164.2589	16.0
.2952	18.0	.3362	19.6	.2617	14.4
.3120	19.2	.3445	20.0	.2648	12.8
.3216	20.0	.3517	20.0	.2675	11.2
.3341	20.4	.3589	20.4	.2700	11.2
.3370	20.8	.3614	20.4	.2727	9.6
.3428	21.3	.3689	20.0	.2747	8.8
.3453	20.0	.3727	20.0	.2770	8.0
.3467	19.2	.3784	18.4	.2789	8.8
.3485	20.0	.3815	16.0	.2803	8.0
.3507	18.7	.3839	13.0	.2831	8.0:
.3521	16.0	.3850	12.0	.2841	9.6
.3533	12.0	.3878	10.0	.2890	10.4
.3543	10.7	.3907	9.6	.3017	12.0
.3557	10.0	.3934	8.0	.3077	13.5
.3578	9.0	.3963	8.0	.3120	16.0
.3600	8.9	.3998	8.9	.3202	17.6
.3617	8.0	.4031	8.9	.3246	18.4
.3634	8.0	.4080	10.0	.3702	20.0
.3656	9.6	.4124	11.0	8165.3948	16.0
.3680	10.4	.4192	14.0	.4098	17.6
.3722	11.2	.4214	13.2	.4192	20.0

TABLE 415. Observations of V. P. Tsesevich

JD hel	s	JD hel	s	JD hel	s
243...		243...		243...	
5585.332	10.1	5747.550	6.4	5758.473	13.2
.340	10.7	.570	6.1	.482	14.1
5592.292	13.8	.593	6.1	.499	15.0
.296	11.4	5748.412	5.9	.511	9.6
.308	6.8	.429	8.5	.527	5.7
.320	6.1	.439	11.4	5761.429	5.4
.329	7.3	.442	13.0	.442	4.4
.336	7.1	.451	14.8	.459	7.4
.344	6.6	.456	11.4	.468	9.9
5593.280	6.1	.462	14 0	.486	10.4
.303	6.1	.471	13.8	.499	11.4
.326	7.9	.485	16.4	.512	15.4
.338	10.4	.492	13.4	.519	10.4
.340	9.8	.506	11.4	.536	8.4
.346	13.3	.512	9.9	.545	5.7
.349	11.2	.517	8.9	.625	13.4
5594.276	7.6	5749.426	10.4	.371	8.4
5599.318	7.1	.434	15.5	.385	5.4
.354	6.2	.444	15.9	.399	8.4
5601.290	9.2	.451	15.4	.407	6.9
.303	12.4	.461	15.0	.415	9.4
.315	12.8	.469	10.4	.426	6.9
.318	10.4	.482	8.4	.432	10.9
5604.294	8.4	.500	6.9	.435	13.3
.312	13.6	.522	8.4	.440	11.4
.315	11.4			.446	13.6
.328	10.4	5753.446	14.7	.451	15.3
5608.317	17.9	.460	9.4	.456	13.0
5746.520	10.4	.471	6.3	.465	14.8
.563	12.8	.489	5.7	.475	14.0
.572	9.4	5758.406	5.7	.489	10.9
.589	7.3	.436	7.3	.496	6.4
5747.532	8.3	.469	9.1		

At Tsesevich's request, V. Satyvaldyev estimated the magnitude from Dushanbe photographs. All observations between 2430407 and 2437760 show a good fit with elements (269). No secular variation of the period was thus discovered.

Another star of this class — *DY Pegasi* — was repeatedly studied at the Odessa Observatory. Grigorevskii and Mandel examined all the old observations and a substantial number of Mandel's visual observations and discovered a secondary period involving a variation in the shape of the light curve. This line of research was continued by Karetnikov and Medvedev in 1963 at Mayaki, a branch of the Odessa Observatory. The observations were carried out with a photoelectric photometer attached to a 200-mm refractor. Although the equation

$$\text{Max hel JD} = 2436071.42469 + 0.0729263727 \cdot E \qquad (270)$$

was on the whole valid, it was impossible to plot a single light curve, since the shape of the curve changed markedly and rapidly. It developed "humps", the entire curve shifted, etc. This is clearly seen from Figure 178, which gives the observations of 15−16 August and 12−13 Spetember 1963 reduced to one period. Detailed examination of the material leads to the conclusion that we are dealing here with something much more complicated than interference of two fluctuations.

323

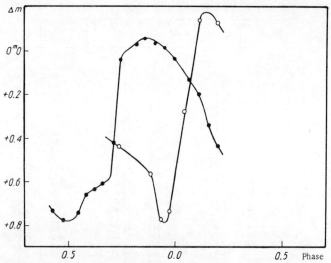

FIGURE 178. Light curves of DY Pegasi:

● 15—16 August, ○ 12—13 September 1963.

STARS WITH OVER-DAY PERIODS

Mandel and Tsesevich studied most of the stars with periods greater than 1^d. Table 416 lists the final improved equations of their periods.

TABLE 416. Stars with over-day periods

Star	Max hel JD	P	ΔE		Period
V 716 Oph	2427959.580	1.1159175		15000	Constant
BF Ser	2428744.097	1.1654393		10500	Ditto
CE Her	{ 2426909.842	1.2094290	before JD 2427000		Changed in
	{ 2426909.846	1.2094365	after JD 2427000		a jump
VX Cap	2425418.648	1.3275616		7900	Constant
XX Vir	2418507.4985	1.3482051		17200	Ditto
VW Mon	2437582.618	1.531893*+			Increases
V 745 Oph	2427636.104	1.595383		6570	Constant
NW Lyr	2436483.323	1.6012330		4800	Ditto
V 632 Sgr	{ 2420304.443	1.631633	{ before J D 2425000		Increases
	{ 2426401.428	1.6317673	{ after J D 2425000		
VZ Aql	2428727.653	1.668235		6800	Constant

* $+0.2879 \cdot 10^{-8} E^2$.

Stars of this subtype occur relatively seldom. A revision of the list has shown that some of the objects were included by mistake. Figures 1? 188 show the light curves of these stars.

Of special interest are NW Lyrae and VZ Aquilae. They have charact istic light curves with a wave on the rise branch. Makarenko studied NW Lyrae from photographs taken in photovisual, photographic, and ultr violet light. Her results completely confirmed our findings (Figure 189) She also obtained the color-variation curves and the shape of the oval described by the representing point in the U—B, B—V diagram (Figure 1

Light curves of this form are probably typical of RR Lyrae stars with
"extreme" periods of about $1.5 - 1^d.6$. Preston's spectral data on four
stars were available to the author:

Star	Sp	ΔS	v_r
V 716 Oph	A7	7	−230
BF Ser	A7	6	−175
CE Her	A8	7	−235
XX Vir	A7	9	− 55

It follows that the average calcium spectral type of these stars is A7
with ΔS ∼ 7, while their radial velocities are very large.

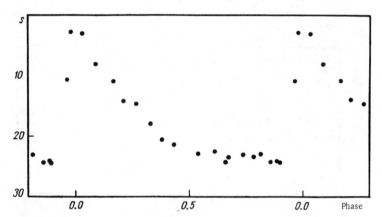

FIGURE 179.   Light curve of V 716 Ophiuchi.

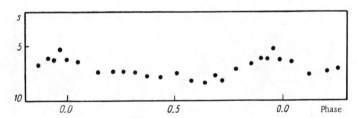

FIGURE 180.   Light curve of BX Delphini.

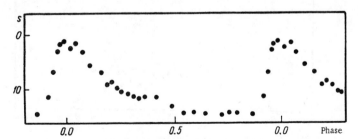

FIGURE 181.   Light curve of BF Serpentis.

325

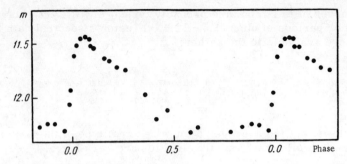

FIGURE 182.   Light curve of CE Herculis.

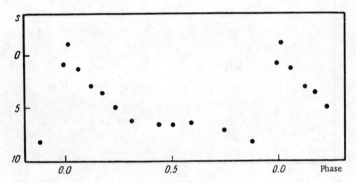

FIGURE 183.   Light curve of VX Capricorni.

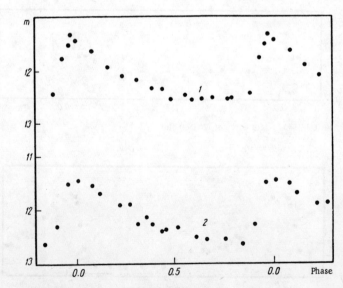

FIGURE 184.   Light curve of XX Virginis:

1) Simeise, 2) Odessa observations.

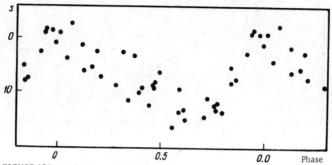

FIGURE 185. Light curve of VW Monocerotis.

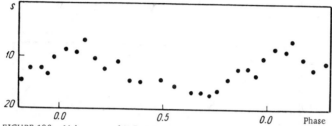

FIGURE 186. Light curve of V 745 Ophiuchi.

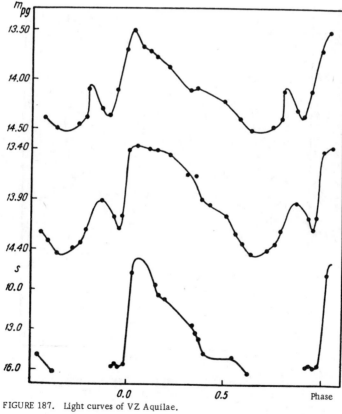

FIGURE 187. Light curves of VZ Aquilae.

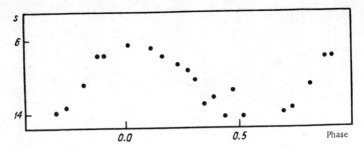

FIGURE 188. Light curve of V 632 Sagittarii.

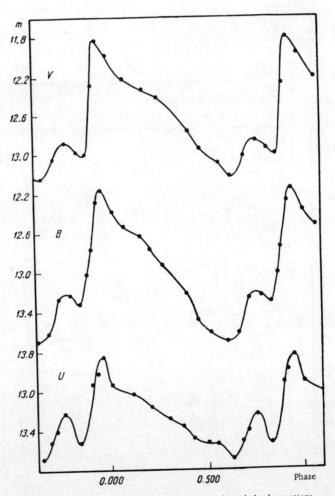

FIGURE 189. Light curves of NW Lyrae from Makarenko's observations.

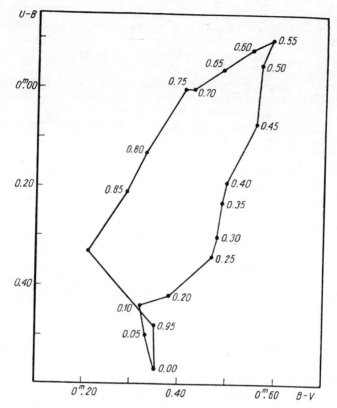

FIGURE 190.   The U—B, B—V "oval" for NW Lyrae.

"STRANGE" STARS

"Strange" stars are a subcategory of RR Lyrae variables, with AC Andromedae as the prototype.   Lange first noted, and Florya then confirmed that this star has two superimposed light periods whose elements according to Lange's latest observations are

$$\text{Max hel JD} = 2437169.353 + 0.525127 \cdot E_1; \qquad (271)$$
$$\text{Max hel JD} = 2437169.353 + 0.711242 \cdot E_2. \qquad (272)$$

The "strangeness" of this star is in that the two fluctuations are superimposed in a highly complex, difficult-to-unravel fashion.   One fluctuation apparently affects the other, so that the resultant pattern is highly complex.

V 567 Ophiuchi is another typical "strange" star which is not readily amenable to analysis. According to Hoffmeister, its elements are

$$\text{Max hel JD} = 2429785.455 + 0.1300729 \cdot E. \tag{273}$$

However, Tsesevich's numerous observations show that these elements are at variance with observations: the light of this star undergoes rapid irregular fluctuations, although on the average the period is close to $0^d.130$. Tsesevich studied other stars of this group, BV Aquarii, AH Camelopardi, BT Aquarii, and AS Virginis.

BV Aquarii

Short-period variable discovered 30 years ago by Hoffmeister /264/, remained uninvestigated to this day. The general behavior of this variable is extremely odd. Sometimes high maxima are observed at intervals of $1.08 - 1^d.11$, and yet no ephemeris with a period of $1.10/n$ could be derived, since the recurrence of these high maxima is irregular and the ephemeris maximum often degenerates into a hardly distinguishable magnitude enhancement.

Having collated a large number of visual observations, Tsesevich found a period of nearly 8^h. Filatov's photographic observations from Dushanbe plates, made at Tsesevich's request, were also used in the search for the true period. Tsesevich's brightness estimates based on Simeise planetary photographs were also amployed.

The approximate equation was applied to construct the seasonal light curves, from which 15 maxima were determined (Table 417). They are fitted with a single equation

$$\text{Max hel JD} = 2437524.218 + 0.3640481 \cdot E; \ P^{-1} = 2.74688977. \tag{274}$$

The O−C residues are plotted in Figure 191. The residues are large, which reveals lack of constancy of the period (a typical feature of RRc stars). Incidentally, visual observations reveal regular variation of these residues.

TABLE 417. Seasonal maxima

Source	Max hel JD	E	O−C
Tsesevich (photogr.)	2420394.33:	− 47054	$+0.031^d$
	7302.47:	− 28078	− .005
	8404.43:	− 25051	− .019
Filatov	30207.490	− 20098	− .089
	2832.386	− 12888	+ .020
	3866.318	− 10048	+ .055
	4656.237	− 7878	− .010
	5371.237	− 5914	− .001
	5742.197	− 4895	− .006
	6430.280	− 3005	+ .027
Tsesevich (visual)	7198.405	− 895	+ .010
	7524.231	0	+ .013
	7549.357	+ 69	+ .020
	7573.337	+ 135	− .027
	7901.344	+ 1036	− .028

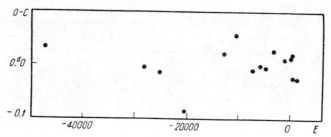

FIGURE 191.   O—C residues for BV Aquarii.

Reduction of all visual observations to one period has shown that the spread of the points at the minimum is not large. It is almost zero at the beginning and the end of the rise branch. At the maxima, on the other hand, the scatter reaches 3/4 of the amplitude! This is a manifestation of the star's "strangeness".

A smooth average light curve was also derived (Figure 192, Table 418).

TABLE 418.   Average light curve

φ	s	n	φ	s	n	φ	s	n
0.008^p	2.2	10	0.308^p	7.6	10	0.691^p	8.9	10
.030	2.2	10	.328	7.2	10	.719	7.9	10
.060	3.2	10	.349	8.7	10	.749	8.1	10
.086	4.5	10	.366	7.6	10	.767	6.6	10
.103	3.3	10	.387	8.0	10	.786	6.7	10
.125	4.6	10	.410	9.3	10	.804	5.5	10
.150	5.2	10	.439	8.2	10	.826	6.1	10
.171	4.6	10	.473	8.6	10	.850	6.7	10
.189	5.8	10	.511	9.0	10	.869	5.5	10
.208	5.2	10	.546	9.1	10	.892	6.6	10
.226	6.4	10	.576	9.1	10	.913	4.2	10
.245	6.2	10	.606	9.4	10	.932	4.8	10
.265	6.1	10	.639	9.3	10	.951	1.6	10
.287	7.5	10	.661	9.6	10	.986	3.3	13

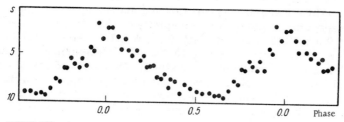

FIGURE 192.   Average visual light curve for BV Aquarii.

Tsesevich searched for the period of the variation of the light curve shape and determined the height of the various maxima. These data (the maxima, the heights M, and the phases ψ calculated from the equation $T_0 = 2437197.3 + 11.56 \cdot n$; $\Pi^{-1} = 0.086505$) are listed in Table 419 (Figure 193).

TABLE 419. Height of maxima

JD	M	ψ	JD	M	ψ	JD	M	ψ
243...			243...			243...		
7197.3	—2.0	0.000	7549.3	5.3	0.450	7902.4	—3.3	0.995
7198.4	—1.5	.095	7557.4	—1.0	.150	7906.4	2.4	.341
7204.2	5.6:	.597	7561.4	3.8	.496	7908.3	5.5	.505
7523.4	1.3	.209	7582.3	0.0	.304	7910.5	3.6	.695
7525.3	5.7	.374	7583.2	4.6:	.382	7912.3	3.6	.851
7526.4	2.4	.469	7873.4	4.8:	.486	7913.4	5.5	.946
7545.4	—2.0	.112	7901.4	—4.1	.908	7943.2	8.5:	.524

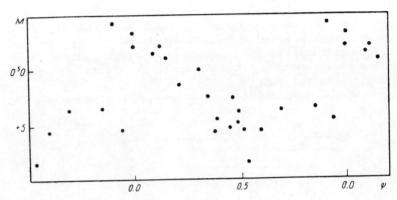

FIGURE 193. Variation of the light range of BV Aquarii.

Figure 193 shows that only one point (the last but one in the table) does not fit the general trend. We therefore conclude that the shape of the light curve varies with a period $\Pi = 11^d.56$. If this is indeed so, we have discovered a Blazhko effect in an RRc star, which is an exception to the general rule.

The second phase ψ was then calculated using the same equation; all the observations were divided into groups according to this phase and the corresponding light curves were plotted. The curves for groups I $(0.1 \leqslant \psi \leqslant 0.3)$, II $(0.3 \leqslant \psi \leqslant 0.5)$, III $(0.5 \leqslant \psi \leqslant 0.7)$, and IV $(0.7 \leqslant \psi \leqslant 0.9)$ are quite reliable, but in group V $(0.9 \leqslant \psi \leqslant 1.1)$ the points fall in two extreme bands with considerable scatter. This seemed to indicate that we were dealing with a phenomenon much more complex than the normal Blazhko effect. The maximum variance is observed for $0.9 < \psi < 1.0$. Photoelectric observations at various longitudes are required to settle this point.

Since the period of the inequality could not be determined, two extreme shapes of the average light curve were plotted, corresponding to $0.98 < \psi < 1.10$ and $0.5 < \psi < 0.7$ (Tables 420, 421, Figure 194).

Comparison stars (Figure 195): visual scale $k = -9^s.8$; $a = 0^s.0$; $b = 13^s.3$; photographic scale $k = 0^s.0$; $a = 9^s.7$; $c = 26^s.6$. All observations are listed in Tables 422—424.

TABLE 420. Average light curve for $\psi \sim 1^n.05$.

φ	s	n	φ	s	n	φ	s	n
$0.\overset{p}{0}14$	−1.5	4	$0.\overset{p}{3}56$	+7.3	5	$0.\overset{p}{8}14$	+3.3	3
.082	+0.3	5	.410	+7.8	4	.907	−0.7	3
.173	+4.8	5	.578	+9.2	4	.950	−2.1	3
.237	+4.5	5	.678	+9.8	4	.976	−2.3	3
.291	+6.9	5	.774	+6.0	4			

TABLE 421. Average light curve for $\psi \sim 0^n.6$

φ	s	n	φ	s	n	φ	s	n
$0.\overset{p}{0}11$	4.8	2	$0.\overset{p}{3}36$	8.4	5	$0.\overset{p}{6}94$	7.5	3
.109	8.4	3	.370	7.9	5	.785	6.9	5
.189	7.7	5	.418	9.4	5	.850	6.4	3
.232	7.5	5	.469	8.6	4	.904	6.0	2
.268	7.7	5	.563	9.8	4	.962	5.1	2
.301	8.3	5	.627	9.1	3			

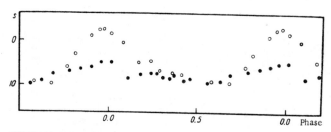

FIGURE 194. Extreme light curves of BV Aquarii:

● $0.0 < \psi < 0.1$; ○ $0.5 < \psi < 0.7$.

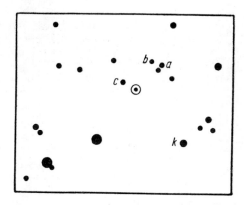

FIGURE 195. Comparison stars of BV Aquarii.

TABLE 422. Simeise photographic observations (Tsesevich)

JD hel	s	JD hel	s	JD hel	s
242...		242...		242...	
0394.328	2.4	6948.432	15.3	8461.300	11.8
	5.3		18.6		20.9
2553.335	19.4	7302.472	3.7	243...	
	18.2	7685.430	5.8	3122.475	15.3
5446.503	11.2		7.3		7.3
	9.7	8398.415	20.9	3178.322	11.4
6186.476	19.4		21.5		5.8
	19.4	8402.434	22.1	3179.313	3.9
6210.391	9.7		20.9		5.8
	7.3	8404.426	5.8	3913.301	7.7
6235.323	16.0		2.4		0.0
	9.7	8460.273	15.3	4623.299	21.5
6236.359	19.8				21.5
	20.9		13.9	4625.335	6.5

TABLE 423. Photographic observations (Filatov, Dushanbe)

JD hel	s	JD hel	s	JD hel	s
242...		243...		243...	
9846.343	11.7	3843.447	4.8	5014.224	21.0 *
9848.300	13.7	3858.382	5.8	5015.226	17.6 *
9868.263	6.8	3861.374	16.3	5019.212	14.2 *
9869.258	12.7	3862.388	13.7	5037.184	15.7 *
9870.289	17.1	3865.377	14.7	5038.154	14.7 *
9871.262	18.1	3866.370	3.2	5042.137	15.2 *
.283	13.7	3868.364	19.1	5043.171	13.7 *
9873.258	13.9	3886.301	14.4	5044.143	17.6 *
9874.273	15.7	3887.301	19.1	5063.090	22.1 *
9875.253	13.9	3889.301	9.7	5360.266	13.7
9879.261	15.7	3891.314	15.7	5366.253	16.3
.297	18.6	3892.313	18.1	5368.241	14.7
9902.156	16.6	3894.294	6.5	5371.255	4.8
.189	18.6	3897.292	6.8	5373.288	19.1
9904.198	12.7	3913.238	4.3	5374.263	12.7
.216	5.8	3970.135	18.7	5394.188	4.3
9907.213	7.5	4223.408	14.7	5395.208	5.4
9923.135	5.8	4225.409	17.6	5396.174	18.1
9925.127	17.6	4244.350	18.1	5397.171	15.0
243...		4245.364	12.9	5400.170	15.7
0207.447	4.8	4248.326	17.6	5695.325	9.7
0282.182	19.6	4254.339	17.1	5700.339	7.8
0284.222	14.7	4255.366	14.7	5717.290	18.1
0288.149	16.0	4270.280	16.6	5719.294	4.8
0931.331	13.7	4272.279	9.7	5721.269	17.2
0940.336	5.8	4279.262	18.1	5722.283	6.2
0967.280	4.8	4280.252	5.4:	5726.257	11.7
1345.249	15.3	4300.204	7.5:	5729.265	14.7
1349.278	14.7	4301.203	14.7	5742.206	4.8
1710.265	6.8	4323.136	9.7	5750.217	9.7
2742.371	17.1	4597.362	17.9	5752.227	16.6
2768.308	9.7	4323.274	12.9	5753.219	17.2
2773.374	12.7	4656.202	4.8	5755.209	13.7
2832.245	7.8	4657.224	18.6	5780.117	12.7
3133.421	13.7	4659.208	12.9	5781.127	15.0
3153.342	19.6	4660.212	3.9	5782.110	13.7
3155.352	12.7	4684.181	9.7	6014.425	15.7
3157.317	18.1	4706.096	19.3 *	6043.377	9.7
3159.356	12.9	4959.391	21.0 *	6045.382	18.6
3174.227	7.5	4985.322	18.1 *	6046.376	16.7
3176.256	18.1	4986.294	22.1 *	6048.387	14.7
3179.269	9.7	4987.306	21.0 *	6082.282	9.7
3183.298	8.7	4988.305	15.7 *	6100.245	17.7
3185.220	14.7	4989.285	16.9 *	6103.210	20.7
3207.184	17.6	4990.273	15.3 *	6377.375	12.7
3213.175	6.8:	5007.238	13.7 *	6397.285	18.7
3244.136	12.7	5008.238	17.6 *	6424.335	19.6
3836.435	15.7	5011.248	14.2 *	6428.354	22.7
3837.431	19.6	5012.223	19.3 *	6430.329	4.3
3842.447	13.4	5013.236	20.6 *	6434.333	5.8
		5013.236	19.8 *	6435.337	4.8
				6460.272	18.7:

* Estimates from panchromatic plates, without yellow filter.

334

TABLE 424. Visual observations (Tsesevich)

JD hel	s	JD hel	s	JD hel	s
243...		243...		243...	
7196.356	8.9	7526.333	8.9	7547.292	7.6
.376	7.3	.346	7.9	.298	7.6
.392	7.8	.353	4.4	.311	8.0
.415	7.8	.372	7.2	.321	7.2
7197.292	−2.0	.380	4.4	.333	7.6
.312	0.0	.392	3.3	.340	7.2
.328	−1.0	.398	2.4	.345	7.8
.340	−1.5	.411	2.7	.356	8.9
.368	0.0	.420	3.7	.365	8.2
.380	2.7	.436	3.5	.368	8.6
.395	2.7	.445	3.7	.382	8.6
.410	5.7	.468	4.4	.385	9.2
7198.275	9.7	.482	4.4	.398	8.0
.288	9.6	.495	4.4	.414	8.6
.309	7.0	.521	4.8	.428	8.2
.315	7.6	7528.311	8.2	.433	8.0
.323	7.3	.330	7.8	7549.280	7.1
.333	5.1	.343	7.9	.290	8.6
.350	2.4	.357	8.6	.344	6.4
.374	0.0	.368	7.3	.356	5.5
.403	−1.5	.391	8.3	.363	5.8
.426	−1.5	7530.318	9.5	.372	5.3
7204.279	5.6	.327	9.2	.386	7.1
.292	7.6	.336	8.6	.395	5.3
.320	7.6	.358	8.9	.404	5.4
.334	8.2	7531.345	9.2	.413	4.6
.335	6.7	.366	9.5	.420	5.7
.358	7.6	.384	9.7	.437	7.1
.370	8.2	.402	9.5	.443	7.1
.380	7.6	.427	9.5	.452	5.6
.401	8.5	.446	8.0	.458	7.1
7497.437	7.8	.456	5.6	7550.260	8.9
.444	7.3	.467	5.6	.276	8.9
7518.355	6.7	.474	3.3	.299	8.9
7521.372	4.9	.480	4.8	7557.256	7.1
.398	4.1	7543.333	10.2	.279	7.1
.416	5.7	.351	9.2	.294	8.5
.440	4.8	.358	7.6	.305	6.6
7522.454	3.3	.374	8.1	.319	5.4
7523.362	10.2	.382	7.1	.325	4.1
.391	9.6	7544.283	7.1	.333	2.7
.409	8.2	.298	7.1	.340	3.0
.434	7.3	.311	8.6	.346	1.9
.450	5.3	.323	7.5	.350	−1.0
.459	1.3	.328	5.3	7561.284	7.2
.468	2.9	.366	7.2	.295	6.6
.492	3.7	.346	7.1	.303	4.8
7524.332	5.1	.364	7.1	.324	5.0
.352	8.2	.373	4.7	.350	7.1
.361	7.2	.378	5.9	.358	5.1
.371	7.6	.396	4.7	.382	6.4
.383	7.3	7545.291	7.4	.392	4.3
.394	8.6	.299	7.0	.397	4.1
.408	8.6	.310	5.3	.404	3.8
.433	9.2	.349	7.2	.413	4.4
.453	9.6	.360	2.9	.426	4.1
.472	8.5	.365	1.3	.435	4.8
.487	8.6	.368	1.9	.443	6.2
.495	8.5	.374	1.2	7570.249	7.9
.505	7.2	.379	1.7	7573.282	7.2
.512	5.1	.391	−1.0	.294	8.6
.521	6.1	.395	−2.0	.306	7.7
.528	7.5	.408	3.1	.313	6.1
.534	6.1	.420	1.9	.331	4.8
.541	4.8	.428	3.6	7575.266	9.2
7525.311	5.7	.434	3.3	.282	7.2
.333	5.8	.440	2.9	.294	7.2
.346	3.7	.446	5.5	.302	6.9
.362	3.3	.454	4.8	.311	7.2
.380	2.7	7547.205	9.8	.317	7.2
7526.313	10.6	.272	8.9	7578.229	10.0
.326	8.9	.281	8.9	.245	11.1

TABLE 424 (continued)

JD hel	s	JD hel	s	JD hel	s
243...		243...		243...	
7578.253	10.2	7901.456	2.9	7910.362	8.2
.275	9.4	7902.276	8.2	.371	6.6
.280	10.0	.296	9.2	.382	5.5
.290	9.2	.310	8.2	.395	7.3
.305	8.7	.328	8.0	.404	6.0
.332	9.2	.345	8.5	.419	5.5
7582.317	9.2	.357	7.8	.437	7.2
.337	10.2	.383	5.7	.450	6.6
.344	7.2	.387	5.5	.468	3.6
.350	6.6	.390	4.8	.476	4.1
.357	5.1	.403	0.7	7911.272	7.2
.360	5.1	.408	2.7	.290	8.2
.365	4.8	.420	2.4	.336	8.9
.377	1.2	.429	0.0	.348	10.9
.382	1.2	.435	—2.2	.380	10.9
.390	0.0	.438	—2.2	.400	11.4
.395	2.9	.444	—3.3	.411	11.4
7583.229	4.8	.450	—2.9	.425	10.9
.245	4.7	.455	—2.9	.436	11.8
.256	4.6	.458	—2.0	.446	10.9
.282	7.2	.464	—2.2	.475	10.0
.299	6.6	.468	—2.4	.488	7.8
.313	8.2	.478	—1.5	7912.268	9.2
.324	7.8	7903.303	8.9	.287	7.3
.332	8.3	.322	8.9	.296	3.6
.382	10.2	.347	9.2	.302	5.2
7857.450	6.4	.384	8.9	.311	4.8
.476	8.9	.396	8.7	.324	5.5
7871.410	14.8	.411	9.5	.334	7.2
.428	7.3	.420	9.7	.345	6.1
.438	6.9	.431	10.9	.355	7.4
.451	6.4	.439	8.9	.368	6.1
.468	6.1	.467	7.5	.377	7.3
.482	7.4	.482	2.1	.389	8.5
.490	7.2	.487	—2.5	.398	9.7
.495	7.4	7904.330	7.0	.408	10.9
7872.366	5.5	.349	6.9	.418	10.9
.394	7.3	.363	7.8	.432	9.7
.417	7.8	.376	8.2	7913.277	8.2
.454	8.9	.408	9.2	.291	8.2
.468	10.0	.454	8.5	.302	7.8
.476	9.7	7906.303	8.2	.326	7.6
.485	9.8	.322	10.0	.339	6.6
.500	10.4	.341	10.0	.349	7.2
.510	8.9	.370	8.9	.367	7.3
7873.369	4.8	.376	8.2	.381	7.3
.398	6.6	.393	8.0	.399	5.5
.416	7.2	.410	7.3	.406	6.6
.432	6.9	.421	5.3	.415	7.8
.443	7.8	.431	7.3	.420	6.1
.470	7.8	.435	6.0	.427	6.6
.479	7.6	.442	3.6	.437	7.8
7899.308	7.8	.450	2.4	.449	7.8
7900.302	7.3	.460	4.1	.458	7.2
.358	5.7	.468	5.9	7926.254	10.9
.364	6.6	.478	3.7	.276	11.6
7901.345	5.8	7908.294	5.5	7942.289	10.0
.366	1.5	.322	8.2	7943.276	8.5
.373	—2.6	.358	7.6	.285	8.5
.385	—4.1	.374	7.3	.303	8.5
.391	—4.1	.392	9.2	.314	9.2
.398	—2.7	.416	9.9	.321	8.2
.407	—2.5	.448	10.9	.333	8.2
.416	—1.4	7910.281	10.0	.340	9.2
.421	1.7	.302	9.8	.351	9.2
.429	4.0	.316	11.6	.357	10.2
.434	2.1	.326	9.2	.370	9.5
.442	1.6	.343	9.2	.384	10.6
.448	2.1	.354	7.8	.396	10.0

AH Camelopardi

Observed by Tsesevich visually and on Odessa photographs in 1956. The elements

$$\text{Max hel JD} = 2433897.519 + 0.3687374 \cdot E; \ P^{-1} = 2.71195707. \tag{275}$$

It was clear from the start, however, that the star has a variable period and a variable light curve. The maximum varies from +11 to -3^s, whereas the minimum is almost constant, $+18^s$. Table 425 lists the epochs, the times of crossing the $11^s.0$ point on the rise branch, the height of the maximum, and the times of the maxima.

TABLE 425. Characteristics of AH Camelopardi

E'	JD hel	O — B	M	Max hel JD	O — A	E
0	2435692.502	d 0.000	2.2	—	—	—
24	5701.373	+ .021	2.2	5701.399	d +0.017	+4892
27	5702.469	+ .011	1.6	5702.489	.000	+4895
59	5714.277	+ .019	2.8	5714.295	+.007	+4927
62	5715.362	— .002	10.1	5715.377	—.017	+4930
65	5716.470	.000	5.2	5716.491	—.010	+4933
70	5718.341	+ .027	6.9	5718.385	+.041	+4938
78	5721.288	+ .023	—1.5	5721.319	+.025	+4946
81	5722.391	+ .021	—2.5	5722.413	+.013	+4949
146	5746.347	+ .009	2.6	5746.373	+.005	+5014
154	5749.301	+ .013	10.6	5749.313	—.005	+5022
179	5758.510	+ .004	2.4	—	—	—
227	5776.226	+ .021	3.2	5776.250	+.014	+5095

The O — B residues were calculated from the equation

$$T \ (11^s0) = 2435692.502 + 0.3687374 \cdot E'. \tag{276}$$

Examination of Table 425 shows that the height of the maximum, as well as the O — A and O — B residues, fluctuate between wide limits. This is possibly a periodic effect, with a period Π close to $46\,P$. This conclusion is somewhat premature, however, especially since the periodicity may prove to be much more complex.

The average light variation curve was calculated from (275). The scatter of the individual points far exceeds the error of observations, and therefore, besides the light curve, the curve of average absolute deviations $\overline{\Delta} = \frac{\Sigma |s - \overline{s}|}{n}$ was computed (Table 426). This curve characterizes the shape variation. It reaches its peak value for phases of $0^P.93$ and $0^P.09$, which is a confirmation of the Blazhko effect.

TABLE 426. Average light variation curve ($\overline{s}$) and the curve of average deviations Δ

φ	$\overline{s}$	Δ	n	φ	$\overline{s}$	Δ	n	φ	$\overline{s}$	Δ	n
p 0.006	5.2	2.6	10	p 0.173	10.5	2.2	10	p 0.887	16.0	3.0	10
.024	3.8	2.3	10	.198	12.2	1.3	10	.910	14.7	2.9	10
.038	4.9	2.7	10	.239	13.0	1.5	9	.928	12.0	2.2	9
.055	4.1	3.0	10	.309	15.3	2.5	10	.949	8.9	2.6	10
.075	6.2	3.9	10	.486	16.6	1.9	10	.965	7.6	3.3	10
.097	8.2	3.8	9	.702	18.1	0.9	10	.978	9.5	2.3	8
.118	7.4	3.3	11	.801	18.1	1.8	10	.992	4.1	2.7	7
.140	10.9	2.7	10	.850	16.9	2.2	10				

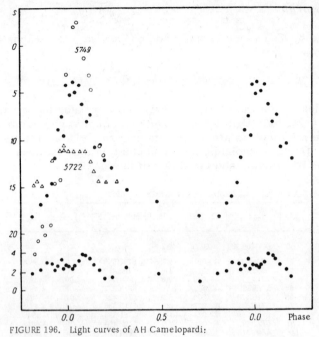

FIGURE 196. Light curves of AH Camelopardi:

● average curve, Δ JD 2435722, O JD 2435749; bottom — the deviations Δ.

Figure 196 shows the extreme curve shapes. Tsesevich's observations are listed in Table 427. Comparison stars (Figure 197): $k = 0^s.0$; $a = 5^s.3$; $b = 15^s.9$; $c = 20^s.7$.

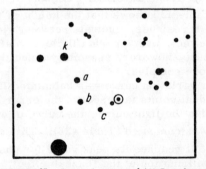

FIGURE 197. Comparison stars of AH Camelopardi.

TABLE 427. Visual observations (Tsesevich)

JD hel	s	JD hel	s	JD hel	s
243...		243...		243...	
5404.275	17.2	5691.490	11.2	5692.396	18.3
.291	18.1	.499	12.7	.430	19.1
.310	19.1	.512	11.4	.446	18.6
5691.437	3.2	.520	12.7	.464	18.5
.452	2.6	.523	12.1	.481	15.2
.468	9.5	5692.355	14.6:	.486	14.0
.478	12.4	.376	17.8	.492	12.4

TABLE 427 (continued)

JD hel	s	JD hel	s	JD hel	s
243...		243...		243...	
5692.499	11.1	5715.359	11.1	5722.385	14.4
.506	10.3	.364	10.6	.391	11.1
.512	2.1	.369	10.1	.399	3.0
.520	2.4	.373	10.1	.411	−2.0
.528	2.9	.378	10.1	.418	−2.5
.534	2.6	.382	9.8	.432	1.3
.540	2.8	.395	10.1	.440	3.2
.544	2.8	.411	11.1	.446	4.7
.546	2.6	.420	11.4	.461	10.9
5693.352	13.5	.431	12.1	.470	11.7
.404	18.3	.460	14.0	5743.306	17.8
.454	18.8	5716.410	18.3	.321	19.5
5694.407	11.4	.432	14.8	5744.347	18.3
.417	12.1	.436	15.9	.454	19.1
.461	14.6	.459	13.8	5746.293	19.5
.512	18.3	.465	11.2	.310	19.7
5700.337	8.5	.469	10.6	.322	19.5
.346	11.1:	.476	10.3	.334	18.3
.372	11.0	.482	8.3	.336	17.1
.416	13.3	.488	2.6	.347	10.2
.453	14.9	.494	5.3	.351	9.2
.515	17.8	.503	6.4	.357	6.8
5701.351	18.9	.511	7.8	.364	3.1
.373	11.1	.517	8.5	.367	3.0
.378	6.4	5718.301	13.8	.374	3.0
.383	6.8	.305	14.0	.379	2.6
.389	3.9	.310	14.0	.384	4.3
.394	2.1	.325	12.1	.390	7.7
.401	2.6	.337	11.1	.399	11.1
.403	2.9	.345	10.6	.405	12.0
.415	3.9	.357	8.8	.415	12.1
.428	3.9	.364	8.3	5748.257	12.4
.436	9.2	.369	8.5	.270	14.0
.453	11.7	.377	7.4	.283	14.9
.460	12.1	.387	6.8	5749.249	14.8
.483	14.0	.397	8.3	.255	14.4
.520	14.8	.413	8.8	.265	14.8
.549	15.9	.422	11.8	.287	12.1
5702.395	19.1	.434	13.2	.303	11.1
.431	19.1	.478	14.2	.309	10.6
.446	19.1	5719.350	19.1	.318	11.1
.454	19.1	.417	14.9	.328	11.1
.460	14.8	.424	14.6	.341	11.1
.463	13.8	.434	13.3	.353	11.1
.468	10.2	5720.435	14.9	.362	12.4
.473	8.2	5721.286	12.1	.368	13.3
.477	7.2	.301	3.9	.377	14.4
.482	4.7	.307	1.8	.392	14.5
.485	−1.0	.313	−0.5	.414	14.4
.492	2.6	.320	−1.5	5758.217	13.3
.499	2.4	.330	1.3	.364	19.4
.505	3.2	.336	1.3	.452	18.5
.512	3.5	.340	2.6	.511	9.2
.517	3.5	.349	1.8	.520	3.5
.531	9.2	.358	5.3	.526	2.6
.540	11.1	.363	8.5	5761.535	13.3
.549	12.4	.374	10.4	.553	11.1
5714.276	11.1	.382	10.6	.559	11.3
.290	3.3	.400	13.4	5776.228	10.1
.292	3.2	.490	17.8	.239	3.5
.295	2.6	5722.288	17.5	.249	4.2
.306	3.7	.321	19.1	.259	3.8
.313	8.1	.333	22.2	.266	5.3
.322	10.6	.339	20.7	.280	7.3
.330	11.1	.349	19.1	.312	10.1
5715.337	14.9	.355	20.0	.330	13.1
.343	17.1	.364	19.1		
.350	17.1	.376	14.7		

A particularly remarkable star. Its period is definitely the true period. It was found to vary, however. Tsesevich estimated its magnitude from all the Simeise photographs, using Kanamori's data and the visual observations of Tsarevskii, Kiperman, and his own. The average seasonal curves gave the maxima listed in Table 428. The O—A residues were calculated from

$$\text{Max hel JD} = 2419625.390 + 0.406816 \cdot E \tag{277}$$

and are plotted in Figure 198. The period fluctuates cyclically, without any apparent periodicity. Comparison stars (Figure 199): visual scale $k = 0^s.0$; $a = 13^s.3$; $b = 18^s.1$; $c = 29^s.1$; photographic scale $k = 0^s.0$; $a = 11^s.2$; $b = 16^s.2$; $c = 26^s.8$.

The large fluctuations of the period preclude the construction of a single average light curve from photographic observations; the visual observations, on the other hand, are too few. The star moreover shows a prominent Blazhko effect. Tsesevich's observations are listed in Tables 429, 430.

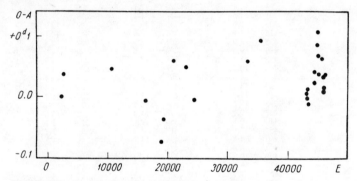

FIGURE 198. O—A residues for BT Aquarii.

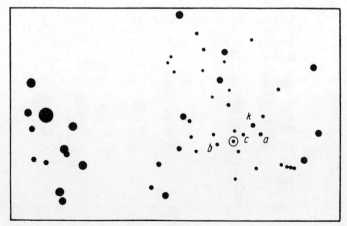

FIGURE 199. Comparison stars for BT Aquarii.

TABLE 428. List of maxima

Source	Max hel JD	E	O — A
			d
Tsesevich (Sim.)	2419625.390	0	0.000
	20741.323	+ 2743	+ .037
	3997.486	+ 10747	+ .044
	6180.412	+ 16113	— .004
Kanamori /292, 293/	7395.906	+ 19101	— .076
	7740.924	+ 19949	— .038
Tsesevich (Sim.)	8020.505	+ 20636	+ .060
	9112.391	+ 23320	+ .052
	9518.336	+ 24318	— .005
	33122.384	+ 33177	+ .060
	4222.451	+ 35881	+ .096
Tsarevskii (visual)	7193.335	+ 43184	+ .003
	7197.400	+ 43194	.000
	7204.330:	+ 43211	+ .014:
Tsesevich (visual)	7204.306	+ 43211	— .010
Tsarevskii (visual)	7549.319	+ 44059	+ .023
	7555.440	+ 44074	+ .042
	7582.289	+ 44140	+ .041
	7872.397	+ 44853	+ .089
	7878.479	+ 44868	+ .069
	7880.513	+ 44873	+ .069
	7881.367	+ 44875	+ .109
	7902.449:	+ 44927	+ .037:
	7907.358	+ 44939	+ .064
Kiperman (visual)	8262.478	+ 45812	+ .034
	8271.432	+ 45834	+ .037
	8286.455	+ 45871	+ .008
	8289.308	+ 45878	+ .014

TABLE 429. Visual observations (Tsesevich)

JD hel	s	JD hel	s	JD hel	s
243...		243...		243...	
7176.321	15.2	7197.344	26.4	7204.295	9.9
.335	15.2	.387	12.1	.308	8.6
.370	16.2	.397	10.7	.310	9.9
.401	16.2	.406	11.2	.319	10.2
7196.258	24.1:	7198.285	15.1	.329	10.4
.302	26.1:	.311	20.3	.341	10.8
.322	26.1:	.342	20.3	.354	11.4
.345	26.1:	.377	22.7	.361	13.3
.372	24.1:	7204.279	21.8	.383	16.2
.388	26.1:	.286	15.7	.401	16.0
7197.268	25.1	.288	11.0		
.294	26.1	.292	10.0		

TABLE 430. Simeise photographic observations (Tsesevich)

JD hel	s	JD hel	s	JD hel	s
241...		242...		242...	
9625.352	9.6	3255.356	21.5	3997.464	8.7
.324	22.5		21.0	.478	6.7
.380	8.4	3283.389	22.1	.493	8.7
242...			22.0	.506	7.5
0715.422	15.5	3620.412	8.1	.521	9.0
0741.337	9.3		6.1	3999.389	19.1
	8.6	3994.492	19.2:	.410	19.1
1070.441	6.1	3997.317	14.5	.431	19.1
	5.6	.331	14.5	.389	19.1
1074.483	6.4	.345	14.5	.410	19.1
	6.5	.358	14.5	.431	19.1
1137.291	21.0	.372	14.5	4362.511	13.7
1431.324	6.4	.386	14.5		14.2
	7.5	.400	14.5	4382.408	13.7
1431.421	13.2	.436	16.2		13.9
	16.2	.450	12.9	4412.318	15.2

TABLE 430 (continued)

JD hel	s	JD hel	s	JD hel	s
242...		242...		242...	
4416.267	19.1	6921.468	17.7	9839.465	8.1
	18.6	6928.402	16.2	243...	
4415.337	15.2		18.3	0224.467	19.1
	13.7	7664.348	19.2	0234.428	10.6
4727.473	18.3		22.0		11.2
5445.482	13.7	7686.329	19.7	0235.447	17.2
	13.2		19.1		17.7
5449.486	8.3	8011.500	16.2	2764.370	6.5
5494.280	17.2		17.2	2768.352	21.0
	14.5	8020.491	5.6		18.6
5495.293	12.9		5.9	2772.400	19.7
	12.9	8024.435	18.3		18.6
5827.491	14.2		20.1	2775.378	6.1
5829.477	10.4	8041.42:	18.1		6.5
5883.370	18.1		16.2	3122.379	6.1
	10.7	8367.480	12.2		8.1
6180.480	8.8		13.2	3154.330	19.7
	6.5	8370.455	12.7		20.1
6187.503	9.8		11.2	3478.471	13.3
6259.324	6.4	8379.483	18.6		13.7
	5.8		12.4	3482.433	6.4
6220.379	10.2	9111.379	15.0		7.8
	10.8	9111.482	21.5	3510.295	17.2
6239.318	20.8		22.0	3838.456	7.5
	21.0	9112.387	7.1		7.1
6240.268	9.7		6.9	4222.454	5.6
	9.3	9132.351	9.8		6.2
6897.483	16.2		8.1	4238.329	7.8
	14.5	9456.477	20.4		7.8
6902.478	18.1	9518.296	9.6	4601.380	18.1
	18.3	9522.284	18.5		19.2
6921.468	17.2		17.4	4607.369	10.2
		9839.465	8.8		14.2

AS Virginis

Discovered by Hoffmeister /262/. First elements determined by Lause /303/. Later observed visually by Tsesevich and Solov'ev /108/. Iwanowska determined the spectral type, Joy the radial velocity.

Tsesevich studied the star from Simeise and Odessa photographs. Reduction to one period and construction of seasonal curves yielded several maxima. The complete list of maxima is given in Table 431.

TABLE 431. List of maxima

Source	Max hel JD	E	O — A
Tsesevich (Sim.)	2420957.319	— 10526	+ 0.032
	5683.467	— 1986	— .037
Lause /303/	6782.567	0	— .032
	6808.565	+ 47	— .045
	6827.378	+ 81	— .048
	6833.460	+ 92	— .054
Tsesevich (visual)	7485.432	+ 1270	— .012
	7490.412	+ 1279	— .013
	7842.400	+ 1915	— .001
Solov'ev /108/	7985.182	+ 2173	— .001
	8299.560	+ 2741	+ .033
Tsesevich (Sim.)	9018.414	+ 4040	— .007
	33031.305	+ 11291	+ .027
Tsesevich (Odessa)	7052.431	+ 18557	— .005
	7808.417	+ 19923	+ .007

The O−A residues were calculated from

$$\text{Max hel } JD = 2426782.599 + 0.5534212 \cdot E.$$ (278)

The star shows the following peculiar features:
1) the period is variable, but its variation is completely irregular (Figure 200);
2) the height of the maximum (and hence the amplitude) is highly variable, and the star possibly displays Blazhko effect (Figure 201).

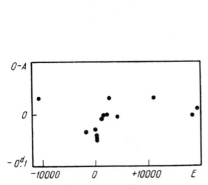

FIGURE 200. O−A residues for AS Virginis.

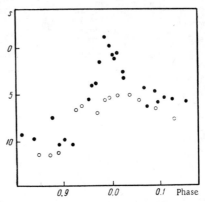

FIGURE 201. Extreme shapes of the light curve of AS Virginis:

● JD 2427485, ○ JD 2427490

Tsesevich's observations are listed in Tables 432−434. Comparison stars (Figure 202): Odessa photographic scale $k = 0^s.0$; $a = 5^s.7$; $b = 12^s.3$; $c = 20^s.3$; Simeise photographic scale $k = 0^s.0$; $a = 9^s.3$; $b = 15^s.5$; $e = 23^s.6$; visual scale $d = 0^s.0$; $b = 8^s.9$; $c = 13^s.4$.

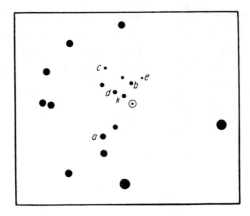

FIGURE 202. Comparison stars for AS Virginis.

343

TABLE 432. Odessa photographic observations (Tsesevich)

JD hel	s	JD hel	s	JD hel	s
243...		243...		243...	
6702.398	10.1	7077.363	3.3	7761.467	9.9
6714.347	4.5	7078.360	14.1	7779.387	11.5
6715.357	6.8	7079.337	12.3	7780.401	9.8
6716.376	14.1	7373.506	7.6	7781.413	9.5
6722.340	10.4	7377.538	12.3	7790.374	9.8
6791.591	8.3	7378.484	4.8	7808.342	8.4
6997.601	12.3	7398.458	7.9	8085.563	12.3
7015.493	9.3	7400.458	12.3	8090.475	10.6
7016.456	4.5	7405.427	12.3	8090.509	12.3
7017.462	11.6	7406.443	12.3	8090.533	9.8
7019.475	12.3	7425.379	8.2	8106.484	13.4
7020.503	9.4	7426.392	11.5	8138.422	10.4
7028.496	13.6	7729.551	12.3	8143.359	11.0
7044.380	10.8	7734.518	12.3	8143.383	9.7
7046.445	10.1	7758.468	13.9	8165.325	7.6
7052.408	2.6	7759.449	13.6		

TABLE 433. Simeise photographic observations (Tsesevich)

JD hel	s	JD hel	s	JD hel	s
242...		242...		242...	
0239.41:	16.9	0988.369	12.8	7535.398	12.4
	17.5	1010.328	19.5	7886.373	16.5
0571.489	19.5		19.1	8245.527	14.9
	20.5	1337.420	13.7		14.8
0580.540	11.4		13.4	8633.356	12.1
0600.336	6.5	3529.416	17.5		12.6
	6.7	3529.416	14.6	9018.377	12.0
0622.334	15.5	4621.287	14.3		12.4
	15.5		13.7	9348.504	13.4
0956.320	8.3	4978.461	6.5		15.5
	11.0		6.5	9366.414	15.0
0957.373	6.6	5347.395	19.5		14.3
	5.3		19.0	9371.348	17.5
0958.347	17.8	5683.458	2.8		17.3
	19.1		1.9	243...	
0959.472	17.8	5716.439	18.2	0107.449	9.3
	15.5		17.5		9.3
0960.326	17.4	6061.457	6.5	2646.428	15.5
	18.5		6.2		14.3
0975.356	18.8	6421.465	14.3	2999.416	12.4
	17.8		13.8		12.8
0977.322	8.8	6799.383	14.1	3031.322	9.3
	7.8	6803.382	16.5		9.3
0978.376	7.2	6803.382	16.2	4125.351	12.4
	8.2	7165.455	12.7		12.4
0984.344	17.5		14.3	4502.338	11.4
	15.5	7213.359	14.0		10.3
0988.369	9.3		13.7		

TABLE 434. Visual observations (Tsesevich)

JD hel	s	JD hel	s	JD hel	s
242...		242...		242...	
7484.320	1.3	7490.412	5.3	7510.376	4.7
.329	3.6	.435	5.9	.400	6.4
.336	5.1	.455	6.9	.418	7.1
.357	6.5	.477	7.9	.450	8.4
.370	6.0	7492.480	11.6	.471	8.5
7485.319	9.4	.501	12.5	7842.324	9.9
.333	9.9	.505	11.9	.332	10.9
.355	7.6	.534	11.9	.343	10.4
.364	10.6	7508.216	7.7	.355	8.4
.377	10.7	.259	8.3	.361	7.7
.396	5.8	.301	8.5	.364	7.5
.401	4.2	.341	10.4	.373	7.7
.405	4.0	7509.330	7.7	.380	7.1
.410	1.6	.365	7.7	.387	5.0
.415	—1.0	.392	8.3	.397	5.0
.420	0.0	.433	10.4	.406	4.6
.426	0.9	7510.215	10.9	.418	4.9
.438	0.6	.224	11.9	.446	5.9
.455	2.8	.240	11.9	.461	7.3
.461	4.2	.276	12.0	.469	7.1
.473	4.9	.287	8.3	.506	8.0
.484	5.4	.292	6.8	7843.335	10.4
.492	5.6	.297	7.0	.361	10.6
.511	5.6	.302	7.0	7861.306	8.4
7490.334	11.9	.306	7.1	.329	8.6
.343	11.9	.313	6.5	.360	8.2
.364	6.9	.319	5.8	.406	9.9
.369	6.5	.334	5.6	7862.241	8.4
.387	7.1	.341	3.8	.334	6.9
.396	5.9	.348	4.1	.437	6.0
.403	5.6	.359	4.7		

CONCLUSION

In summing, we come to the following conclusions.

1. The assembly of RR Lyrae stars is inhomogeneous. It comprises the following distinct groups:

a) ultrashort-period stars with periods of $0.05 - 0^d.178$, which as a rule display light curves of highly variable shape (Blazhko effect), but are immune to any noticeable secular variation of the period;

b) short-period cepheids with almost symmetric light curves between 0.104 and $0^d.411$. This group is in its turn divided into two subgroups: c_1 for stars with secular variation of the period (the majority) and c_2 for stars with widely variable periods (RZ Cep, RU Psc, SX UMa, TV Boo, RV CrB);

c) ordinary RR Lyrae stars with asymmetric light curves, which are classified as RRa and RRb according to the curve shape. Their periods lie between 0.270 and $0^d.838$. They fall into three subgroups: a_1 for stars without secular variation of the period, a_2 for stars with period varying in proportion to time, a_3 stars with jump-like or irregular variation of the period during the observations;

d) long-period stars with over-day periods. These, like the ultrashort-period stars, constitute a homogeneous groups.

The different groups slightly overlap in terms of the periods. For example, the upper bound value for the period of stars in groups c_1, c_2 is $0^d.411$, the lower bound value for groups $a_1 - a_3$ is $0^d.270$, etc.

2. In subgroup a_3 stars with periods close to $0^d.36$ stand out. Their periods are particularly unstable. This non-stability, however, is not entirely determined by the above critical value of the period, since subgroup a_1 also contains stars with precisely the same period which have retained a constant pulsation rhythm for 50 years.

3. The frequency of periods of RR Lyrae stars is close to a normal distribution peaked between $P = 0^d.50$ and $P = 0^d.55$. The distribution curve for subgroup a_1 is bimodal. It splits into two separate distributions with maxima near 0.40 and $0^d.50$. We may thus distinguish between two further subgroups a_{11} and a_{12}.

4. Stars of subgroup a_{11} have constant periods (or periods varying in proportion to time). Their kinematic characteristics (small radial velocities) assign them to Population I. Stars of subgroup a_{12} possibly do not differ from group a_3 stars, but the observations are insufficient to reveal period variation.

5. Of particular interest is the distribution of stars over their spectral types. In Chapter 1 we described in some detail the spectral classification of RR Lyrae variables. Preston has shown that the hydrogen spectral types of all the stars in the minimum are identical, being F5.5, whereas the spectral types in the minimum of the same stars determined from the

intensity of the Ca II lines are different. Figure 203 plots the dependence of the spectral types Sp Ca II on the period P according to Preston's data (circles — stars without secular variation of the period, subgroup a_1; triangles — stars with periods varying in proportion to time, subgroup a_2; dots — stars with variable periods, subgroup a_3).

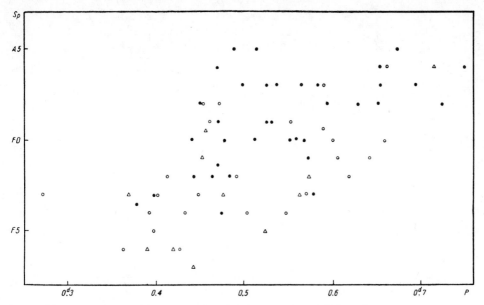

FIGURE 203. Spectral types of RR Lyrae stars from Ca II lines vs. the period (Preston).

Stars of subgroups a_1 and a_2 thus generally have more normal spectra. Stars of subgroup a_3 are conversely depleted in calcium. This difference is particularly pronounced for stars with periods of $0.4 - 0^d.55$; for stars with large periods this difference is inconspicuous.

The last point probably deserves special attention, as it may prove to be of importance in the theory of cepheids and stellar evolution. In my opinion, non-stability of the period is associated with spontaneous modification of stellar structure. This should give rise to convection, which may affect the structure and the chemical composition of the star's shell.

The observed features also admit of an alternative interpretation. The normal spectra of stars of subgroups a_{11} and a_2 may be attributed to their being Population I stars. Hence the conclusion that Population I stars are more stable with regard to pulsation: they more readily retain a steady pulsation mode and better resist external disturbances. Stars of subgroup a_3, on the other hand, are most probably Population II stars. This is evident from the low content of metals, their galactic distribution, and the kinematic properties. These stars are much less stable and are liable to spontaneously change their structure. If these changes occurred more than once during the observations, the star invariably returned to the original state which it had occupied before the first change. If this is indeed so, then considering the age difference between stars of different populations, we conclude that a_{11} and a_2 stars reached the RR Lyrae

347

stage on their way from the main sequence, whereas a_3 stars entered this stage in very advanced stages of their evolution.

6. The Blazhko effect is observed in stars of all subgroups, but mostly in stars of subgroup a_3. There are reasons to believe that in stars of subgroup a_3 this effect follows a different course than in stars of subgroups a_{11} and a_2.

LIST OF ABBREVIATIONS USED IN THE BIBLIOGRAPHY

Publications in Russian

ATs	Astronomicheskii Tsirkulyar [Astronomical Circular]
AZh	Astronomicheskii Zhurnal [Astronomical Journal]
Eng. Bull.	Byulleten' Astronomicheskoi Observatorii im. V.P. Engel'gardta pri Kazanskom Gosudarstvennom Universitete im. V.I. Ul'yanova-Lenina [Bulletin of the Engel'gardt Astronomical Observatory, the Ul'yanov-Lenin Kazan State University]
Eng. Izv.	Uchenye Zapiski Kazanskogo Gosudarstvennogo Universiteta im. V.I. Ul'yanova-Lenina. Izvestiya Astronomicheskoi Observatorii im. V.P. Engel'gardta [Scientific Notes of the Ul'yanov-Lenin Kazan State University. Transactions of the Engel'gardt Astronomical Observatory)
Len. Bull.	Byulleten' Astronomicheskoi Observatorii Leningradskogo Gosudarstvennogo Universiteta [Bulletin of the Astronomical Observatory of the Leningrad State University]
Len. Eph.	Catalogue and Ephemeris of Short-Period Cepheids published by the Astronomical Insitute and Astronomical Observatory of the Leningrad State University
L'vov Circ.	Tsirkulyar Astronomicheskoi Observatorii L'vovskogo Gosudarstvennogo Universiteta [Circular of the Astronomical Observatory of the L'vov State University]
Mirov. Bull.	Mirovedenie [The Universe: Bulletin of the Soviet Society of the Observers of Variable Stars]
Odessa Izv.	Izvestiya Astronomicheskoi Observatorii Odesskogo Gosudarstvennogo Universiteta [Transactions of the Astronomical Observatory of the Odessa State University]
Pulkovo Bull.	Byulleten' Glavnoi Astronomicheskoi Observatorii v Pulkove [Bulletin of the Main Astronomical Observatory in Pulkovo]
PZ	Peremennye Zvezdy [Variable Stars]
Rostov Bull.	Byulleten' Astronomicheskoi Observatorii Rostovskogo Gosudarstvennogo Universiteta [Bulletin of the Astronomical Observatory of the Rostov State University]
Tadzhik Circ.	Tsirkulyar Tadzhikskoi Astronomicheskoi Observatorii, Dushanbe [Circular of the Tadzhik Astronomical Observatory, Dushanbe]
Tadzhik Eph.	Catalogue and Ephemeris of Short-Period Cepheids published by the Tadzhik Astronomical Observatory, Dushanbe
Tadzhik Trudy	Trudy Tadzhikskoi Astronomicheskoi Observatorii, Dushanbe [Proceedings of the Tadzhik Astronomical Observatory, Dushanbe]
Tashk. Circ.	Tsirkulyar Tashkentskoi Astronomicheskoi Observatorii [Circular of the Tashkent Astronomical Observatory]
Tashk. Trudy	Trudy Tashkentskoi Astronomicheskoi Observatorii [Proceedings of the Tashkent Astronomical Observatory]
Trudy II, III, IV	Proceedings of II, III, IV Astronomical Congresses (1920—1928)
Trudy GAISh	Trudy Gosudarstvennogo Astronomicheskogo Instituta im. P.K. Shternberga [Proceedings of the Shternberg State Astronomical Institute]
Vil'nyus Bull.	Byulleten' Astronomicheskoi Observatorii Vil'nyusskogo Gosudarstevennogo Universiteta [Bulletin of the Astronomical Observatory of the Vil'nyus State University]

Publications in Other Languages

AAc	Acta Astronomica. Ser. c
AAS	Acta Astronomica. Supplementa
AJ	The Astronomical Journal

349

All. Pb.	Publications of the Allegheny Observatory of the University of Pittsburgh
AN	Astronomische Nachrichten
ApJ	The Astrophysical Journal
Asiago Contr.	Contributi dell'Osservatorio Astrofisico dell'Universita di Padova in Asiago
BA	Bulletin Astronomique
BAF	Bulletin de l'Association Française d'Observateurs d'Étoiles Variables
Bamb. Ver.	Veröffentlichungen der Remeis-Sternwarte zu Bamberg
BAN	Bulletin of the Astronomical Institute of the Netherlands
Berg. Mitt.	Mitteilungen der Hamburger Sternwarte in Bergedorf
Bol. Boll.	Bolletin dell'Osservatorio Astronomico Universiterio di Bologna
Bud. Mitt.	Mitteilungen der Sternwarte Budapest — Svábhegy
BZ	Beobachtungs-Zirkular der Astronomischen Nachrichten
Coelum	Coelum
Enebo	Beobachtungen veränderlicher Sterne angestellt auf Dombas (Norwegen) von Sigurd Einbu (Enebo)
Erg. AN	Astronomische Abhandlungen. Ergänzungshefte zu den Astronomischen Nachrichten
HA	Annals of the Astronomical Observatory of Harvard College
Hamb.-Berg. Abh.	Abhandlungen der Hamburger Sternwarte in Bergedorf
HB	Bulletin of the Harvard College Observatory
HC	Harvard College Observatory Circular
Heid. Ver.	Veröffentlichungen der Staatlichen Sternwarte Heidelberg — Königstuhl
IAU Bull. Var. St.	Information Bulletin on Variable Stars. Commission 27 IAU
JO	Journal des Observateurs
KVB	Kleine Veröffentlichungen der Remeis-Sternwarte Bamberg
KVBB	Kleinere Veröffentlichungen der Universitats-Sternwarte zu Berlin — Babelsberg
Laws Bull.	Laws Observatory Bulletin
Lyon Bull.	Bulletin de l'Observatoire de Lyon
Mem. SAItal.	Memorie della Società Astronomica Italiano
Mil. Mer. Contr.	Contributi dell'Osservatorio Astronomico di Milano — Meratte (Nuova serie)
MN	Monthly Notices
MVS	Mitteilungen über veränderliche Sterne, Berlin — Babelsberg und Sonneberg
NBI	Nachrichtenblatt der Astronomischen Zentralstelle
ROB	Royal Observatory Bulletin
Torun Bull.	Bulletin of the Astronomical Observatory of N. Copernicus University in Torun
Tokyo Bull.	Tokyo Astronomical Bulletin
Utrecht Rech.	Recherches Astronomiques de l'Observatoire d'Utrecht
Valk. Ver.	Sternwarte des Ignatiuskollegs, Valkenburg (L) Niederlande Veröffentlichungen
Vilno Bull.	Bulletin de l'Observatoire Astronomique de Vilno
VJS	Vierteljahrsschrift des Astronomischen Gesellschaft
VSS	Veröffentlichungen der Sternwarte der Deutschen Akademie der Wissenschaften zu Berlin in Sonneberg
Yale Trans.	Transactions of the Astronomical Observatory of Yale University
Zs Ap	Zeitschrift für Astrophysik

BIBLIOGRAPHY

Publications in Russian
(in Cyrillic Alphabetical Order)

1. ALANIYA, I.F.— ATs, 1954, 146.
2. ALANIYA, I.F.— ATs, 1956, 173.
3. BATYREV, A.A.— PZ, 1950, **7**, 243.
4. BATYREV, A.A.— PZ, 1950, **7**, 246.
5. BATYREV, A.A.— PZ, 1950, **7**, 247.
6. BATYREV, A.A.— PZ, 1951, **8**, 152.
7. BATYREV, A.A.— PZ, 1951, **8**, 157.
8. BATYREV, A.A.— PZ, 1951, **8**, 163.
9. BATYREV, A.A.— PZ, 1951, **8**, 284.
10. BATYREV, A.A.— PZ, 1952, **9**, 48.
11. BATYREV, A.A.— PZ, 1952, **9**, 52.
12. BATYREV, A.A.— PZ, 1952, **9**, 49.
13. BATYREV, A.A.— PZ, 1953, **9**, 217.
14. BATYREV, A.A.— PZ, 1953, **9**, 298.
15. BATYREV, A.A.— PZ, 1953, **9**, 336.
16. BATYREV, A.A.— PZ, 1955, **10**, 116.
17. BATYREV, A.A.— PZ, 1955, **10**, 192.
18. BATYREV, A.A.— PZ, 1957, **12**, 142.
19. BATYREV, A.A.— PZ, 1957, **12**, 141.
20. BATYREV, A.A.— PZ, 1957, **12**, 137.
21. BELYAVSKII, S.I.— PZ, 1935, **4**, 372.
22. BELYAVSKII, S.I.—Pulkovo Bull., 1915, 6, 111.
23. BLAZHKO, S.N.— Tadzhik Eph., 1935, 37.
24. BLAZHKO, S.N.— Tadzhik Circ., 1935, **5**, 6.
25. BLAZHKO, S.N.— Len. Bull., 1933, 3, 19.
26. GRIGOR'EVA, N.B.— PZ, 1938, **5**, 177.
27. GUR'EV, N.I.— Tadzhik Trudy, 1958, 7.
28. GUR'EV, N.I.— Tadzhik Circ., 1938, 33.
29. GUR'EV, N.I.— Tadzhik Circ., 1934, 26.
30. GUR'EV, N.I.— Tadzhik Circ., 1938, 36.
31. GUR'EV, N.I.— Tadzhik Circ., 1938, 32.
32. DZIGVASHVILI, R.M.— PZ, 1951, **7**, 337.
33. DOMBROVSKII, V.A.— Len. Bull., 1934, 4.
34. DOMBROVSKII, V.A.— Tadzhik Circ., 1935, 13.
35. DOMBROVSKII, V.A.— Tadzhik Circ., 1936, 14.
36. DOMBROVSKII, V.A. and N.F. FLORYA. — Len. Bull., 1934, **4**, 29.
37. ERLEKSOVA, T.E.— PZ, 1953, **9**, 219.
38. ZVEREV, M.S.— PZ, 1937, **5**, 109.
39. ZVEREV, M.S. and V.P. TSESEVICH.— PZ, 1953, **9**, 68.
40. IVANOV, N.I.— Trudy II, III, IV, Leningrad, 1930, 151.
41. ISHCHENKO, I.M.— PZ, 1939, **5**, 243.

42. KLEPIKOVA, L.A.— PZ, 1958, **12**, 164.
43. KOSHUBA, P.— Vil'nyus Bull., 1961, 3.
44. KUKARKIN, B.V.— PZ, 1930, **2**, 49.
45. KUKARKIN, B.V.— PZ, 1938, **5**, 196.
46. KUKARKIN, B.V.— PZ, 1946, **6**, 6.
47. KUKARKIN, B.V.— Trudy GAISh, 1948, **16**, 133.
48. KUKARKIN, B.V.— PZ, 1933, **4**, 67.
49. KUKARKIN, B.V.— PZ, 1933, **4**, 68.
50. KUKARKIN, B.V.— PZ, 1940, **5**, 295.
51. KUKARKIN, B.V.— PZ, 1933, **4**, 205.
52. KUKARKIN, B.V.— PZ, 1929, 1, 39.
53. KULIKOVSKII, P.G.— PZ, 1934, **4**, 295.
54. LAVROV, M.I.— ATs, 1950, 100.
55. LAVROVA, N.— ATs, 1959, 203.
56. LANGE, G.A.— Len. Bull., 1934, 4, 28.
57. LANGE, G.A.— Tadzhik Eph., 1935, 36.
58. LANGE, G.A.— Tadzhik Trudy, 1938, 1, 2.
59. LANGE, G.A.— PZ, 1932, 3, 136.
60. LANGE, G.A.— Tadzhik Circ., 1935, 5.
61. LANGE, G.Z.— Tadzhik Circ., 1935, 4.
62. LANGE, G.A.— Tadzhik Circ., 1935, 11.
63. LANGE, G.A.— Tadzhik Eph., 1935, 23.
64. LANGE, G.A.— ATs, 1959, 201.
65. LANGE, G.A.— ATs, 1960, 216.
66. LANGE, G.A.— ATs, 1961, 219.
67. LANGE, G.A.— ATs, 1961, 223.
68. LANGE, G.A.— ATs, 1961, 225.
69. LANGE, G.A.— ATs, 1959, 207.
70. LANGE, G.A.— ATs, 1960, 215.
71. LANGE, G.A. and R.K. KANISHCHEVA.— ATs, 1961, 219.
72. MANDEL', O.E.— ATs, 1960, 209.
73. MARTYNOV, D.Ya.— Eng. Izv., 1938, 20.
74. MARTYNOV, D.Ya.— Eng. Bull., 1934, 1, 8.
75. MARTYNOV, D.Ya.— Eng. Izv., 1951, **26**, 39.
76. MOKHNACH, D.O.— Len. Eph., 1932, 24.
77. MUSTEL, E.R.— PZ, 1934, **4**, 339.
78. MUSTEL, E.R.— PZ, 1934, **4**, 277.
79. NACHAPKIN, V.A.— Tadzhik Circ., 1939, 41.
80. NACHAPKIN, V.A.— Tadzhik Circ., 1940, 44.
81. ODYNSKAYA, O.K.— PZ, 1947, **6**, 191.
82. ODYNSKAYA, O.K. and B.A. USTINOV.— PZ, 1951, **7**, 334.
83. ODYNSKAYA, O.K. and V.P. TSESEVICH.— Odessa Izv., 1954, 2, 2, 93.
84. OLIINYK, G.T.— L'vov Circ., 1962, 37/38, 29.

85. PARENAGO, P.P.— PZ, 1930, 3, 7.
86. PARENAGO, P.P.— PZ, 1933, 4, 83.
87. PARENAGO, P.P.— PZ, 1933, 4, 150.
88. PARENAGO, P.P.— PZ, 1934, 4, 311.
89. PARENAGO, P.P.— PZ, 1940, 5, 330.
90. PARENAGO, P.P.— PZ, 1947, 6, 214.
91. POPOV, M.V.— PZ, 1963, 14, 425.
92. RADLOVA, L.N.— Len. Bull., 1934, 4, 28.
93. RAZGULYAEVA, M.— PZ, 1951, 8, 254.
94. SATANOVA, E.A.— ATs, 1961, 218.
95. SELIVANOV, S.M.— Tadzhik Eph., 1935/36.
96. SELIVANOV, S.M.— Tashk. Circ., 1934, 33.
97. SELIVANOV, S.M.— Tashk. Circ., 1935, 43.
98. SELIVANOV, S.M.— PZ, 1936, 5, 90.
99. SELIVANOV, S.M.— PZ, 1936, 5, 92.
100. SOLOV'EV, A.V.— PZ, 1940, 5, 313.
101. SOLOV'EV, A.V.— Tadzh. Circ., 1936, 21.
102. SOLOV'EV, A.V.— PZ, 1938, 5, 199.
103. SOLOV'EV, A.V.— PZ, 1953, 9, 99.
104. SOLOV'EV, A.V.— Tadzhik Circ., 1936, 18.
105. SOLOV'EV, A.V.— Tadzhik Circ., 1938, 40.
106. SOLOV'EV, A.V.— Tadzhik Circ., 1943, 52.
107. SOLOV'EV, A.V.— Tadzhik Trudy, 1938, 1,2.
108. SOLOV'EV, A.V.— Tadzhik Trudy, 1941, 1,3.
109. SOLOV'EV, A.V.— ATs, 1943, 24.
110. SOLOV'EV, A.V.— PZ, 1938, 5, 168.
111. SOLOV'EV, A.V.— PZ, 1938, 5, 199.
112. SOLOV'EV, A.V.— Tadzhik Circ., 1935, 3.
113. SOLOV'EV, A.V.— Tadzhik Eph., 1935/36.
114. STRASHNYI, I.I.— AZh, 1932, 9, 200.
115. STRELKOVA, E.P.— ATs, 1960, 209.
116. TORONDZHADZE, A.— PZ, 1948, 6, 328.
117. USTINOV, B.A. and V.P. TSESEVICH.—
     Trudy GAISh, 1953, 23, 62—249.
118. FILIN, G.Ya.— Tadzhik Circ., 1950, 85/86.
119. FLORYA, N.F.— Len. Bull., 1933, 3, 19.
120. FLORYA, N.F.— Len. Eph., 1932.
121. FLORYA, N.F.— Len. Bull., 1934, 4, 9.
122. FLORYA, N.F.— Tashk. Circ., 1934, 27.
123. FLORYA, N.F.— Tashk. Circ., 1934, 34.
124. FLORYA, N.F.— Tashk. Trudy, 1933, 4, 2,
     35.
125. FLORYA, N.F.— Len. Bull., 1934, 4, 9.
126. FLORYA, N.F.— PZ, 1932, 4, 21.
127. FLORYA, N.F.— PZ, 1932, 4, 38.
128. FLORYA, N.F.— PZ, 1933, 4, 73.
129. FLORYA, N.F.— PZ, 1933, 4, 126.
130. FLORYA, N.F.— PZ, 1933, 4, 204.
131. FLORYA, N.F.— PZ, 1933, 4, 216.
132. FLORYA, N.F.— PZ, 1932, 4, 36.
133. FLORYA, N.F.— Tashk. Circ., 1932, 2.
134. FLORYA, N.F.— Len. Bull., 1933, 3, 4.
135. FLORYA, N.F.— Tadzhik Eph., 1935.
136. FLORYA, N.F.— PZ, 1940, 5, 262.
137. FLORYA, N.F.— Tadzhik Circ., 1935, 7.
138. FLORYA, N.F.— PZ, 1933, 4, 198.
139. FRIDEL', Yu.V.— L'vov Circ., 1963, 39/40,
     73.
140. FRIDEL', Yu.V.— L'vov Circ., 1962, 37/38,
     53.
141. TSAREVSKII (Shteiman), G.S.— ATs, 1958,
     190.
142. TSAREVSKII (Shteiman), G.S.— PZ, 1957,
     12, 128.
143. TSESEVICH, V.P.— PZ, 1929, 2, 13.
144. TSESEVICH, V.P.— PZ, 1931, 3, 91.
145. TSESEVICH, V.P.— PZ, 1948, 6, 174.
146. TSESEVICH, V.P.— PZ, 1947, 6, 184.
147. TSESEVICH, V.P.— PZ, 1949, 7, 12.
148. TSESEVICH, V.P.— Mirov. Bull., 1925, 4.
149. TSESEVICH, V.P.— Len. Bull., 1933, 3.
150. TSESEVICH, V.P.— Len. Bull., 1934, 4.
151. TSESEVICH, V.P.— Tadzhik Eph., 1935.
152. TSESEVICH, V.P.— Len. Eph., 1932, 24.
153. TSESEVICH, V.P.— Tadzhik Circ., 1934, 2.
154. TSESEVICH, V.P.— Tadzhik Eph., 1935/36.
155. TSESEVICH, V.P.— Tadzhik Circ., 1935, 5.
156. TSESEVICH, V.P.— Tashk. Trudy, 1930, 3,
     58.
157. TSESEVICH, V.P.— Odessa Izv., 1948, 1, 2.
158. TSESEVICH, V.P.— Odessa Izv., 1952, 2, 2,
     111.
159. TSESEVICH, V.P.— ATs, 1949, 86.
160. TSESEVICH, V.P.— ATs, 1956, 170.
161. TSESEVICH, V.P.— AZh, 1961, 38, 293.
162. TSESEVICH, V.P.— ATs, 1958, 188.
163. CHUDOVICHEV, N.I.— PZ, 1930, 2, 58.
164. CHUPRINA, R.I.— PZ, 1958, 12, 228.
165. SHARONOV, V.V.— Tashk. Trudy, 1930, 3,
166. YUDKINA, V.P.— Rostov Bull., 1960, 1, 25.
167. YUDKINA, V.P.— PZ, 1954, 9, 315.
168. YUDKINA, V.P.— PZ, 1951, 8, 289.
169. YUDKINA, V.P.— PZ, 1954, 9, 321.

Publications in Other Languages
(in Latin Alphabetical Order)

170. AHNERT, P.— MVS, 1959, 397.
171. AHNERT, P.— MVS, 1959, 419.
172. AHNERT, P.— MVS, 1959, 420.
173. AHNERT, P.— MVS, 1959, 426.
174. AHNERT, P.— MVS, 1960, 514.
175. AHNERT, p.— MVS, 1960, 520.
176. AHNERT, P.— MVS, 1961, 546.
177. AHNERT, P.— KVBB, 1941, 24.
178. AHNERT, P.— MVS, 1959, 425.
179. ALBITZKY, V.— AN, 1929, 5633.
180. ALMAR, J.— Bud. Mitt., 1961, 51.
181. ASHBROOK, J.— HA, 1942, 109, 7.
182. ASHBROOK, J.— AJ, 1946, 52, 58.
183. BAADE, W.— Berg. Mitt., 1921, 14.
184. BAADE, W.— AJ, 1961, 66, 7.
185. BALAZS, J.— AN, 1938, 6341.
186. BELJAWSKY, S.— BZ, 1923, 5, 24.
187. BELJAWSKY, S.— AN, 1927, 5482.

188. BELJAWSKY, S.— AN, 1929, 5677.
189. BELJAWSKY, S.— AN, 1923, 5261.
190. BELJAWSKY, S.— AN, 1925, 5346.
191. BELJAWSKY, S.— AN, 1926, 5457.
192. BELJAWSKY, S.— AN, 1928, 5595.
193. BEYER, M.— AN, 1934, 6035.
194. BEYER, M.— NBl, 1955, 9, 5.
195. BLAŽKO, S.— BZ, 1925, 7, 28.
196. BLAŽKO, S.— AN, 1910, 4456.
197. BLAŽKO, S.— AN, 1913, 4626.
198. BLAŽKO, S.— AN, 1922, 5167.
199. BLAŽKO, S.— AN, 1926, 5467.
200. BLAŽKO, S.— AN, 1928, 5545.
201. BOHLIN, K.— AN, 1924, 5292.
202. BOHLIN, K.— AN, 1925, 5375.
203. BORN, F. and H. SORFONIEVITSCH.— AN,
     1953, 281, 116.
204. BORN, F. and H. SOFRONIEVITSCH.— AN,
     1955, 282, 235.
205. BOTTLINGER, K. and K. GRAFF.— AN,
     1913, 4687.
206. BOYCE, E.H.— HB, 1939, 911.
207. BROGLIA, P.— Mil. Mer. Contr., 1958, 125.
208. BROGLIA, P.— Mil. Mer. Contr., 1959, 142.
209. BUGOSLAWSKI, N.— AN, 1926, 5484.
210. BUSCH, H., J. BARTHEL, and K. HAUSSLER.—
     MVS, 1959, 429.
211. CANNON, A.— HC, 1907, 129.
212. CANNON, A.— AN, 1907, 4186.
213. CANNON, A.— HC, 1910, 159.
214. CANNON, A.— AN, 1910, 4432
215. CANNON, A.— HC, 1919, 218.
216. CERASKI, W.— AN, 1902, 3860.
217. CERASKI, W.— AN, 1907, 4207.
218. CERASKI, W.— AN, 1909, 4328.
219. CERASKI, W.— AN, 1910, 4387,
220. CERASKI, W.— AN, 1914, 4717.
221. DETRE, L.— AN, 1934, 6074.
222. DETRE, L.— KVB, 1962, 34.
223. DOMKE, K. and E. POHL.— AN, 1953, 281,
     116.
224. DUBIAGO, A.— AN, 1930, 5713.
225. DZIEWULSKI, W.— Torun Bull., 1953, 12, 26.
226. DZIEWULSKI, W.— Torun Bull., 1952, 10, 15.
227. DZIEWULSKI, W.— Vilno Bull., 1938, 21.
228. ENEBO, S.— Enebo, 1911, 5, 35; Archiv for
     Mathematik og Naturvidenskab XXXI, 1911,
     10.
229. EROPKIN, D. and B. OKUNEV.— AN, 1930,
     5742.
230. ESCH, M.— AN, 1918, 4969.
231. ESCH, M.— AN, 1921, 5094.
232. ESCH, M.— Valk. Ver., 1937, 2, 42.
233. FARQUHAR, A. V.— ApJ, 1948, 107, 276.
234. FLORJA, N.— AN, 1930, 5729.
235. FRINGANT, A. M.— JO, 1961, 44, 9—10,
     165—232.
236. FURUJELM, R.— AN, 1922, 5161.

237. GAPOSCHKIN, S.— HC, 1934, 392.
238. GAPOSCHKIN, S.— HA, 1950, 115, 25.
239. GEYER, E.— Zs Ap, 1961, 52, 229.
240. GEYER, E., R. KIPPENHAN, and W.
     STROMEYER.— KVB, 1955, 9.
241. GINORI, N. V.— Mem. SA Ital., 1912, 2, 1, 89.
242. GRAFF, K.— BZ, 1922, 4, 8.
243. GRAFF, K.— AN, 1919, 4992,
244. GRAFF, K.— AN, 1922, 5198.
245. GUTHNICK, P.— AN, 1908, 4284.
246. GUTHNICK, P.— AN, 1929, 5619.
247. GUTHNICK, P. and R. PRAGER.— KVBB,
     1929, 6, 16.
248. GUTHNICK, P. and R. PRAGER.— BZ, 1929,
     11, 32.
249. HAAS, J.— AN, 1926, 5453.
250. HARTWIG, G.— Bamb. Ver., 1932, 1, 489.
251. HARTWIG, G.— Bamb. Ver., 1932, 1, 538.
252. HARTWIG, G.— AN, 1908, 4277.
253. HERBIG, G. H.— ApJ, 1948, 107, 276.
254. HOFFMEISTER, C.— AN, 1916, 4839.
255. HOFFMEISTER, C.— AN, 1916, 4843.
256. HOFFMEISTER, C.— AN, 1919, 4985.
257. HOFFMEISTER, C.— AN, 1923, 5228.
258. HOFFMEISTER, C.— AN, 1927, 5503.
259. HOFFMEISTER, C.— AN, 1928, 5595.
260. HOFFMEISTER, C.— BZ, 1932, 14, 76.
261. HOFFMEISTER, C.— AN, 1929, 5655.
262. HOFFMEISTER, C.— AN, 1930, 5748.
263. HOFFMEISTER, C.— AN, 1931, 5791.
264. HOFFMEISTER, C.— AN, 1933, 5919.
265. HOFFMEISTER, C.— AN, 1934, 6002.
266. HOFFMEISTER, C.— AN, 1934, 6058.
267. HOFFMEISTER, C.— AN, 1935, 6118.
268. HOFFMEISTER, C.— AN, 1936, 6171.
269. HOFFMEISTER, C.— MVS, 1950, 115.
270. HOFFMEISTER, C.— VSS, 1951, 1, 5.
271. HOFFMEISTER, C.— VSS, 1947, 1, 2.
272. HOFFMEISTER, C.— Erg. AN, 1949, 12, 1.
273. HOFFMEISTER, C.— AN, 1963, 287, 1/2, 55.
274. HUGHES, E. M.— HB, 1931, 883.
275. HUGHES-BOYCE, E.— HA, 1942, 109, 2.
276. HUTH, H.— MVS, 1963, 742.
277. ICHINOHE, H.— AN, 1908, 179, 18, 279.
278. IVANOV, N.— AN, 1925, 5346.
279. IVANOV, N.— AN, 1926, 5457.
280. IVANOV, N.— AN, 1926, 5432.
281. IWANOWSKA, W.— Torun Bull., 1953, 11, 1.
282. JACCHIA, L.— AN, 1931, 5783.
283. JACCHIA, L.— AN, 1936, 6251.
284. JACCHIA, L.— BZ, 1936, 18, 55.
285. JACCHIA, L.— HB, 1940, 912.
286. JACCHIA, L.— Bol. Bull., 1935, 3, 43.
287. JENSCH, A.— AN, 1934, 6058.
288. JENSCH, A.— AN, 1935, 6119.
289. JENSCH, A.— AN, 1935, 6135.
290. JORDAN, F.— All. Pb., 1929, 7.
291. KAHO, S.— Tokyo Bull., 1956, Ser. 2, 87.

292. KANDA, S.— BZ, 1934, 16, 27.
293. KANDA, S.— BZ, 1935, 17, 34.
294. KINMAN, T.D.— ROB, 1961, 37.
295. KINMAN, T.D. and C.A. WIRTANEN.— ApJ, 1963, 137, 698.
296. KIPPENHAM, R.— AN, 1953, 281, 4.
297. KLEISSEN, E.— AN, 1938, 6393/6394.
298. KORDYLEWSKI, K.— AAS, 1959, 3, 139.
299. KOWALCZEWSKI, M. and F. KEPINSKI.— AAc, 1934, 2, 91.
300. LAUSE, F.— AN, 1930, 5721.
301. LAUSE, F.— AN, 1932, 5854.
302. LAUSE, F.— AN, 1932, 5902.
303. LAUSE, F.— AN, 1934, 6003.
304. LAUSE, F.— AAc, 1930, 1, 98.
305. LEINER, E.— AN, 1920, 5077.
306. LEINER, E.— BZ, 1921, 2, 17.
307. LEINER, E.— AN, 1921, 5094.
308. LEINER, E.— AN, 1931, 5788.
309. LUIZET, M.— BA, 1907, 24, 342.
310. LUIZET, M.— AN, 1910, 4456.
311. LUIZET, M.— Lyon Bull., 1923, 5, 133.
312. LUIZET, M.— Lyon Bull., 1930, 12, 94.
313. LUYTEN, W.J.— AN, 1937, 6263.
314. MANNINO, G.— Asiago Contr., 1950. 19.
315. McKNELLY, R.D.— AJ, 1948, 53, 215.
316. MARTIN, C. and H. PLUMMER.— MN, 1913, 73, 166.
317. MARTIN, C. and H. PLUMMER.— MN, 1913, 73, 440.
318. MARTIN, C. and H. PLUMMER.— MN, 1913, 73, 654.
319. MARTIN, C. and H. PLUMMER.— MN, 1919, 79, 190.
320. MICZAIKA, G.R.— Heid. Ver., 1946, 14, 12.
321. MORGENROTH, O.— AN, 1935, 6119.
322. MORGENROTH, O.— AN, 1935, 6135.
323. NIJLAND, A.— Utrecht Rech., 1923, 8, 156.
324. NIJLAND, A.— Utrecht Rech., 1923, 8, 229.
325. NIJLAND, A.— BAN, 1935, 7, 249—250.
326. NOTNY, P.— IAU Bull. Var. St., 1962, 7.
327. OPPOLZER, Th.— AN, 1906, 173, 2, 43.
328. PARENAGO, P.— AN, 1930, 5755.
329. PAYNE-GAPOSCHKIN, C.— HA, 1954, 113, 3.
330. PAYNE-GAPOSCHKIN, C.— HA, 1952, 118, 9.
331. PICKERING, E.— HC, 1907, 127.
332. PICKERING, E.— AN, 1907, 4181.
333. PICKERING, E.— HC, 1908, 135.
334. PICKERING, E.— AN, 1908, 4258.
335. PICKERING, E.— HC, 1909, 142.
336. PICKERING, E.— HC, 1918, 201.
337. PICKERING, E.— AN, 1918, 4963.
338. POHL, E.— AN, 1955, 282, 5, 235.
339. PRAČKA, J.— AN, 1909, 4335.
340. PRAGER, R.— HB, 1939, 911.
341. PRESTON, G.W.— ApJ, 1959, 130, 2.
342. RICHTER, G.— MVS, 1960, 493.
343. RICHTER, G.— VSS, 1961, 4, 6.
344. ROBINSON, V.— HB, 1930, 876.
345. ROBINSON, V.— HA, 1933, 90, 48—62—71.
346. ROBINSON, V.— HA, 1933, 90, 46—63—72.
347. ROMANO, G.— Coelum, 1958, 26, 3—4, 37.
348. SHAJN, P.— AN, 1929, 5639.
349. SHAPLEY, H.— Laws Bull., 1911, 17.
350. SHAPLEY, H. and E.M. HUGHES.— HA, 1934, 90, 4.
351. SILVA, E.C.— BAF, 1951, 1, 3—9.
352. SOLOVIEV, A.— AN, 1936, 6232.
353. SPERRA, H.— AN, 1909, 4407.
354. SPINRAD, H.— ApJ, 1961, 133, 479.
355. STEARNS, C.L.— Yale Trans., 1923, 3, 2, 61.
356. STROMEYER, W., R. KNIGGE and H. OTT.— Bamb. Ver., 1962, 5, 14.
357. SUBBOTIN, M.T.— AN, 1927, 5529.
358. SWOPE H.— AJ, 1961, 66, 7.
359. VAN SCHEWICK, H.— BZ, 1940, 22, 81.
360. VAN SCHEWICK, H.— AN, 1942, 272, 198.
361. VOGT, H.— AN, 1920, 5070.
362. VORONTSOV-VELYAMINOV, B.— AN, 1925, 5457.
363. WACHMANN, A.A.— Hamb.-Berg Abh., 1961, 6, 1.
364. WATERFIELD, R.L.— HB, 1927, 848, 25.
365. WENZEL, W.— MVS, 1962, 653—654.
366. WENZEL, W.— MVS, 1962, 656—657.
367. WILLIAMS, S.— MN, 1902, 62, 200.
368. WILLIAMS, S.— MN, 1903, 63, 304.
369. WILLIAMS, S.— MN, 1905, 65, 586—588.
370. WOLF, M.— AN, 1920, 5043.
371. ZESSEWITSCH, W.— Lyon Bull., 1929, 11, 49A.
372. ZESSEWITSCH, W.— AN, 1930, 5754.
373. ZESSEWITSCH, W.— AN, 1931, 5767.
374. ZESSEWITSCH, W.— AN, 1931, 5771.
375. ZINNER, E.— AN, 1913, 4679.
376. ZINNER, E.— AN, 1916, 4839.
377. ZINNER, E.— VJS, 1916, 51, 262. 338.
378. ZINNER, E.— AN, 1931, 5772.

Index of Stars

Oph V731 94
      V734 173—175, 208
      V745 324, 327
      V765 94
      V768 94
      V777 94
      375.1934 94
      HV 10913 94
      HV 10539 315—317, 318
Ori   CM 93
Peg VV 143—144, 208
      AE 245—246, 288
      AO 66—68, 95
      AV 82—84, 96
      BF 288
      BH 92, 95
      BP 13
      CG 92, 95
      CS 94
      DH 318
      DY 323—324,
      DZ 263—267, 288
      ES 93
      ET 63—65, 95
Per   TU 267—271, 287
      AR 51—54, 95
Phe SX 13, 290
Psc RU 318, 346
      RY 155—157, 208
      SS 318
      SW 247—250, 288
      SY 93
Sgr   V632 324, 328
      V756 93
      V1176 311—312, 318
Sco V559 36—38, 95
Ser   VY 92, 95
      AN 86—88, 96
      AT 284—287, 288
      AV 61—63, 95
      BF 324, 325

Ser   CF 94
      CG 94
      CO 94
      CS 152—155, 208
      CW 298—299, 318
Sex T 318
Tau SS 32—36, 287
      YZ 318
      AI 93
Tri   U 56—57, 95
UMa RV 211—212, 287
      SX 318, 346
      TU 81, 220—221, 287
Vel   AI 13, 290
Vir   ST 226—231, 287
      UU 136—138, 208, 209
      VX 94
      WW 93
      XX 324, 325, 326
      AF 93
      AM 93
      AS 342—345
      AT 253—256, 288
      AU 308—311, 318
      AV 79—80, 95
      BB 212—214, 287
      BI 306—308, 318
      BL 94
      BM 94
      BN 114—115, 208
      BO 147—149, 208
      BQ 94
      BS 313—314, 318
      BU 94
      BV 145—146, 208
      BW 133—134, 208
      BX 94
      SVS 122 93
      SVS 123 93
Vul   BC 93
      BN 186—187, 208